Springer Complexity

Springer Complexity is an interdisciplinary program publishing the best research and academic-level teaching on both fundamental and applied aspects of complex systems – cutting across all traditional disciplines of the natural and life sciences, engineering, economics, medicine, neuroscience, social and computer science.

Complex Systems are systems that comprise many interacting parts with the ability to generate a new quality of macroscopic collective behavior the manifestations of which are the spontaneous formation of distinctive temporal, spatial or functional structures. Models of such systems can be successfully mapped onto quite diverse "real-life" situations like the climate, the coherent emission of light from lasers, chemical reaction-diffusion systems, biological cellular networks, the dynamics of stock markets and of the internet, earthquake statistics and prediction, freeway traffic, the human brain, or the formation of opinions in social systems, to name just some of the popular applications.

Although their scope and methodologies overlap somewhat, one can distinguish the following main concepts and tools: self-organization, nonlinear dynamics, synergetics, turbulence, dynamical systems, catastrophes, instabilities, stochastic processes, chaos, graphs and networks, cellular automata, adaptive systems, genetic algorithms and computational intelligence.

The two major book publication platforms of the Springer Complexity program are the monograph series "Understanding Complex Systems" focusing on the various applications of complexity, and the "Springer Series in Synergetics", which is devoted to the quantitative theoretical and methodological foundations. In addition to the books in these two core series, the program also incorporates individual titles ranging from textbooks to major reference works.

Editorial and Programme Advisory Board

Understanding Complex Systems

Founding Editor: J.A. Scott Kelso

Future scientific and technological developments in many fields will necessarily depend upon coming to grips with complex systems. Such systems are complex in both their composition (typically many different kinds of components interacting with each other and their environments on multiple levels) and in the rich diversity of behavior of which they are capable. The Springer Series in Understanding Complex Systems series (UCS) promotes new strategies and paradigms for understanding and realizing applications of complex systems research in a wide variety of fields and endeavors. UCS is explicitly transdisciplinary. It has three main goals: First, to elaborate the concepts, methods and tools of self-organizing dynamical systems at all levels of description and in all scientific fields, especially newly emerging areas within the Life, Social, Behavioral, Economic, Neuro- and Cognitive Sciences (and derivatives thereof); second, to encourage novel applications of these ideas in various fields of Engineering and Computation such as robotics, nano-technology and informatics; third, to provide a single forum within which commonalities and differences in the workings of complex systems may be discerned, hence leading to deeper insight and understanding. UCS will publish monographs and selected edited contributions from specialized conferences and workshops aimed at communicating new findings to a large multidisciplinary audience.

NECSI

New England Complex Systems Institute Book Series

Series Editor

Dan Braha

New England Complex Systems Institute
24 Mt. Auburn St.
Cambridge, MA 02138, USA

New England Complex Systems Institute Book Series

The world around is full of the wonderful interplay of relationships and emergent behaviors. The beautiful and mysterious way that atoms form biological and social systems inspires us to new efforts in science. As our society becomes more concerned with how people are connected to each other than how they work independently, so science has become interested in the nature of relationships and relatedness. Through relationships elements act together to become systems, and systems achieve function and purpose. The study of complex systems is remarkable in the closeness of basic ideas and practical implications. Advances in our understanding of complex systems give new opportunities for insight in science and improvement of society. This is manifest in the relevance to engineering, medicine, management and education. We devote this book series to the communication of recent advances and reviews of revolutionary ideas and their application to practical concerns.

H.Qudrat-Ullah J.M. Spector P.I. Davidsen (Eds.)

Complex Decision Making

Theory and Practice

With 132 Figures and 22 Tables

Springer

NECSI

Hassan Qudrat-Ullah
York University
School of Administrative Studies
4700 Keele St.,
Toronto, ON M3J IP3
Canada
hassanq@yorku.ca

J. Michael Spector
Florida State University
Learning Systems Institute
Instructional Systems Program
C 4622 University Center
Tallahassee, FL 32306-2540
USA
mspector@lsi.fsu.edu

Pål I. Davidsen
University of Bergen
Department of Geography
System Dynamics Program
Fosswinkelsgate 6
5007 Bergen
Norway
davidsen@ifi.uib.no

This volume is part of the NECSI Studies on Complexity collection.

Library of Congress Control Number: 2007932775

ISSN 1860-0832
ISBN 978-3-540-73664-6 Springer Berlin Heidelberg New York

Springer is a part of Springer Science+Business Media
springer.com

Typesetting: by the authors and Integra, India
Cover design: WMX Design, Heidelberg

Printed on acid-free paper SPIN: 12077991 54/3180/Integra 5 4 3 2 1

New England Complex Systems Institute

President
Yaneer Bar-Yam
New England Complex Systems Institute
24 Mt. Auburn St.
Cambridge, MA 02138, USA

For over 10 years, The New England Complex Systems Institute (NECSI) has been instrumental in the development of complex systems science and its applications. NECSI conducts research, education, knowledge dissemination, and community development around the world for the promotion of the study of complex systems and its application for the betterment of society.

NECSI was founded by faculty of New England area academic institutions in 1996 to further international research and understanding of complex systems. Complex systems is a growing field of science that aims to understand how parts of a system give rise to the system's collective behaviors, and how it interacts with its environment. These questions can be studied in general, and they are also relevant to all traditional fields of science.

Social systems formed (in part) out of people, the brain formed out of neurons, molecules formed out of atoms, and the weather formed from air flows are all examples of complex systems. The field of complex systems intersects all traditional disciplines of physical, biological and social sciences, as well as engineering, management, and medicine. Advanced education in complex systems attracts professionals, as complex systems science provides practical approaches to health care, social networks, ethnic violence, marketing, military conflict, education, systems engineering, international development and terrorism.

The study of complex systems is about understanding indirect effects. Problems we find difficult to solve have causes and effects that are not obviously related. Pushing on a complex system "here" often has effects "over there" because the parts are interdependent. This has become more and more apparent in our efforts to solve societal problems or avoid ecological disasters caused by our own actions. The field of complex systems provides a number of sophisticated tools, some of them conceptual helping us think about these systems, some of them analytical for studying these systems in greater depth, and some of them computer based for describing, modeling or simulating them.

NECSI research develops basic concepts and formal approaches as well as their applications to real world problems. Contributions of NECSI researchers include studies of networks, agent-based modeling, multiscale analysis and complexity, chaos and predictability, evolution, ecology, biodiversity, altruism, systems biology, cellular response, health care, systems engineering, negotiation, military conflict, ethnic violence, and international development.

NECSI uses many modes of education to further the investigation of complex systems. Throughout the year, classes, seminars, conferences and other programs assist students and professionals alike in their understanding of complex systems. Courses have been taught all over the world: Australia, Canada, China, Colombia, France, Italy, Japan, Korea, Portugal, Russia and many states of the U.S. NECSI also sponsors postdoctoral fellows, provides research resources, and hosts the International Conference on Complex Systems, discussion groups and web resources.

The New England Complex Systems Institute is comprised of a general staff, a faculty of associated professors, students, postdoctoral fellows, a planning board, affiliates and sponsors. Formed to coordinate research programs that transcend departmental and institutional boundaries, NECSI works closely with faculty of MIT, Harvard and Brandeis Universities. Affiliated external faculty teach and work at many other national and international locations. NECSI promotes the international community of researchers and welcomes broad participation in its activities and programs.

To my parents, Safdar Khan and Fazeelat Begum, for their unconditional love and sacrifice for my education, and to my loving wife Tahira Qudrat, who always been there for me, and my wonderful children Anam, Ali, Umer, and Umael, who never cease to amaze and delight me but all that with innocent complexities.
I promise to keep my office more tidy now.

----Hassan Qudrat-Ullah

To my wife, ChanMin Kim Spector, who listens to my ramblings about complex systems and tolerates the time I spend trying to write sensibly about them, and to my children, Randy, Julia, Sam, Eli and Miriam, who represent a unique and beautiful kind of complexity.

----J. Michael Spector

To my wife, Inger-Lise, and my wonderful children, Robert Heuthe and Jon Leander, they are all very patient and encouraging.

---- Pål I. Davidsen

Preface and Acknowledgements

The environments in which we live and work are becoming increasingly complex. The technological innovations, globalization, and political upheavals hugely impact our decision making capabilities. Myriad interactions between available but limited physical and intellectual resources and the objectives and goals we, as a society in 21st century, want to realize severely impinge on the decision makers, be it in public or private domain. In the face of this increasing complexity of the decision making environments, the decision maker must become better at understanding and coping with the dynamic, complex problems. However, at the core of the decision-making process is the need for quality information that allows the decision makers to better understand the impact of e.g., feedback processes, non-linear relationships between variables, and time delays on the performance of the complex system. One of the most important sources of such information is the outcome of the both the model building process and the application of the model of the complex system. Modeling supports decision making by providing specific "what-ifs" scenario analysis opportunity to the decision makers in a "non-threatening" manner.

Thus the primary aim of this book is to disseminate the roles and applications of various modeling approaches. The key focus is on the development and the applications of system dynamics and agent-based modeling approaches in service of dynamic decision making. Invitations to contribute to this book were sent via special-interest electronic list servers around the globe. Several renowned authors were also specially invited to contribute. Each prospective contributor was initially asked to prepare a two-to three-page proposal on his/her contribution. These proposals were reviewed by the editors and suggestions were made to prepare the full papers. The submitted papers were then reviewed by independent reviewer panels. Each panel consisted of three members, one from the editors and two independent experts in the filed. The final acceptance/rejection decisions were made by the editors based on the revised papers submitted by the contributors.

The book contains five parts. Part I, Complex decision making: Concepts, theories and empirical evidence, has two chapters. It examines the nature and the dimensions of decision making and learning in dynamic task, based on published literature. Part II of the book, Tools and techniques for decision making in complex systems, consists of six chapters and deals with the range to tools, methods, and technologies that support decision making in complex, dynamic environments including Method of Dynamic Handling of Uncertainty, System Dynamics, System Dynamics and Multiple Objective Optimization based hybrid approach, Object Oriented Programming based simulation methods, Data Envelop Analysis, Group Model Building, and Agent-based Modeling that can help people make better decisions in complex systems. Part III of the book, System dynamics and agent-based modeling in action has five chapters and provides the empirical evidence to the application of both modeling approaches to variety of complex decision making situations covering both individual and group decision making perspectives. In part IV of the book, Methodological issues in research on complex decision making, the two chapters deals with the two key issues: (i) understanding of "goal dynamics" and, (ii) measure of change in the mental models of the decision makers in dynamic tasks. Part V of the book, Future directions on complex decision making, discusses the emerging trends and challenges in using modeling and model based tools for improved decision making in complex, dynamic environments.

We are grateful to the authors of various chapters for their contributions. It had been a long process from the initial outlines to developing the full chapters and then revising them in the light of reviewers' comments. We sincerely acknowledge the authors' willingness to go through this long process. We also acknowledge the work and knowledge of various reviewers of the chapters, many of which had to be done at short notice.

Thanks to all the people at NECSI especially Dan Braha, and Springer-Verlag with whom we corresponded for their advice and facilitation in the production of this book.

Ms. Hemalatha prepared camera-ready copy of the manuscript with her usual professionalism and cooperation and we wish to record our thanks to her.

Finally, we are grateful to our families for their support all the way through.

H. Qudrat-Ullah
Toronto, Canada

J. M. Spector
Florida, USA

P. I. Davidsen
Bergen, Norway

Contents

PART I

Complex Decision Making: Concepts, Theories, and Empirical Evidence

Chapter 1

How to Improve Dynamic Decision Making? Practice and Promise

Mustafa Karakul
Hassan Qudrat-Ullah
School of Administrative Studies
York University, Canada
mkarakul@yorku.ca
hassanq@yorku.ca

1. Introduction

Decision makers today face problems that are increasingly complex, and interrelated (Diehl & Sterman, 1995; Moxnes, 2000; Sterman, 1989b). Many important decisions routinely made are dynamic in nature - a number of decisions are required rather than a single decision, decisions are interdependent, and the environment in which decision is set changes over time (Brehemer, 1990; Edwards, 1962; Hogarth, 1981; Sterman, 1989b). The development of managerial capability to cope with dynamic tasks is ever in high demand. However, the acquisition of managerial capability of decision making in dynamic tasks has many barriers (Bakken, 1993). On the one hand, corporate and economic systems do not lend themselves well to real-world experimentation. On the other hand, most of the real-world "decisions and their outcomes" are not closely related in both time and space.

Recent advances in computer technology together with the availability of new simulation tools provide a potential solution to this managerial need. Computer simulation-based interactive learning environments (ILEs) allow the compression of time and space. Additionally, ILEs provide an opportunity for practicing managerial decision making in a non-threatening way (Issacs & Senge, 1994). In an ILE, decision makers can test out their assumptions, practice exerting control over a business situation, and learn from the "immediate" feedback of their decisions. However, the empirical evidence to the effectiveness of ILEs is inconclusive (Bakken, 1993; Diehl & Sterman, 1995; Keys & Wolf, 1990; Lane, 1995; Moxnes, 2000).

One way to improve the efficacy of a learning tool such as ILE is to incorporate insights and knowledge about how the learners interact with peers, with teachers and facilitators, and with tools, models, and simulations (Spector, 2000). This knowledge about the nature and extent of human-human and human-simulation interactions can be used to help create a typology of ILEs based on how conducive they are to improving decision making in dynamic tasks. While an extensive research on dynamic decision making (DDM) (Funke, 1995; Hsiao, 1999; Kerstholt & Raaijmakers, 1997; Kleinmuntz & Thomas, 1987; Rapoport, 1975; Sterman, 1989a, b) and learning in ILEs (Bakken, 1993; Homer & Hirsch, 2006; Keys & Wolf, 1990; Kriz, 3003; Langley & Morecroft, 1995; Lane, 1995) exists, knowledge about interactions between decisions, decision-makers, and decision-making processes is extremely fragmented. The aims, therefore, of this chapter are (i) to present a systematic integration of the literature on DDM and learning in ILEs, and (ii) to develop a theoretical model for improving DDM.

This chapter starts with a brief description of some background concepts and terminology. Next, the review of experimental research on DDM and learning in ILEs is presented. After review and assessment of prior research on DDM, the development of the proposed conceptual model for decision making in dynamic tasks follows. Finally, some suggestions for future research are presented.

2. Review of Experimental Research

2.1. Background Concepts

We use "ILEs" as a term sufficiently general to include microworlds, management flight simulators, learning laboratories, and any other computer simulation-based environment – the domain of these terms is all forms of actions whose general goal is the facilitation of decision making and learning. Motivated by the on-going work in the system dynamics discipline (Davidsen, 2000; Lane 1995; Moxnes, 2000; Qudrat-Ullah, 2005), this conception of ILE embodies learning as the main purpose of an ILE. Likewise, computer-simulation models, human intervention, and decision making are considered as the fundamental components of an ILE (Bakken, 1993; Cox, 1992; Crookall, Martin, Saunders, & Coote, 1987; Davidsen, 1996; Davidsen & Spector, 1997; Goodyear, 1992; Lane, 1995; Sterman, 1994).

Under this definition of ILE, learning goals are made explicit to the decision-makers. A computer-simulation model is built to represent adequately the domain or issue under study with which the decision makers can experience and induce real world-like responses (Lane, 1995). Human intervention refers to active keying in of the decisions by the decision makers into the computer-simulation model via decision-making environment or interface. Hence, in the remainder of this chapter, the terms "game," "simulation," "micro worlds," and "simulators" are used in the senses described above.

What is dynamic decision making? Dynamic decision-making situations differ from those traditionally studied in static decision theory in at least three ways: a number of decisions are required rather than a single decision, decisions are interdependent, and the environment changes, either as a result of decisions made or independently of them or both (Edwards, 1962). Recent research in system dynamics has characterized such tasks by feedback processes, time delays, and non-linearities in the relationships between decision task variables (Hsiao, 1999; Sengupta & Abdel-Hamid, 1993; Sterman, 1994). In dynamic tasks, contrary to static tasks such as gambling, locating a park on a city map, and counting money, multiple and interactive decisions are made over several periods whereby these decisions change the environment (Forrester, 1961; Sterman, 1989a, 1994).

Task performance. Researchers have operationalized the construct "task performance" in many ways. Maximizing, minimizing, predicting, achieving, controlling, and performing with task goals are the common measures for task performance. Examples of these measures are provided in Tables 2, 3, and 4.

Task knowledge. The task knowledge category primarily concerns how well the learners in an ILE acquire the knowledge. Declarative – procedural knowledge distinction is the most commonly employed typology in surveyed studies. Declarative knowledge pertains to the knowledge about principles, concepts, and facts about the task – designer's logic or structural knowledge. Structural knowledge is measured through written or verbal questions about the precise nature of relationships among various system components or the nature of decision-induced causal variations in the output variables. Procedural knowledge concerns how decision makers actually control or manage the task- operator's logic or heuristics knowledge. To evaluate the task knowledge, a pre-task and / or post-task questionnaire is often used.

Transfer learning. Transfer learning is the measure that has been used to assess how well the decision makers learn from the previous task by making them play another task either in the same domain (Huber, 1995) or in a different domain (Bakken, Gould & Kim, 1994).

2.2. Characteristics of the Existing Research

Numerous studies, listed in Table 2, on DDM and learning in ILEs use decision-task factors as an integral part of larger manipulations. Relatively few studies exist, however, where the nature of facilitation manipulation is such that the effects of the form of facilitation and level of facilitation can be determined clearly. A moderate number of studies, listed in Table 1 and Table 3, empirically examine the influences of learner characteristics and features of decision-making environment on task performance and learning.

Thirty-nine experimental studies provide enough information about the nature of predictor manipulations to be considered here. In most of the studies, task performance is a major dependent variable, while in a few cases, task knowledge and transfer learning are the outcome variables. These studies are listed in Tables 1, 2, and 3. For each study, types of predictor factors, task structure, decisions required, measurements of performance, as well as, a short summary of major results are provided.

Table 1. Summary of the Reviewed Empirical Studies on Learner Factors

Predictor Variables	Decisions Required	Measurement of Task Performance and Learning	Results; Source
Task Structure encompasses : Delay, nonlinearity, and positive loops			
Cognitive styles (CS)	Single decision in capital investment game STRATEGUM-2	Minimize the cost (relative to the benchmark), over time	Positive effect of CS (Abstract component of Gregoric test) on TP; Trees, Doyle, and Radzicki (1996)
Heuristic strategies: schema-driven acquisition or random acquisition	A medical decision making task with a single decision	Maximize the proportions of patients cured	Schema-driven acquisition was superior to random acquisition on TP; Kleinmuntz (1985)
Prior Knowledge about the domain (PK)	Control a system in control theory domain by applying controllers	Mean scores on the pre-task / post-task knowledge test	Higher the PK the higher the posttest scores; Njoo and de Jong (1993)
Prior professional experience (PPE)	Three decisions in a simulator (PEOPLE EXPRESS)	Mean scores on the relationship between variables test	Subjects with no PPE transferred insights better than those with PPE; Bakken, Gould, and Kim (1994)
Prior task knowledge (PTK) through a two day training	Six decisions in a social welfare game-JOBS	Mean scores on the relationship between variables in the game	No effect of PTK on SK; Maxwell (1995)
Prior task knowledge (PTK) in terms of academic background	Two forecasts and two decisions in a stock management task, over time	Mean scores on relationship between variables test	Positive effect of PTK on structural knowledge (SK); Bakken (1993)
Specific exploratory strategies (SES) or global exploratory strategies (GES)	Two decisions in a business management simulated system JEANSFABRIK	Maximize the profit, over time	Subjects with SES were more effective than those with GES in TP; Grubler, Renkel, Mandal, and Reiter (1993)
Systematic-elaboration instruction (SEI) and goal-planning instruction (GPI)	One-to-twenty five decisions in a dynamic task system MORO	Achieve the rank-ordered states of goal	Subjects in both groups improved more than those in control group on TP; Jansson (1995)
Systematic-elaboration instruction (SEI) and goal-planning instruction (GPI)	One-to-twenty five decisions in a dynamic task system MORO	Mean scores on post-experimental questionnaire	Both groups possessed better system model (HK) than the control group; Jansson (1995)
Systematic-elaboration instruction (SEI) and goal-planning instruction (GPI)	One-to-twenty five decisions in a dynamic task system MORO	Total amount of time spent in minutes and seconds	Both groups showed increase in decision time as compared to the control group; Jansson (1995)

Task experience (TE) in terms of academic background	Two forecasts and two decisions in a stock management task	The accumulation of profit	Positive effect of TE on task performance (TP); Bakken (1993)
Task experience (TE) in terms of repeated trials	Single decision in a DESSY	Amount of decision times	TE decreased the decision time; Brehmer and Allard (1991)
Task experience in terms of repeated trials	Single decision in management task	Minimize total costs	Positive effect of TE on TP; Diehl and Sterman (1995)
Task experience through a two day training session	Six decisions in a social welfare game-JOBS	Minimize the cost	No effect of TE on TP; Maxwell (1995)
Task experience through repeated trials	Single decision in a real time fire-fighting game	Minimize the percentage area lost (to fire)	TE decreased the decision time in terms of average trial time; Brehmer and Svenmark (1995)
Task Structure encompasses : Delay and nonlinearity			
Task experience in terms of repeated trials	Two decisions in a strategy game	Maximize profits relative to benchmark	Positive effect of TE on TP; Paich and Sterman (1993)
Task Structure encompasses : Delay			
Extended length of training(ELT) and decision aid: history graph	A decision in Sugar Production Factory task	Meet (the production) goals	ELT alone did not improve TP but ELT with history graph did; Gibson (2000)
Task experience (TE) in terms of repeated trials	A decision in a sugar production game; A decision in computer person game	Mean scores of test of direct and crossed relationships between variables	Positive effect of TE on TP; Berry and Broadbent (1987)
Task experience (TE) in terms of repeated trials	Three decisions in a game simulating British economy (ECONEX)	Minimize the difference of TP compared with a benchmark	Positive effect of TE on TP; Broadbent and Aston (1978)
Task experience through selective mode(SM) and unselective mode (UM)	Single decision in a computer person game	Median scores of verbal protocols and mean scores on knowledge test	TE through SM is positively related to SK and UM induces HK; Hayes and Broadbent (1988)
Task experience through task practice (TE)	A decision in a sugar production game; A decision in computer person game	Mean scores of test direct and crossed relationships between variables	TE has positive effect on test direct knowledge and negative on crossed relationships; Berry and Broadbent (1987)
Task knowledge (TK) about salient and non-salient relationships	A decision in a sugar production game; A decision in computer person game	Number of attempts and number of errors of direction	Positive effect of TK on TP; Berry and Broadbent (1987)

Table 2. Summary of the Reviewed Empirical Studies on Decision Task Factors

Predictor Variables	Decisions Required	Measurement of Task Performance and Learning	Results; Source
Task Structure encompasses : Delay, nonlinearity, and feedback loops			
Frequency of oscillations	Two forecasts and two decisions in a stock management	The accumulation of profit relative to benchmark profit	Low frequency leads to better task performance (TP); Bakken (1993)
Task transparency (TT)	Multiple decisions (25) in a business simulator SITMECOM	Mean scores of pre-game and post-game questionnaire	Positive effect of TT on SK; Gonzalez, Machuca, and Castillo (2000)
Task transparency (TT)	Multiple decisions in LEARN!	Maximize profits	TT has a positive influence on TP; Gröbler, Maier, and Milling (2000)
Task transparency (TT)	Multiple decisions (25) in a business simulator SITMECOM	Maximize profits	TT has positive influence on acquisition of SK; Machuca, Ruiz, Domingo, and Gonzalez (1998)
Task transparency (TT) in terms of full task information (TI)	Two decisions in a renewable-resource-management flight simulator	Maximize net present value of the profits over time plus resource value in the final year	Subjects with full TI performed worse than ones with imperfect information; Moxnes (1998)
Task transparency in terms of goal conditions	Two decisions in a simulated ecosystem game	Number of times to reach the target in a given number of trials	Subjects with number goal performed better than those with ratio goal; Yang (1997)
Task Transparency (TT)	Multiple decisions (25) in a business simulator SITMECOM	Maximize profits	Positive effect of TT on TP; Gonzalez, Machuca, and Castillo (2000)
Time delay (TD) and feedback gains (FG)	Single decision in a generic stock management task	Minimize total costs relative to the benchmark	Effects of TD and FG on decisions are insignificant; Diehl and Sterman (1995)
Time delays (TD), positive feedback and gains built in the task model (PG)	Single decision in stock management game Beer Game	Minimize the total company cost	Detrimental effect of TD and PG on TP; Sterman (1989a)
Time delays (TD), positive feedback and gains built in the task model (PG)	Single decision in a capital investment simulation game STRATEGUM-2	Minimize the average absolute deviation	Detrimental effect of TD and PG on TP; Sterman (1989b)
Time delays (TD): in terms strength, feedback gains (FG)	Single decision in a generic stock management task	Minimize total costs relative to the benchmark	Increase in TD decreases subjects TP; A stronger FG deteriorate TP; Diehl and Sterman (1995)

Time delays in terms strength (TD) and feedback gains (FG)	Two decisions in a market strategy simulation game	Maximize profits relative to benchmark	Subjects perform poorly as complexity (TD and FG) increases; Paich and Sterman (1993)
Total number of variables (TNV), interaction between subsystems (IBS), and Random variation (RV)	Three decisions in a welfare administration simulation game	Minimize the percent difference (in cost) relative to the benchmark	Detrimental effect of TNV and RV on TP but positive effect of IBS on TP; Mackinnon and Weaing (1980)
Uncontrollable positive feedback loops (UPFL)	Single decision in capital investment STRATEGUM-2	Minimize the deviation between "backlog" and "capital stock"	Detrimental effect of UPFL on TP; Young, Chen, Wang, and Chen (1997)
Uncontrollable positive feedback loops (UPFL)	Single decision in capital investment STRATEGUM-2	Minimize the deviation between "backlog" and "capital stock"	UPFL has significant influence on decision scope; Young, Chen, Wang, and Chen (1997)
Task Structure encompasses feedback loops			
Task semantic: semantically embedded version (S^+) and abstract version (S^-)	Single decision in a complex dynamic system with multiple goals	Average Euclidean distance between subject and the optimal intervention target	Subjects in S^- group outperformed those in S^+ in terms of TP; Beckmann and Guthke (1995)
Task semantic: semantically embedded (S^+) and abstract vesrion (S^-)	Single decision in a complex system with multiple goals	Number of items completed and number of prompts required	S^+ has no relation but S^- has positive relation to SK; Beckmann and Guthke (1995)
Task semantic embedding (TSE)	Single decision in an investment task	Maximize the capital	No effect of TSE on TP; Huber (1995)
Task Structure encompasses delay and nonlinearity			
Time delay (TD): in terms of strength	Single decision in FIRE FIGHTING	Minimize the percentage	Detrimental effect of TD on TP; Brehmer (1995)
Task Structure encompasses delay			
Task transparency (TT)	A decision in a sugar production game; A decision in computer person game	Mean scores of test direct and crossed relationships between variables	TI improves SK but not about indirect relationships; Berry and Broadbent (1987)

Table 3. Summary of the Reviewed Empirical Studies on Decision-making environment Factors

Predictor Variables	Decisions Required	Measurement of Task Performance and Learning	Results; Source
Task Structure encompasses : Delay, nonlinearity, and positive loops			
Feedback in terms of feed forward (FF), outcome feedback (OF) and cognitive feedback (CF)	Single decision in a software project management game	Percentage deviation of two goal variables from the initial stated estimates	Subjects in CF group had the best TP, followed by FF and OF groups; Sengupta and Abdel-Hamid (1993)
FF, OF, and CF	Single decision in a software project management game	Whether subjects used the feedback and how long did they view it	More use of CF and not more information per se improved TP; Sengupta and Abdel-Hamid (1993)
FF, OF, and CF	Single decision in a software project management game	Average time spent by subject making staffing decisions in each interval	CF group used longer decision time, followed by FF and OF; Sengupta and Abdel-Hamid (1993)
FF, OF, and CF	Single decision in a software project management game	Average amount of fluctuations and reversals in direction	Subjects in OF group fluctuated the most, followed by FF and CF; Sengupta and Abdel-Hamid (1993)
Graphical feedback (GF)	One-to-twenty five decisions in a dynamic task system MORO	Achieve the rank-ordered states of five goal variables	GF alone is ineffective in improving TP; Putz-Osterloh, Bott, and Koster, (1990)
Learning by doing (LBD)	-Delay, nonlinearity and positive loops; One-to-twenty five decisions in a dynamic task system MORO	Achieve the rank-ordered states of five goal variables	LBD alone is ineffective in improving TP; Putz-Osterloh, Bott, and Koster, (1990)
Outcome feedback in terms of knowledge of results (KOR) and benchmark outcome (BO)	Single decision in stock management game Beer Game	Minimize the total company cost	HK more effectively contributes to TP than SK; Hsiao (2000)
Learning conditions in terms of process-learning (PL) and content-learning (CL)	Multiple decisions in various computer-based simulations games	Scaled score on pre-task and post-task test	Subjects in PL condition exhibited improved cognitive strategies; Breuer and Kummer (1990)
Outcome feedback: knowledge of results (KOR) and benchmark outcome (BO)	Single decision in stock management game Beer Game	Minimize the total company cost	Providing BO improves subjects TP; Hsiao (2000)

Outcome feedback: knowledge of results (KOR) and benchmark outcome (BO)	Single decision in stock management game Beer Game	Mean scores on post-gaming questionnaire	Providing BO improves subjects SK and HK; Hsiao (2000)
Simulators with different cover stories but identical structures of the task model	Three decisions in a simulator (PEOPLE EXPRESS)	Maximize profits, over time	Subjects transferred insights from the first to second simulator; Bakken, Gould, and Kim (1994)
System trends condition- unstable graphs	One-to-twenty five decisions in a dynamic task system MORO	Achieve the rank-ordered states of five goal variables	Unstable graphs with experience cause an improvement in TP; Putz-Osterloh, Bott, and Koster, (1990)
System trends conditions- stable graphs	One-to-twenty five decisions in a dynamic task system MORO	Achieve the rank-ordered states of five goal variables	GF alone is ineffective in improving TP; Putz-Osterloh, Bott, and Koster, (1990)
Task Structure encompasses judgmental tasks			
Field-dependence (FD) or independence and color: mono-color or multi-color	Three decisions in a marketing decision making simulator Brand Manager's Allocation Problem	Maximize profits	TP of FD with color-enhanced reports was 73 % better than those without such reports; Benbasat and Dexter (1985)
Information system: graphical or tabular	Three decisions in a marketing decision making simulator Brand Manager's Allocation Problem	Maximize profits	No differences on TP were found; Benbasat and Dexter (1985)
Task Structure encompasses delay			
Explicit feedback (subjects' performance vs. rule's)	Two decisions in a production planning task	Minimize aggregate total costs	Explicit feedback improved subjects' task performance (TP); Davis and Kottermann (1995)
Explicit description of benchmark rule	Two decisions in a production planning task	Minimize aggregate total costs	Explicit description of rule had no significant effect on subjects' TP; Davis and Kottermann (1995)
Learning conditions in term of "observation" (LBO) and "active intervention"(LBI)	A decision in a sugar production game; A decision in computer person game	Number of trails on target	Subjects with LBO condition were not able to show improvement in TP; Berry (1991)
Learning conditions in term of "observation" (LBO) and "active intervention"(LBI)	A decision in a sugar production game; A decision in computer person game	Mean scores on post-task questionnaire	Subjects in LBO condition did not improved SK; Berry (1991)
What-if-analysis	Two decisions in a production planning task	Minimize aggregate total costs	Provision of what-if-analysis had no significant effect on subjects' TP; Davis and Kottermann (1994)

2.3. A Model of the Effects of Learner Factors

Figure 1 depicts the key variables determining the effects of individual differences on task performance and learning.

DDM researchers have shown a reasonable recognition to prior knowledge. Bakken (1993) reports that subjects with business management backgrounds perform better in a real estate game, which presumably requires more management expertise, than in an oil tanker game. However, Maxwell's study (1995), a two-day session on simulation techniques and task knowledge, shows no effect of training on task performance. This is not surprising: even an extended training session alone is not sufficient because subjects appear not to take feedback delays into account unless they are provided with additional decisional aids (Gibson, 2000). Among various decisional aids, human facilitation is crucial to the development of better understanding of the dynamic tasks (Spector, 2000). In the next section we present a model that incorporates human facilitation based decisional aids.

Task experience explores the relationships of decision inputs and outputs by trial and error, enhances their causal understanding of task structure, establishes reliable decision rules, and as a result, improves task performance (Hsiao, 1999). Broadbent & Aston (1978) establish the fact that subjects can learn through task practice to make better decisions than they can when they are new to the task. Yet, the same subjects might not improve in their ability to answer verbal questions. Conversely, verbal instructions can improve subjects' question answering ability but not their task control performance (Berry & Broadbent, 1987). Berry and Broadbent (1988) argue that subjects may learn about how to handle tasks through an implicit manner. For instance, airline pilots maneuver air planes well even though relatively few of them can explain the structure and design of air planes. This "insight without awareness" (Stanley, Mathews, & Kolter-Cope, 1989) often works in decision processes and can be identified through a systematic research design.

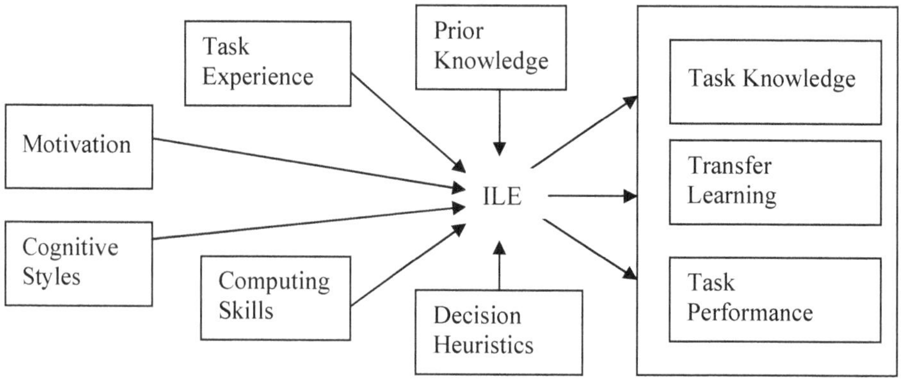

Figure 1. A Model of the Effects of Learner Factors

Motivation has a positive influence on subjects' task performance. Dörner, Reusing, Reiter & Staudel (1983) in their LOHHAUSEN study showed that DDM performance was related to motivational factors to a greater extent than intellectual factors. However, Beckmann and Guthke (1995) suspected that LOHHAUSEN findings might have been because subjects' interactions with the system were mediated by the experimenter. Drawing on the seminal work of Gagné (1985), where motivation is regarded as a fundamental factor in achieving optimal learning in complex tasks, we would expect support for LOHHAUSEN results. Consequently, a subsequent LOHHAUSEN study where experimenter's role is explicitly manipulated might help empirically resolve this inconclusive finding.

Computing skills have been demonstrated to be helpful for familiarization with the task systems but not in task performance (Trees, Doyle, & Radzicki, 1996). The irrelevance of computing skills to task performance seems predictable as the subjects in DDM studies are allowed to spend sufficient time familiarizing themselves with computer-simulation interfaces (Hsiao, 1999).

Trees et al. (1996) investigate the extent to which cognitive styles help explain individual differences in DDM. They reported that scoring higher on the Abstract component of the Gregoric test has marginal explanatory power for task performance. Nevertheless, when learning is associated with the acquisition of the prototypic knowledge, future studies need to explore: does the efficacy of decision making in dynamic tasks depend on cognitive styles?

In ILEs, decision makers may be provided with decision heuristics. Yang's (1997) empirical study confirms that subjects, given the heuristic- focus on goals - are able to achieve better control and understanding of the tasks that they are subjected to. Likewise, Putz-Osterloh, Bott, and Koster (1990), using the DYNAMIS micro world, found significant improvements in structural knowledge for subjects using efficient strategies for intervention.

The results of Jansson's (1995) study show that the performance of both groups who received heuristic instructions is significantly better than that of the control group. These findings are in sharp contrast to a large amount of research that documents people's problems dealing with complex systems (Berry & Broadbent, 1988; Brehmer, 1990, 1995; Dörner, 1980; Paich & Sterman, 1993; Sterman, 1989a, b). However, the nature of the tasks, relatively simple task in Jansson (1995) versus relatively complex task studies (Paich & Sterman, 1993; Sterman, 1989a, b), does not warrant a credible comparison. Nevertheless, these results provide impetus for future research on how the heuristics strategies help improve decision making in dynamic tasks.

2.4. A Model of the Effects of Decision Task Factors

Figure 2 shows the major decision task factors influencing decision making and learning in ILEs.

Several researchers (e.g., Berry & Broadbent, 1987; Gröbler, 1998; Gröbler, Maier, & Milling, 2000; Gonzalez, Machuca, & Castillo, 2000; Jansson, 1995; Machuca, Ruiz, Domingo, & Gonzalez, 1998) have explored the issue of task transparency. Berry and Broadbent (1987) finnd that providing subjects with task information only improves their understanding of the direct relationships rather than the indirect relationships between the task system variables.

Gröbler et al., (2000) performed an experiment to evaluate the relevance and effects of structural transparency. The results showed that a presentation about the structure of the system had a positive influence on subjects' task performance. Moreover, Jansson (1995) has shown that the provision of task structure information increases subjects' decision time and information use.

Thus, regarding the impact of the task transparency, studies conclude that subjects improve their task performance but are unable to improve on verbal knowledge test. A reasonable explanation to this puzzling conclusion would be that subject's inability to verbalize the acquired knowledge might be an issue of the instrument used for evaluation rather than the effect of task transparency.

Semantic embedding of the task refers to whether or not the task is couched within a well-understood and familiar context. Beckmann and Guthke (1995) compare two semantic embeddings (CHERRY TREE vs. MACHINE) of the same system structure with respect to subjects' knowledge acquisition strategies. They report that the semantically rich embedding seem to prevent the problem solvers from using efficient analytic knowledge acquisition strategies. Thus, task semantics influence subjects' decision making process. It would be interesting to explore whether an ecological embedding of a task promotes decision making and learning in dynamic tasks.

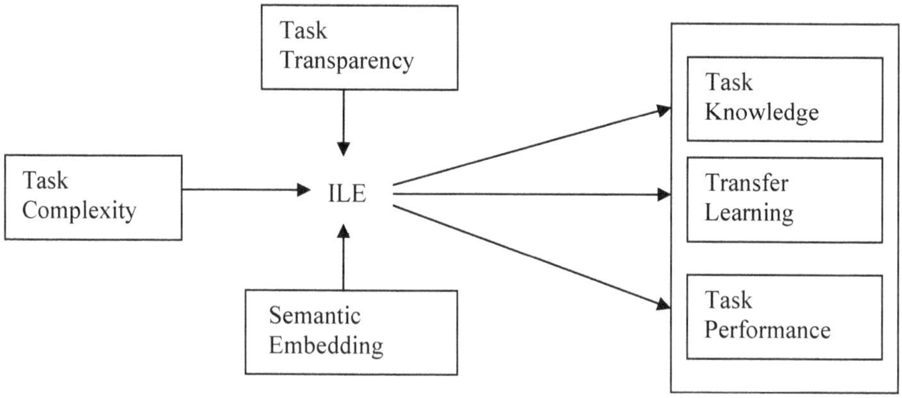

Figure 2. A Model of the Effects of Decision Task Factors

Some indicators of task complexity include total number of variables, interaction between subsystems, random variation, miscellaneous task characteristics, positive feedback and gains, lagged effects, decision effectiveness, and frequency of oscillations (Diehl & Sterman, 1995; Hsiao, 2000; Moxnes, 1998; Paich & Sterman, 1993; Sterman, 1989a)

Mackinnon and Wearing (1980), using a welfare administration model, examine the impact of total number of variables, interaction between subsystems, and random variation on task performance. Their results show that an increase in the total number of variables and random variation deteriorates the subjects' task performance. However, contrary to their hypothesis, subjects performed better when interaction between subsystems existed. On the other hand, research on system dynamics (e.g., see in Paich & Sterman, 1993) suggests that negative feedback loops can stabilize system behavior through the interaction between subsystems.

The pioneering work of Sterman (1989a, 1989b), the misperception of feedback (MOF) hypothesis, attributes the decision makers, failure to manage dynamic tasks to their inability to identify endogenous positive feedback gains enlarging apparently tiny decision errors and side effects. Many researchers (Diehl & Sterman, 1995; Paich & Sterman, 1993; Young, Chen, Wang, & Chen, 1997) have confirmed this hypothesis. It has also been shown that decision time does not increase in proportion to the increasing strength of positive gains. Young et al. (1997), using the microworld STRATEGEM-2, tested whether the decision scope was reduced when decision makers triggered some uncontrollable positive feedback loops. They reported strong evidence for the hypothesis.

Sterman (1989a) reports two facets of subjects' failure to appreciate time delays. Firstly, they ignore the time lag between the initiation of a control action and its full effect. Secondly, they are overly aggressive for correcting the discrepancies between desired and actual state of the variable of interest. Logically, the same failure to appreciate the delayed effect of decisions also applies for counter-correction because subjects fail to understand the full effect of their previous discrepancy correction. Substantial confirmatory evidence to the detrimental effect of time delays on task performance comes from empirical studies adopting various experimental settings (e.g., Berry and Broadbent, 1988; Brehmer, 1990, 1995; Diehl & Sterman, 1995; Moxnes, 1998; Paich & Sterman, 1993; Sterman, 1989a).

In summary, Sterman's famous MOF hypothesis – human decision makers, generally, are dynamically deficient in dealing with decision tasks-lends support mainly from the inability of decision makers to identify endogenous positive feedback gains and to appreciate the delayed effects of decisions. How to deal with the misperception of feedback becomes the key question to be addressed in the proposed conceptual model to be presented in the next section.

2.5. A Model of the Effects of Decision-making Environment Factors

Figure 3 shows the major factors of a decision-making environment influencing decision makers' performance in ILEs.

Subjects with active intervention perform better in task control performance but poorly on knowledge verbalization measure (Funke and Muller, 1988). Interestingly, the passive observers, who are poor in control performance, show improved performance on system knowledge. Berry (1991) finds that learning by observation, both knowledge acquisition and control performance, is possible when the task is changed from a task with non-salient relations to a task with salient relations among the system variables. Thus, besides the heuristic strategies, task salience appears to be another way that might help reduce MOF and consequently improve subjects' decision making and learning in ILEs. Our proposed model, to be presented in the next section, will take into account these findings.

On the effect of time pressure on DDM behavior, a general speedup of information processing and a deterioration of task performance is reported (Kerstholt, 1994). When people face deteriorating system performance, depending upon information costs, they make suboptimal strategy choices (Kerstholt, 1996). Thus, Kerstholt suggests that subjects' reactions to time pressure depend on the absolute costs of information.

DDM literature identifies three types of information feedback - feed forward, outcome feedback, and cognitive feedback: Feed forward refers to a set of pre-task heuristics, available to the decision-makers, for effectively performing the task (Bjorkman, 1972); outcome feedback pertains to the provision of past decisions and outcomes to the subjects (Sterman, 1989a, b); and cognitive feedback is conceptualized as information reflecting task structure and the decision-making environment (Benbasat & Dexter, 1985).

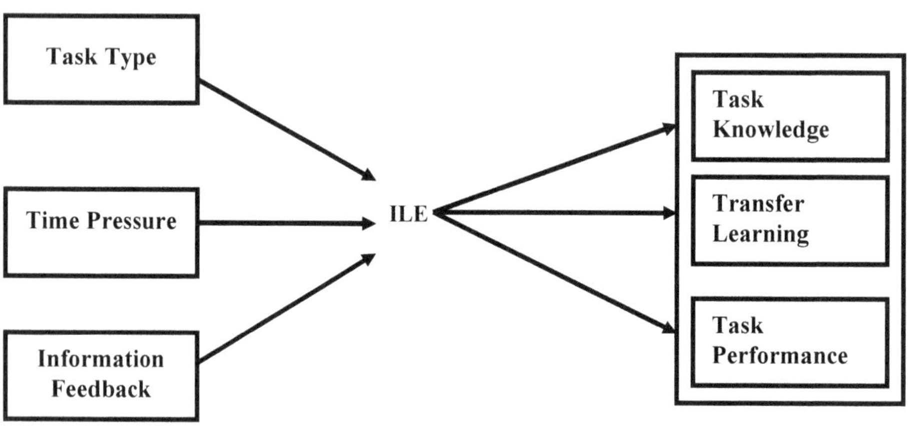

Figure 3. A Model of Effects of Decision-making Environment Factors

It has been argued that outcome feedback permits the decision makers to adjust to the general direction of judgment through judgment-action-feedback loop (Hogarth, 1981). Kleinmuntz (1985) argues that availability of Bayesian probability helps subjects with task performance. Sanderson (1989), on the other hand, supports that making previous decisions and outcomes available to subjects would prevent them from developing correct task knowledge. Other studies show similar dysfunctionalities in performance when subjects are exposed to repeated trials even with minimal delays in feedback (Brehmer, 1990) and are presented with complete decisions and outcomes (Sterman, 1989a,b).

In the context of a software development project task, the incremental efficacy of cognitive feedback and feed forward over outcome feedback in improving task performance is demonstrated in the experimental work by Sengupta and Abdel-Hamid (1993). The subjects receiving outcome feedback alone have shown inferior task performance while addition of cognitive feedback has improved their task performance in the complex software project task. It has also been shown that the provision of benchmark outcome improves task performance (Hsiao, 2000).

Contrary to MOF hypothesis, decision makers with decision aids in the form of feedback may develop adequate task system model (Conant & Ashby, 1970) and hence perform better in dynamic tasks. Our proposed model, to be presented in the next section, will further expand on this argument.

2.6. Studies on Facilitation

When learning is considered as a progression towards expertise (e.g., as in Sternberg 1995), the human facilitation becomes critical (Cox, 1992; Crookall et al., 1987; Davidsen & Spector, 1997; Goodyear, 1992; Wolf, 1990). Davidsen and Spector (1997) analyze the successful uses of system dynamics based learning environments. They find that many of the successful ILEs depend on the effective pre-task preparations and instructions by the human facilitator. More importantly, learning effects appear highly dependent on debriefing discussions. Elsom-Cook (1993) argues that the key role of the facilitator is to facilitate the "institutionalization of knowledge". Learners can have many experiences with ILEs. Initially, they have no way to know which experiences are important and useful for real-world situations. Similar concerns have been echoed in the *"assimilation paradox"* (Briggs, 1990). Assimilation paradox asserts that self-directed learners, in the absence of help and guidance, face difficulties in assimilating the new knowledge with the existing knowledge and mental models. Debriefing reviews appear to help learners overcome these difficulties and update their mental models (Ledrman, 1992).

Using the business simulator LEARN!, Gröbler et al. (2000) conducted an experiments to investigate the impact of transparency on learning in complex dynamic environment. They operationalized transparency in terms of provision of structural information about the underlying task system. They report a support for presentation by the facilitator- a pre-task level support.

Qudrat-Ullah (2002), using FISHBANKILE, investigated the effects of human facilitation on task performance and transfer learning. He finds that the subjects provided with post-task facilitation perform the best, followed by those provided with in-task facilitation. Subjects provided with pre-task facilitation perform the least.

In summary, learners, in ILEs and in the absence of human facilitation face difficulties in assimilating the new knowledge with the existing mental models. Debriefing sessions and exercises have been shown to induce transferable learning effects. Reflection on the experiences gained during the simulation session (s) may help the learners to avoid the so-called "video arcade syndrome": people can win the game but without having any clue about how (Bakken, 1993).

3. The Promise of the Future

Despite an overwhelming support for MOF hypothesis, prior research on DDM provides a substantial evidence to the effectiveness of various decisional aids in improving DDM. The misperception of feedback perspective concludes that people perform poorly in dynamic tasks because they ignore time delays and are insensitive to feedback structure of the task system. Contrary to MOF hypothesis, an objective scan of real world decisions would suggest that experts can deal efficiently with highly complex dynamic systems in real life, for example, maneuvering a ship through restricted waterways (Kerstholt & Raaijmakers, 1997). This example suggests that people are not inherently incapable of better performance in dynamic tasks. Instead, decision makers need to acquire the requisite expertise.

Turning to the future, the most fundamental research question for DDM research seems to be: How to acquire prototypic expertise in dynamic tasks? A solution to this question would effectively provide a way to improve DDM. We propose a potential solution as shown in Figure 4. Our conceptual model presents an integrated approach to improve DDM. Specifically, it draws on (i) decisional aids, and (ii) human computer interface (HCI) design principles.

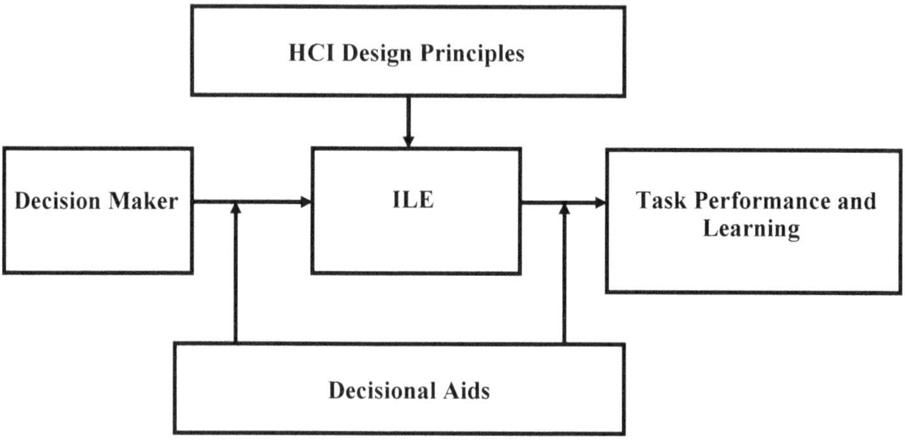

Figure 4. The Conceptual Model: How to Improve Dynamic Decision Making?

3.1. Decisional Aids

The surveyed studies provide a substantial evidence to the effectiveness of various decisional aids (e.g., task experience, task salience, task transparency, feedback, decision heuristics, debriefing, and human facilitation) in promoting DDM in ILEs. Moreover, all factors except task experience appear to be either a direct (e.g., feedback, decision heuristics, debriefing, and human facilitation) or an indirect (e.g., task salience and task transparency) function of human decisional aids. For example, a presentation by the human facilitator about the structure of the task system results in task salience (Gröbler et al., 2000). Therefore, decisional aid in the form of human facilitation constitutes the distinguishing and fundamental component of our model. In an ILE session, decisional aids can be provided at three levels: pre-, in-, and post-task levels. Pre-task level decisional aids can be conceptualized as information provided by the human facilitator to a decision maker about the model of the task prior to performing the task (Davidsen & Spector, 1997; Grobler et al., 2000). In-task decisional aids attempt to improve the individuals' decision-making performance by (i) making the task goals explicit at early stages of learning, (ii) helping them keep track of goals during the task, and (iii) providing them with 'diagnostic information' (Cox, 1992). Post-task level decisional aids aim at improving performance by providing the decision-makers an opportunity to reflect on their experiences with the task (Cox, 1992; Davidsen & Spector, 1997). Qudrat-Ullah (2002) found that the subjects provided with post-task facilitation perform the best, followed by those provided with in-task facilitation. Pre-task facilitation subjects perform the least.

3.2. HCI Design Principles

An interest in and the successes of the HCI design in disciplines such as management information system, information science, and psychology, all sharing the common goal of improving the organizational decision making, are huge (Carey et al., 2004). However, despite the fact that all the ILEs have to have an HCI element, the role of HCI design in improving DDM has received little attention from the DDM researchers. Only recently, Howie et al. (2000) and Qudrat-Ullah (2001) have directed DDM research in this dimension.

Howie et al. (2000) has emprically investigated the impact of an HCI design-based on human factors guidelines on DDM. The results reveal that the new interface designs based on human-computer interaction principles lead to improved performance in dynamic tasks as compared to the original interfaces. Qudrat-Ullah (2001) studied the effects of an HCI design based on learning principles (Gagné, 1995) on DDM. The results show that the ILE with HCI design based on learning principles versus the traditional ILE-used in studies supporting MOF hypothesis, is effective on improving the subjects, task performance and learning in the dynamic task. We suspect the lack of focus on HCI design principles are the root cause of the subjects' poor performance in the earlier studies on DDM supporting MOF hypothesis.

4. Concluding Remarks

DDM research is highly relevant to the managerial practice (Huber, 1984). Improving decision making and learning in dynamic environments presents a major challenge to researchers. This chapter provides a first thrust towards the accomplishment of this objective by theoretically integrating DDM and learning in ILEs literature. Additionally, this chapter makes a case for the inclusion of (i) decisional aids in the form of human facilitation and (ii) HCI design principles in any ILE model aimed at improving decision making in dynamic tasks. The model can serve as the foundation for future experimental-based empirical research. Departing from traditional DDM research focus on how poorly subjects perform in dynamic tasks, our model attempts to increase our understanding of the way in which expertise on DDM are acquired.

Contrary to MOF hypotheses, we believe that the lack of emphasis on human facilitation and HCI design principles resulted, at least in part, in poor performance in dynamic tasks by people. In future studies, by utilizing our model, one can compare results, build a cumulative knowledge base of effective decisional aids and HCI design principles, and study the trade-offs among various kinds of decisional aids to DDM. Future studies should also be designed to re-assess the earlier studies on DDM supporting MOF hypotheses. Pursued vigorously and systematically, research on decisional aids such as human facilitation and HCI design principles should be beneficial to both DDM researchers and practitioners.

References

Bakken, B. E. (1993). *Learning and transfer of understanding in dynamic decision environments.* Unpublished doctoral dissertation. MIT, Boston.

Bakken, B., Gould, J. & Kim, D. (1994). Experimentation in learning organizations: A management flight simulator approach. In J. Morecroft & J. Sterman (Eds.) *Modeling for learning organizations* (pp. 243–266). Portland, Or.: Productivity Press.

Beckmann, J. F. & Guthke, J. (1995). Complex problem solving, intelligence, and learning ability. In P. Frensch & J. Funke (Eds.), *Complex problem solving: The European perspective* (pp. 177–200). NJ: Lawrence Erlbaum Associates Publishers.

Benbasat, I. & Dexter, A. S. (1985). An experimental evaluation of graphical and color-enhanced information presentation. *Management Science*, ***31***, 1348–1364.

Berry, D. C. & Broadbent, D. E. (1987). The combination of explicit and implicit learning processes in task control. *Psychological Research*, ***49***, 7–16.

Berry, D. C. & Broadbent, D. E. (1988). Interactive tasks and the implicit-explicit distinction. *British Journal of Psychology*, ***79***, 251–271.

Berry, D. C. (1991). The role of action in implicit learning. *Quarterly Journal of Experimental Psychology*, ***43A***, 8–906.

Bjorkman, M. (1972). Feed forward and feedback as determinants of knowledge and policy: Notes on neglected issue. *Scandinavian J. Psychology*, ***13***, 152–158.

Brehmer, B. & Allard, R. (1991). Dynamic decision making: The effects of task complexity and feedback delay. In J. Rasmussen, B. Brehmer & J. Leplat (Eds.) *Distributed decision making: Cognitive models for cooperative work* (pp. 319–334). New York: Wiley.

Brehmer, B. (1990). Strategies in real-time dynamic decision *making*. In R. M. Hogarth (Ed.), *Insights in decision making* (pp. 262–279). Chicago: University of Chicago Press.

Brehmer, B. (1995). Feedback delays in complex dynamic decision tasks. In P. Frensch & J. Funke (Eds.) *Complex problem solving: The European perspective* (pp. 103–130). NJ: Lawrence Erlbaum Associates Publishers.
Breuer, K. & Kummer, R. (1990). Cognitive effects from process learning with computer-based simulations. *Computers in Human Behavior*, **6**, 69–81.
Briggs, P. (1990). Do they know what they are doing? An evaluation of word-processor user's implicit and explicit task-relevant knowledge, and its role in self-directed learning. *International Journal of Man-Machine Studies*, **32**, 385–298.
Broadbent, B. & Aston, B. (1978). Human control of a simulated economic system. *Ergonomics*, ***21***, 1035–1043.
Carey, J., Galletta, D., Kim, J., Te'eni, D., Wildemuth, B., & Zhang, P. (2004). The role of human-computer interaction in management information systems curricula: A call to action. *Communications of the Association for Information Systems*, ***13***, 357–379.
Collins, A. (1991). Cognitive apprenticeship and instructional technology. In L. Idol & B. F. Jones (Eds.) *Educational values and cognitive instruction: Implication for reform* (pp. 11–139). New York: Elsevier Science.
Conant, R. & W. Ashby. (1970). Every good regulator of a system must be a model of the system. *International Journal of System Science*, ***1***, 89–97.
Cox, R. J. (1992). Exploratory learning from computer-based systems. In S. Dijkstra, H. P. M. Krammer, & J. J. G. van Merrienboer (Eds.) *Instructional models in computer-based learning environments* (pp. 405–419). Berlin, Heidelberg: Springer-Verlag.
Crookall, D., Martin, A., Saunders, D., & Coote, A. (1987). Human and computer involvement in simulation. *Simulation & Games: An International Journal of Theory, Design, and Research*, **17**(3), 345–375.
Davidsen, P. I. & Spector, J. M. (1997). *Cognitive complexity in system dynamics based learning environments*. International system dynamics conference. Istanbul, Turkey: Bogacizi University Printing Office.
Davidsen, P. I. (1996). Educational features of the system dynamics approach to modelling and simulation. *Journal of Structural Learning*, ***12***(4), 269–290.
Davidsen, P. I. (2000). Issues in the design and use of system-dynamics-based interactive learning environments. *Simulation & Gaming*, ***31***(2), 170–177.
Davis, F. D. & Kottermann, J . E. (1994). User perceptions of decision support effectiveness: Two productions planning experiment. *Decision Sciences*, ***25***(1), 57–78.
Davis, F. D. & Kottermann, J . E. (1994). Determinants of decision rule use in a production planning task. *Organizational Behavior and Human Decision Processes*, ***63***, 145–159.
Diehl, E. & Sterman, J. D. (1995). Effects of feedback complexity on dynamic decision making. *Organizational Behavior and Human Decision Processes*, ***62***(2), 198–215.
Dörner, D. & Pfeifer, E. (1992). Strategic thinking, strategic errors, stress, and intelligence. *Sprache & Kognition*, ***11***, 75–90.
Dörner, D. (1980). On the difficulties people have in dealing with complexity. *Simulations and Games*, ***11***, 8–106.
Dörner, D., Kreuzig, H. W., Reither, F., & Staudel, T. (Eds.). (1983). *Lohhausen: On dealing with uncertainty and complexity*. Bern, Switzerland: Hans Huber.
Edwards, W. (1962). Dynamic decision theory and probabilistic information processing. *Human Factors*, ***4***, 59–73.
Elsom-Cook, M. T. (1993). Environment design and teaching intervention. In D. M. Town, T. de Jong, & H. Spada. (Eds.) *Simulation-based experiential learning* (pp. 165–176). Berlin: Springer-Verlag.
Forrester, J. W. (1961). *Industrial dynamics*. Cambridge, MA: Productivity Press.

Funke, J. (1995). Experimental Research on Complex Problem Solving. In P. Frensch & J. Funke (Eds.) *Complex problem solving: The European perspective* (pp. 3–25). NJ: Lawrence Erlbaum Associates Publishers.

Funke, J. & Muller, H. (1988). Active control and prediction as determinants of system identification and system control. *Sprache & Kognition, 7*, 176–186.

Gagné, R. M. (1985). *The conditions of learning* (4th edition). New York: Holt, Rinehart, & Winston.

Gibson, F. P. (2000). Feedback delays: How can decision makers learn not to buy a new car every time the garage is empty? *Organizational Behavior and Human Decision Processes,* ***83***, 141–166.

Gonzalez, M., Machuca, J., & Castillo, J. (2000). A transparent-box multifunctional simulator of competing companies. *Simulation & Gaming,* ***31***(2), 240–256.

Goodyear, P. (1992). The provision of tutorial support for learning with computer-based simulations. In E. Corte, M. Lin, H. Mandal, & L. Verschaffel. (Eds.) *Computer-based learning environments and problem solving* (pp. 391–409). Berlin: Springer-Verlag.

Gröbler, A, Maier, F. H., & Milling, P. M. (2000). Enhancing learning capabilities by providing transparency in transparency. *Simulation & Gaming,* ***31***(2), 257–278.

Gröbler, A. (1998). *Structural transparency as an element of business simulators.* Paper presented at 16th International System Dynamics Conference, Quebec City, Canada.

Grubler H., Renkal A., Mandal H., & Reiter, W. (1993). Exploration strategies in an economics simulation game. In D. M. Town, T. de Jong, & H. Spada. (Eds.) *Simulation-based experiential learning* (pp. 225–233). Berlin: Springer-Verlag.

Hayes, N. A. & Broadbent, D. E. (1988). Two modes of learning for interactive tasks. *Cognition,* ***28***, 249–276.

Hogarth, R. M. (1981). Beyond discrete biases: Functional and dysfunctional aspects of judgmental heuristics. *Psychological Bulletin,* ***9***(2), 197–217.

Homer, J. B., Hirsch, G.B., 2006. System dynamics modeling for public health: Background and opportunities. *American Journal of Public Health,* ***96*** (3), 452–458.

Hsiao, N. (1999). *In search of theories of dynamic decision making: A literature review*. Paper presented at the International System Dynamics Conference, Wellington, New Zealand.

Hsiao, N. (2000). *Exploration of outcome feedback for dynamic decision making.* Unpublished doctoral dissertation, State University of New York at Albany, Albany.

Huber, G. P. (1984). The nature and design of post-industrial organization. *Management Science,* ***30***, 928–951.

Huber, O. (1995). Complex problem solving as multistage decision making. In P. Frensch & J. Funke (Eds.) *Complex problem solving: The European perspective* (pp. 151–173). NJ: Lawrence Erlbaum Associates Publishers.

Issacs, W. & Senge, P. (1994). Overcoming limits to learning in computer-based learning environments. In J. Morecroft and J. Sterman (Eds.) *Modeling for learning organizations* (pp. 267–287). Portland, Or.: Productivity Press.

Jansson, A. (1995). Strategies in Dynamic Decision making: Does Teaching Heuristic Strategies By Instructors Affect Performance? In J. Caverni, M. Bar-Hillel, F.Barron, & H. Jungermann (Eds.) *Contributions to decision making-I* (pp. 213–253). Amsterdam: Elsevier.

Kerstholt, J. H. (1994). The effect of time pressure on decision-making behavior in a dynamic task environment. *Acta Psychologica,* ***86***, 89–104.

Kerstholt, J. H. (1996). The effect of information costs on strategy selection in dynamic tasks. *Acta Psychologica,* ***94***, 273–290.

Kerstholt, J. H. & Raaijmakers, J. G. W. (1997). Decision making in dynamic task environments. In R. Ranyard, R. W. Crozier & O. Svenson (Eds.) *Decision making: Cognitive models and explanations* (pp. 205–217). New York, NY: Routledge.

Keys, J. B. & Wolfe, J. (1990). The role of management games and simulations in education and research. Journal of Management, ***16***, 307–336.

Kleinmuntz, D. (1985). Cognitive heuristics and feedback in a dynamic decision environment. *Management Science*, *31*, 680–701.

Kleinmuntz, D. & Thomas, J. (1987). The value of action and inference in dynamic decision making. *Organizational Behavior and Human Decision Processes*, ***39***, 341–364.

Kriz, W. C., 2003. Creating effective learning environments and learning organizations through gaming simulation design. *Simulation & Gaming*, ***34***(4), 495–511.

Lane, D. C. (1995). On a resurgence of management simulations and games. *Journal of the Operational Research Society*, *46*, 604–625.

Langley, P. A. & Morecroft, J. D. W. (1995). Learning from microworlds environments: A summary of the research issues. In G. P. Richardson & J. D. Sterman (Eds.) *System Dynamics' 96*. Cambridge, MA: System Dynamics Society.

Ledrman, L. C. (1992). Debriefing: towards a systematic assessment of theory and practice. *Simulations & Gaming*, ***23***(2), 145–160.

Machuca, J. A. D., Ruiz, J. C., Domingo, M. A., & Gonzalez, M. M. (1998). 10 years of work on transparent-box business simulation. Paper presented at 16th International System Dynamics Conference, Quebec City, Canada.

Mackinnon, A. J. & A.J. Wearing. (1980). Complexity and decision making. *Behavioral Science*, ***25***(4), 285–292.

Maxwell, T. A. (1995). *Decisions: Cognitive styles, mental models, and task performance.* Unpublished doctoral dissertation, State University of New York at Albany, Albany.

Moxnes, E. (2000). Not only the tragedy of the commons: Misperceptions of feedback and policies for sustainable development. *System Dynamics Review*, ***16*** (4): 325–348.

Moxnes, E. (1998). Not only the tragedy of the commons: Misperceptions of bioeconomics. *Management Science*, ***44***, 1234–1248.

Njoo, M. & de Jong, T. (1993). Supporting exploratory learning by offering structured overviews of hypotheses. In D. M. Town, T. de Jong, & H. Spada. (Eds.) *Simulation-based experiential learning* (pp. 207–223). Berlin: Springer-Verlag.

Paich, M & J. D. Sterman. (1993). Boom, bust, and failures to learn in experimental markets. *Management Science*, ***39***, 1439–1458.

Putz-Osterloh, W. Bott, B., & Koster, K. (1990). Modes of learning in problem solving – Are they transferable to tutorial systems. *Computers in Human Behaviors*, ***6***, 83–96.

Qudrat-Ullah, H. 2005a. "Improving Dynamic Decision Making through HCI Design Principles", In C. Ghaoui, (Ed.), *The Encyclopedia of Human Computer Interaction*, USA: Information Science Publishing, pp. 311–316.

Qudrat-Ullah, H., 2005b. MDESRAP: a model for understanding the dynamics of electricity supply, resources, and pollution. *International Journal of Global Energy Issues*, ***23***(1), 1–14.

Qudrat-Ullah, H. (2002). *Decision making and learning in complex dynamic environments*. Unpublished doctoral dissertation, National University of Singapore, Singapore.

Qudrat-Ullah, H. (2001). *Improving Decision Making and Learning in Dynamic Task: An Experimental Investigation.* Paper presented at Annual Faculty Research Conference, NUS Business School, National University of Singapore, Singapore.

Rapoport, A. (1975). Research paradigms for studying dynamic decision behavior. In D. Wendt & C. A. J. Vlek (Eds.) *Utility, probability, and human decision making* (pp. 349–375). Dordrecht, Holland: Reidel.

Sanderson, P. M. (1989). Verbalizable knowledge and skilled task performance: Association, dissociation, and mental Model. *Journal of Experimental Psychology: Learning Memory and Cognition*, ***15***, 729–739.

Sengupta, K & Abdel-Hamid, T. (1993). Alternative concepts of feedback in dynamic decision environments: An experimental investigation. *Management Science*, ***39***, 411–428.

Spector, J. M. (2000). System dynamics and interactive learning environments: Lessons learned and implications for the future. *Simulation & Gaming*, ***31***(4), 528–535.

Stanley, W. B., R. C. Mathews, R. R. Buss, & S. Kolter-Cope. (1989). Insight without awareness: On the interaction of verbalization, instruction and practice in a simulated process control task. *The Quarterly Journal of Experimental Psychology*, ***41A***(3), 553–577.

Stienwachs, B. (1992). How to facilitate a debriefing. *Simulation & Gaming*, ***23***(2), 186–195.

Sterman, J. D. (1989a). Modeling managerial behavior: Misperceptions of feedback in a dynamic decision making experiment. *Management Science* , ***35***, 321–339.

Sterman, J. D. (1989b). Misperceptions of feedback in dynamic decision making. *Organizational Behavior and Human Decision Processes*, ***43***, 301–335.

Sterman, J. D. (1994). Learning in and around complex systems. *System Dynamics Review*, ***10*** (2–3), 291–323.

Sternberg, R. J. (1995). Expertise in complex problem solving: A comparison of alternative conceptions. In P. Frensch & J. Funke (Eds.) *Complex problem solving: The European perspective* (pp. 3–25). NJ: Lawrence Erlbaum Associates Publishers.

Trees, W. S., Doyle, J. K., & Radzicki, M. J. (1996). *Using cognitive styles typology to explain differences in dynamic decision making in a computer simulation game environment*. Paper presented at the International System Dynamics Conference.

Wolf, J. (1990). The evaluation of computer-based business games. In J. Gentry (Ed.) *Guide to business gaming and experiential learning* (pp. 279–300). London: Nichols.

Yang, J. (1997). Give me the right goals, I will be a good dynamic decision maker. *International System Dynamics Conference* (pp. 709–712). Istanbul, Turkey: Bogacizi University Printing Office.

Young, S. H., Chen, C. P., Wang, S. & Chen, C. H. (1997). An experiment to study the relationship between decision scope and uncontrollable positive feedback loops. *International System Dynamics Conference* (pp. 15–20). Istanbul, Turkey: Bogacizi University Printing Office.

Chapter 2

Expertise and Dynamic Tasks

J. Michael Spector
Learning Systems Institute
Florida State University
mspector@lsi.fsu.edu

1. Introduction

Life in the information age is becoming increasingly complex. New technologies are introduced into the workplace on a daily basis. Local economies are increasingly affected by global market situations. People are moving about and interacting with others in ways never before seen. It is no surprise that serious and challenging problems have arisen unlike any witnessed before. To sustain progress on this fragile planet, society must design new systems to manufacture safe and useful products, promote sustainable agricultural development, ensure that public health standards are adequate, protect the environment, resolve political disputes, respond to crises, educate people for life in the 21st century, and so much more. In short, people must become better at understanding and coping with complex, dynamic problems. This chapter presents an overview of the requirements of some of the most challenging problems, introduces the concept of developing expertise in solving complex and dynamic problems, and discusses the challenges of assessing progress of learning and performance in these areas.

The specific problem addressed herein is the difficulty of determining progress of learning and development of expertise in domains involving dynamic, ill-structured problems. Ill-structured problems are those lacking well-defined and complete specification of desired outcome states, input conditions, problem constraints, and processes involved in transforming input conditions into desired output states (Rittel & Weber, 1973; Simon, 1973). Such problems often involve many interrelated factors with time-dependent and non-linear relationships (e.g., delayed effects, singularities, tipping points, etc.), such that the problem situation changes over time in ways that are not easily predicted or anticipated (Dörner, 1996; Forrester, 1987; Sterman, 1994).

Examples of such problems exist in many domains, ranging from engineering design to public health policy. Engineering examples include: (a) designing heat shielding for the space shuttle to allow safer re-entries with greater payloads; (b) developing new transportation systems that reduce pollution, improve access and maintain public safety; and (c) designing Web-accessible databases that provide security, redundancy and ease of use. Examples in the area of public health include: (a) controlling the spread of viruses, (b) responding to medical emergencies involving multiple victims, and, (c) helping patients with histories of chronic pain and allergies.

In many ill-structured problem solving situations, the desired outcome state may not be well-defined and is likely to change as the problem situation evolves. In some cases, current conditions and problem constraints are not completely specified or known, although these are likely to become better understood as the situation evolves. The effects of specific actions and decisions are rarely understood in advance; unanticipated side effects are common when trying to resolve complex problems. Key aspects of such problems include:

- *dynamics* – the problem situation changes as circumstances change.
- *expertise* – what is perceived to be challenging and difficult by a person with little training and experience may be perceived as easy and straightforward by a highly experienced person.
- *collaboration* – complex, ill-structured problems are typically resolved by teams rather than by individuals; the collective and changing knowledge and experience of team members is relevant to team performance.
- *criticality* – time is often critical to successful resolution of a complex problem; in addition to time criticality, there may also be critical life and death issues or other critical aspects to success or failure.

Complex, ill-structured problems may be considered *wild* and *wicked*; they are quite different from typical classroom and textbook problems. Professional training programs may be aimed at such problems. Designing effective training requires: (a) a method to assess learning and performance that recognizes alternative problem approaches and conceptualizations since standard solutions do not exist; (b) metrics to determine levels of performance and expertise that provide a basis for assessing problem-solving abilities, designing training programs and providing meaningful feedback; and, (c) support for collaborative teams engaged in critical problem-solving tasks.

After a discussion of the first two of these aspects of ill-structured problems – dynamics and expertise, a methodology based on causal concept maps to assess progress of learning and relative levels of performance will be presented. The chapter will conclude with a discussion of significant challenges and open issues.

2. Dynamic Tasks

The kinds of problems described above as wild and wicked can be collectively considered *dynamic tasks*. The circumstances surrounding such problem situations are subject to change as the problem evolves, so there are dynamics associated with time-dependent time factors. For example, in a situation involving an epidemic, whether or not an individual or group who may be infected display symptoms will depend in part on the time since infection. At time T_1 it may be the case that X% show symptoms of infection. At time T_2 this percentage will change even if no new persons become infected. Other aspects of a problem situation are also subject to change, making developing and implementing a solution particularly challenging.

However, it is not only the circumstances that impact a problem situation that are subject to change. Changing circumstances may be such that the nature of the problem situation is itself changed. That is to say that the structure of the problem may change or the system in which the problem is set may change in significant ways. For example, consider the use of genetically modified seeds in agriculture. Research suggests that there is a possibility for a modified gene to jump to natural relatives growing nearby by pollination and other means. Quist and Chapela (2001) report and analyze widespread existence of transgenic DNA *introgressed* into traditional maize in Mexico. There is a heated debate about the benefits and risks associated with genetically modified crop cultivation. As a dynamic system, major factors involved in the problem include the genetically modified crop, the natural crop, the introgressed crop, natural weeds, (potential) transgenic weeds, natural pests, and (potential) transgenic pests. It is known that introgression is possible from modified seeds to natural seeds. If a new transgenic seed emerges in the wild, it is ecologically plausible that in time it would stimulate emergence of its own pest. There is also a small possibility that a gene jump from genetically modified actors can act in a nonlinear fashion and involve multiple feedbacks and delays. This system can potentially produce complex dynamic behaviors, including the invasion of the entire field by super-resistant transgenic weed or the inability of the transgenic crop to survive in the wild in the long run or the domination of the entire field by the transgenic crop.

A final aspect of problem and task dynamics involves the perceptions of those involved in the problem situation. What people know and are able to do in response to challenging problems changes as well. Typically, as a dynamic problem situation evolves, those involved in responding to the problem will learn new and relevant things to take into consideration. For example, in responding to an emergency medical situation, the caregiver may be treating a person for bleeding and a broken bone, and then discover a new and potentially more serious problem. During a crisis situation, those managing the crisis situation frequently uncover new problems requiring immediate response. Moreover, they gain information and problem understanding that may change how they think about a solution. Finally, problem solvers also learn what to expect from their team members, which may also affect how the problem situation is approached. For example, if it becomes known that one individual or group cannot manage a particular task approach very well, that task may be directed to others ,or teams may re-composed, or perhaps new approaches or tools may be introduced.

In summary, the complex, ill-structured tasks that are the focus of this discussion may be considered dynamic as a result of changing circumstances, evolving systems and structural considerations, altered understanding and perceptions of situations and those involved, or any combination of these. It is clear that the large problems mentioned at the outset (agricultural development, environmental planning, public health, etc.) are subject to much change and variation, and that effective responses should take these changes into account.

3. The Nature of Expertise

3.1. Development of Expertise

Studies pertaining to the development of expertise suggest that it takes a great deal of focused and deliberate practice to achieve superior levels of performance (Ericsson, 2004). Practice does not always make one better; moreover, people often require ten or more years of experience before reaching the highest levels of performance. This finding occurs in a number of domains ranging from chess to medical diagnosis. It is noteworthy that these domains have recognized means to establish superior performance and that they are what Jonassen (1997; 2004) would characterize as planning or diagnosis problems.

Researchers argue that deliberate practice can account for differences in long term working memory and superior performance (Ericsson, 2004; Ericsson & Simon, 1993; Ericsson & Smith, 1991). Deliberate practice consists of individuals identifying difficult tasks or aspects of specific tasks that they do not believe they have adequately mastered; they then devote years of practice to improving performance on these tasks; upon mastery, they move on to other challenging tasks. Such deliberate practice requires high levels of motivation, dedication and awareness as well as skill and self-discipline. Years of experience alone will not necessarily result in the development of expertise, as evidenced by the many amateurs who devote years to a game or sport without significant improvement.

The empirical research that does exist in this area has mostly involved individual planning and diagnosis problems that do not require immediate responses nor communication or coordination with others in response to a representative task problem. The finding that is especially relevant to the approach described below concerns expert representations. Namely, it is possible to develop an assessment

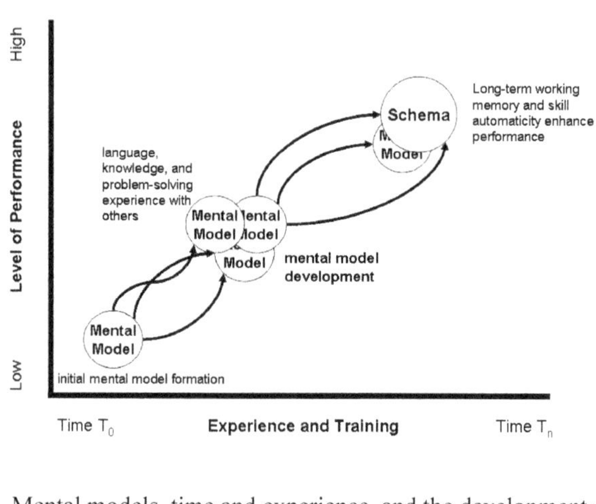

Figure 1. Mental models, time and experience, and the development of expertise

methodology to determine progress of learning and improvement in performance using expert representations along with an individual's history of problem conceptualizations as points of reference (Christensen, Spector, Sioutine, & McCormack, 2000; Davidsen & Spector, 1997; Spector, 2000; Spector, Christensen, Sioutine, & McCormack, 2001; Spector, Dennen, & Koszalka, 2005; Spector & Koszalka, 2004).

3.2. Expert Performance

The challenge to be confronted with regard to dynamic tasks of the kind described above is that measures of performance are inherently problematic. Learning in ill-structured, complex problem-solving situations is not directly measurable by a particular solution or test score or knowledge item. Rather, learning is better treated as an ongoing process in which problem-solvers become increasingly expert-like in their problem-solving activities (Bransford, Brown, & Cocking, 2000; Bruner, 1985; Carin & Bass, 2001). Measures of learning in dynamic task domains should be directed at this process of gradual improvement. The general problem with regard to evaluating learning and performance in dynamic problem-solving task domains is that there is no well-established and reliable assessment methodology.

As a consequence, previous evaluations have focused on: (1) an analysis of efforts and resources required to successfully implement a particular approach (e.g., cost-benefits); and/or, (2) an analysis of the immediate and easily observed effects reported by those involved in such efforts (e.g., student reactions). The former are relevant to the issue of scalability; moreover, they are of particular interest to those interested in larger issues of systemic change and educational reform and innovation (Chang et al., 1998; Dede, 1998). Scalability is a legitimate issue but is not the focus of the assessment methodology outlined below, which focuses on observable effects relevant to and predictive of problem solving ability. There is evidence to suggest that problem-centered approaches to training and performance improvement have a positive effect on learner attitudes and interests and contribute to improved motivation (Merrill, 2002; Novak, 1998; Seel, 2003). However, there has been only indirect or suggestive evidence of improved learning outcomes associated with problem-centered approaches.

Learning outcome measures typically collected for simpler domains (e.g., scores on knowledge tests or performance on highly algorithmic procedural demonstrations) are only indirectly relevant to higher-order problem solving and understanding of complex problems. The limitations of these kinds of evaluation are well known (Herl et al., 1999). One may perform very well on a knowledge test about disease indicators but still not do well with regard to diagnosing particular cases. Attitude, change agency, cost-benefits, and knowledge measures are relevant in an overall evaluation. However, the critical aspect of improved higher order learning is often overlooked because it is difficult to determine what progress students are making in understanding the complexities involved in more advanced problem-solving areas of science and engineering domains (Romiszowski & LeBlanc, 1991).

The central problem remains how to reliably determine improvement in higher order learning and problem solving with regard to dynamic problems., especially with regard to transfer of learning. Such learning is typically supported by a variety of tools and technologies consistent with the learning-with-technology paradigm advocated by many educational researchers (Dede, 1998; Jonassen 2004). Such tools and technologies include, among others, interactive simulations and system dynamics based learning environments. The issue is not which particular technology or approach to use in a given situation; nor is the focus on how to integrate technology effectively in learning and performance improvement. Rather, the issue is how to establish that significant improvement in higher order learning and improved understanding of dynamic task domains has occurred. This determination is vital to assessing progress of learning, and it is necessary for evaluating instructional and performance support systems.

4. Measuring Dynamic Task Performance

Because the dynamic problem tasks typically lend themselves to multiple solutions and solution approaches, it is not possible to directly assess knowledge and performance in comparison with a standard solution. An alternative approach to measurement is to first determine if experts think about the same kinds of thing when approaching a complex problem, and, further, how experts think about the relationships among critical problem factors (Grotzer & Perkins, 2000; Novak, 1998; Schvaneveldt, 1990; Spiro Coulson Feltovich & Anderson, 1998). Figure 2 depicts one way of gathering such information using annotated causal influence diagrams.

The method of presenting individuals with a complex problem situation and asking them to develop an annotated concept to show they were thinking about the problem was developed independently in Germany by Seel and colleagues (Seel, Al-Diban, Blumschein, 2000), in Norway by Spector and colleagues (Davidsen & Spector, 1997; Milrad, Spector, & Davidsen, 2002; Spector, Christensen, Sioutine & McCormack, 2001), and by other researchers (O'Connor & Johnson, 2004; Stoyana & Kommers, 2002; Taricani & Clariana, 2006). These efforts have demonstrated that it is in principle possible to use the problem representations and conceptualizations of experts as a point of reference in determining relative levels of understanding in novice (non-expert) problem solvers.

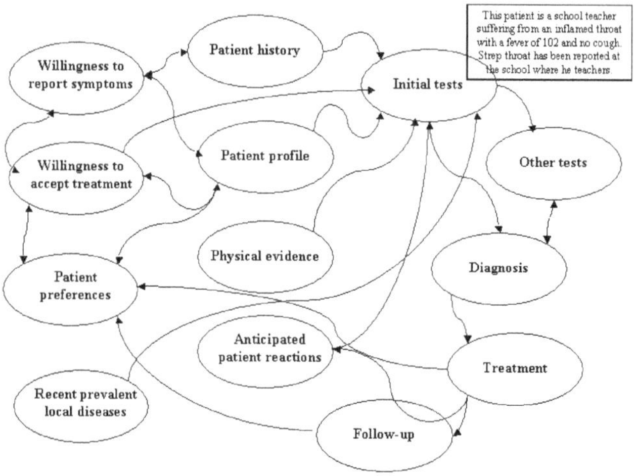

Figure 2. Annotated causal influence diagrams (only one node is annotated in this example)

A study at Syracuse University (Spector & Koszalka, 2004) further developed this approach to measuring relative levels of expertise in complex domains. The study, is called the "Enhanced Evaluation of Learning in Complex Domains (DEEP)" (NSF EREC 03-542), established the feasibility of using annotated concept diagrams to assess progress of learning in complex domains. The DEEP study explored expert and novice problem conceptualizations in environmental planning, engineering design and medical diagnosis. Annotated concept maps, inspired by the causal influence diagrams used by system dynamicists, were used to gather problem conceptualizations from problem respondents. The basic concept was to confirm that experts (those with extensive experience in a domain and strong academic backgrounds) would exhibit clearly recognizable patterns in their problem conceptualizations, although experts did develop a variety of problem responses. Inexperienced persons exhibited recognizably different patterns from those of the experts. Finally, a preliminary set of metrics was developed to determine how far from an expert-like problem conceptualization a non-expert was.

The DEEP study involved the selection of two representative problems for each of three complex domains. Subjects (both expert and non-expert respondents) were provided with a problem scenario (typically about one page in length) and asked to indicate things that may be relevant to a solution. Subjects were asked to document these items and provide a short description of each item along with a brief explanation of how and why it is relevant. They were also asked to indicate and document other assumptions about the problem situation that they are making (initially and again at the end of the activity). Subjects were asked to develop a solution approach to the problem. Required parts of this representation included: (a) a list of key facts and factors influencing the problem situation; (b) documentation of each causal factor – what it is and how it influences the problem; (c) a graphical depiction of the problem situation (see Figure 3; see also http://deep.lsi.fsu.edu/DMVS/jsp/index.htm for online access to the DEEP tool); (d) annotations on the graphical depiction (descriptions of each item - linked to factors already described when possible); (e) a solution approach based on the representations already provided; and, (f) an indication of other possible solution approaches. Figure 3 depicts a screen from DEEP showing how problem conceptualizations were captured. In DEEP, respondents are asked to annotate each node and also each link. The open-ended nature of these annotations makes analysis difficult but it allows respondents to think in any direction and use any vocabulary they desire.

There are three levels of analysis in DEEP. A surface analysis is performed on those things that are easily counted and automated, such as the number of nodes and links, words per annotation, and so on. Table 1 shows a summary of this surface analysis for the problems in the DEEP study. These data suggest that biology experts tend to identify more nodes (critical problem factors) than biology non-experts; however, this pattern does not hold true for medicine and engineering where there were no significant differences in average. Likewise, medical experts tended to provide much lengthier annotations per node than medical non-expert; again, this particular pattern does not appear in the other two domains. In short, analysis of surface features of the problem conceptualizations are useful in distinguishing experts from non-experts within a domain, and they also suggest that experts in different domains tend to have somewhat domain-specific ways of thinking about problems.

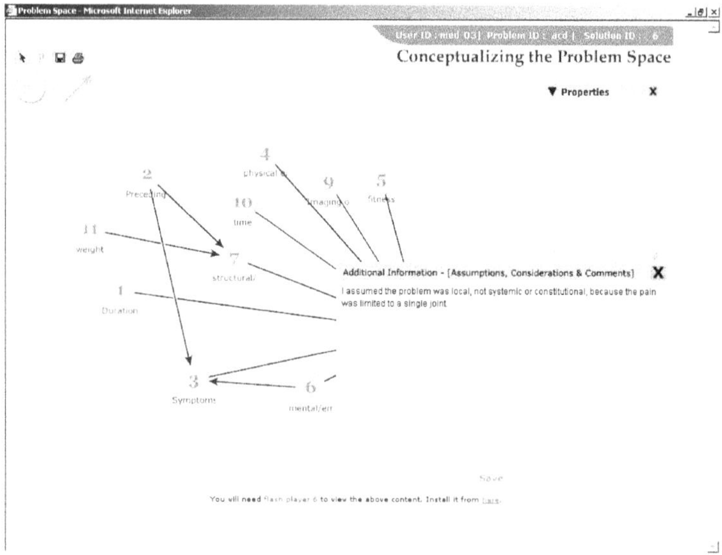

Figure 3. Capturing problem conceptualizations in DEEP

One would hardly be satisfied with such simple metrics, however. In determining how expert a respondent actually is, one would probably want to know not just how many nodes, links and words were used but also how the nodes were linked together and what was said about each node and link. The former kind of analysis is called a structural analysis and the latter a semantic analysis in DEEP. The goal is eventually to be able to automate relevant aspects of the analysis so that dynamic feedback can be provided to learners, instructors and those performing a problem-solving task.

A critical decision was made in the DEEP study not to separate the structural and semantic analyses since this was a validation study only. In order to compare the structures of two different responses, one first has to determine if the nodes in each one are the same or different, regardless of having the same or different node names. In other words, a semantic analysis of the content of each node and link was required in order to understand just what that node or link represented. Then it could be determined to be the same as or similar to a node or link in an expert's problem conceptualization.

The semantic analysis revealed that links, which could be of any type, actually into one of five categories: cause-effect, example-non-example, correlation, process-step, and formula-equation. Table 2 shows the types of links identified by medical experts and non-experts in response to one of the medical scenario. Not surprisingly, no formulas or equations were identified by the medical respondents. At first glance, there may be no noticeable differences, but a close examination shows that medical experts identified a high percentage of process links in comparison with non-experts. In interviews after the study, the reason became obvious. Experienced medical practitioners were familiar with standard diagnostic procedures and used them when appropriate. Non-experts had to reason through the biology and chemistry of the human body to arrive at their representations. In the other domains, experts tended to identify more causal links than did non-experts, so this finding is specific to medical diagnosis.

Table 1. Surface analysis by scenario by domain

Prob. #	Domain	Group	Nodes	One-way links	Two-way links	Words per node	Words per node-name	Words per link
1	Medicine							
		Novice	13.07	16	6.86	15.86	1.83	11.57
		Expert	10.6	17.6	4	26.6	2.24	21.2
	Eng'ing							
		Novice	9	12	2	22.9	2.6	7.25
		Expert	8	8	1.5	11.15	2.55	16
	Biology							
		Novice	7.36	10.29	1.14	39	2	24.86
		Expert	16	20.6	0.6	12.8	1.8	8.8
2	Medicine							
		Novice	12.21	16.36	5	21.5	2.06	15.36
		Expert	12.25	15.5	1.5	31.75	2.35	14.63
	Eng'ing							
		Novice	8.5	8.5	1.5	27.5	3.05	18.05
		Expert	10.5	15.5	2	16.4	2.6	11.75
	Biology							
		Novice	7.13	24.5	1.13	10	4.25	23.94
		Expert	16.8	25.8	1.4	12.2	1.6	12

Table 2. Types of links identified on the second medical scenario

Experts (4)			Novices (14)		
Type of link	No. of links	Percent	Type of link	No. of links	Percent
cause/effect	28	41.2%	cause/effect	137	46.6%
example	0	0.0%	example	27	9.2%
correlation	10	14.7%	correlation	123	41.8%
process	30	44.1%	process	7	2.4%
total links	68	100%	total links	294	100%

Table 3. Most critical nodes for medical novices (Scenario 1; N = 14; Links = 318)

Node	Links	From / To	Percentage
History of present illness	67	31 / 36	21.07%
Hypothesis testing	55	17 / 38	17.30%
Chief complaint	49	14 / 35	15.41%
Patient's background/ social history	45	27 / 18	14.15%
Stress	44	23 / 21	13.84%

Table 4. Most critical nodes for medical experts (Scenario 1; N = 5; Links = 106)

Node	Links	From / To	Percentage
Tests	19	8 / 11	17.93%
Differential diagnosis	16	10 / 6	15.09%
Diagnosis	14	8 / 6	13.21%
History	10	8 / 2	9.43%
Post-visit additional information	10	3 / 7	9.43%
Clinical knowledge	10	7 / 3	9.43%

The structural analysis also demonstrated differences between experts and non-experts and patterns specific to the different problem domains.

Tables 3 and 4 provide the five most crucial nodes identified by novices and experts in response to one of the medical scenarios. Both experts and novices had evidence as critical, but the quantified results of medical tests were the most critical node for experts whereas the qualitative analysis of the patient's illness was the most critical node for non-experts.

It should be noted that the medical novices (non-experts) were interns. In the other domains, the novices or non-experts were upper division university students majoring in engineering or biology. The experts each had more than 10 years work experience and substantial publications in addition to holding a doctorate degree.

The scenarios that were used in this study represented actual problems whose solution was known to the investigators but not to problem respondents (with one exception as one of the biology experts recruited to the study had actually worked on the environmental problem that had been selected for one of the scenarios). Using actual complex problems allowed the investigators to determine how well problem conceptualizations adequately captured factors that proved relevant to the actual solution (Healey, Undre, & Vincent, 2004).

The DEEP methodology appears quite useful and robust, although the software certainly requires refinement. To be scalable and useful in dynamic problem solving situations, it will be necessary to de-couple the structural and semantic analyses. One way of doing this is to provide a list of pre-defined nodes and links, as has been done in the analysis of concept maps (Herl et al., 1999). This approach may provide too much direction to problem respondents, however. A compromise is to allow respondents to drag and drop words and phrases from the problem scenario onto a causal diagram and then also allow open-ended naming of nodes. In any case, it does appear possible to distinguish the relative level of expertise and understanding in complex problem solving domains based on the quality of problem conceptualizations developed by respondents to representative dynamic problem solving tasks.

5. Concluding Remarks

In conclusion, if we wish to make sustained progress in planning and implementing training programs and performance support environments appropriate for the dynamic problems of the 21st century, it is imperative that we take seriously the issue of helping people learn to cope with dynamic problems. Dynamic problem tasks involve ill-structured aspects and individuals with different academic, cultural, and experiential backgrounds. Supporting individuals and organizations in resolving the complex problems that confront society is a significant challenge for educators.

Within the context of learning in and about complex domains, peoples are required to develop flexible processes and techniques for approaching a variety of problem situations. Critical to success in these endeavors is a methodology for assessing progress of learning and improvement in performance. The DEEP methodology presents a promising way to determine and predict relative levels of expertise in ill-structured problem solving situations.

Many significant research questions remain to be explored. Identifying different patterns of responses to different types of problem scenarios is one open task. Determining how experts in different domains thing different and similarly is another area worth investigating. Developing the means to automatically and dynamically analyze problem conceptualizations is a matter of the utmost urgency. One shares what one has learned, and one hopes that with the help of others, improved methods and tools will be developed.

References

Bransford, J. D., Brown, A. L., & Cocking, R. R. (eds.), 2000, *How people learn: Brain, mind, experience and school* (expanded edition), National Academy Press (Washington).

Bruner, J. S., 1985, Models of the learner. *Educational Researcher*, ***14***, 6, 5–8.

Carin, A. A., & Bass, J. E., 2001, *Teaching science as inquiry* (9th ed.), Prentice-Hall (Upper Saddle River).

Chang, H., Honey, M., Light, D., Moeller, B., & Ross, N., 1998, *The Union City story: Education reform and technology: Students' performance on standardized tests*. Education Development Center/Center for Children & Technology (New York).

Christensen, D. L., Spector, J. M., Sioutine, A., & McCormack, D., 2000, *Evaluating the impact of system dynamics based learning environments: A preliminary study*, International System Dynamics Society Conference (Bergen).

Davidsen, P. I., & Spector, J. M., 1997, Cognitive complexity in system dynamics based learning environments. In Y. Barlas, V. G. Diker, & S. Polat (eds.), *Systems Dynamics Proceedings: Systems approach to learning and education in the 21st Century* (Vol. 2) (pp. 757–760), Bogaziçi University (Istanbul).

Dede, C., 1998, *Learning with technology*, ASCD (Alexandria).

Dörner, D., 1996, *The logic of failure: Why things go wrong and what we can do to make them right* (R. Kimber & R. Kimber, Trans.), Metropolitan Books (New York).

Ericsson, K. A., 2004, Deliberate practice and the acquisition and maintenance of expert performance in medicine and related domains, *Academic Medicine,* ***10,*** S1-S12.

Ericsson, K. A., & Simon, H. A., 1993, *Protocol analysis: Verbal reports as data (Revised edition),* MIT Press (New York).

Ericsson, K. A., & Smith, J. (eds.), 1991, *Toward a general theory of expertise: Prospects and limits*, Cambridge University Press (New York).

Forrester, J. W., 1987, Nonlinearity in high-order models of social systems, *European Journal of Operational Research,* ***30***(2), 104–109.

Grotzer. T. A., & Perkins, D. N., 2000, *A taxonomy of causal models: The conceptual leaps between models and students' reflections on them,* The National Association of Research in Science Teaching (New Orleans).

Healey, A. N., Undre, S., & Vincent, C. A., 2004, Developing observational measures of performance in surgical teams, *Quality & Safety in Health Care,* ***13***, 33–40.

Herl, H. E., O'Neil, H. F., Jr., Chung, G. L. W. K., Bianchi, C., Wang, S., Mayer, R., Lee, C. Y., Choi, A., Suen, T., & Tu, A., 1999, *Final report for validation of problem solving measures*, CSE Technical Report 501, CRESST (Los Angeles).

Jonassen, D. H., 1997, Instructional design models for well-structured and ill-structured problem-solving learning outcomes, *Educational Technology Research &Development,* ***45***(1), 1042–1629.

Jonassen, D. H., 2004, Learning to solve problems: An instructional design guide, Pfeiffer (San Francisco).

Lave, J., 1988, *Cognition in Practice: Mind, mathematics, and culture in everyday life*, Cambridge University Press (Cambridge).

Merrill, M. D., 2002, First principles of instruction. *Educational Technology Research & Development,* ***50*** (3), 43–59.

Milrad, M., Spector, J. M., & Davidsen, P. I., 2002, Model facilitated learning. In S. Naidu (ed.), *Investigating the effects of information and communications technology: What is working, where and how*, Kogan Page (London).

Novak, J. D., 1998, *Learning, creating and using knowledge: Concept maps™ as facilitative tools in schools and corporation,* Erlbaum (Mahwah).

O'Conner, D. L., & Johnson, T. E., 2004, Analysis-constructed shared mental model methodology: Using concept maps as data for the measurement of shared understanding in teams. First International Conference on Concept Mapping (Pamplona).

Quist, D., & Chapela, I. H., 2001, Transgenic DNA introgressed into traditional maize landraces in Oaxaca, Mexcio, *Nature,* ***414***, 541–543.

Rittel, H., & Webber, M., 1973, Dilemmas in a general theory of planning, *Policy Sciences,* ***4***, 155–169.

Romiszowski, A. J., & LeBlanc, G., 1991, Assessing higher-order thinking skills in computer-based education, In D. Saunders & P. Race (eds.), *Aspects of Educational Technology, Volume XXV,* Kogan Page (London).

Russell, T. L., 1999, *The no significant difference phenomenon*, North Carolina State University (Raleigh).

Schvaneveldt, R. W. (ed.), 1990, *Pathfinder associative networks: Studies in knowledge organization*, Ablex (Norwood).

Seel, N. M. (2003). Model-centered learning and instruction, *Technology, Instruction, Cognition and Learning, 1*(1), 59–85.

Seel, N., Al-Diban, S., & Blumschein, P., 2000, Mental models and instructional planning, In J. M. Spector & T. M. Anderson (eds.), *Integrated & Holistic Perspectives on Learning, Instruction & Technolog* (pp. 129–158), Kluwer (Dordrecht).

Simon, H., 1973, The structure of ill-structured problems, *Artificial Intelligence, **4***, 181–201.

Spector, J. M., 2000, System dynamics and interactive learning environments: Lessons learned and implications for the future, *Simulation and Gaming, **31***(3), 457–464.

Spector, J. M., Christensen, D. L., Sioutine, A. V., & McCormack, D., 2001, Models and simulations for learning in complex domains: Using causal loop diagrams for assessment and evaluation, *Computers in Human Behavior, **17***, 517–545.

Spector, J. M., & Koszalka, T. A., 2004, *The DEEP methodology for assessing learning in complex domains* (Final report to the National Science Foundation Evaluative Research and Evaluation Capacity Building), Syracuse University (Syracuse).

Spiro, R. J., Coulson, R. L., Feltovich, P. J., & Anderson, D., 1988, Cognitive flexibility theory: Advanced knowledge acquisition in ill-structured domains, In Patel, V. (ed.), *Proceedings of the 10th Annual Conference of the Cognitive Science Society*, Erlbaum (Hillsdale).

Sterman, J. D., 1994, Learning in and about complex systems, *System Dynamics Review, **10***(2–3), 291–330.

Stoyanova, N., & Kommers, P., 2002, Concept mapping as a medium of shared cognition in computer-supported collaborative problem solving, *Journal of Interactive Learning Research, **13***(1/2), 111–133.

Taricani, E. M., & Clariana, R. B., 2006, A technique for automatically scoring open-ended concept maps, *Educational Technology Research &Development, **54***(1), 65–82.

PART II

Tools and Techniques for Decision Making in Complex Systems

Chapter 3

Dynamic Handling of Uncertainty for Decision Making

S.C. Lenny Koh
Management School
University of Sheffield, UK
S.C.L.Koh@sheffield.ac.uk

1. Introduction

Off-shoring and outsourcing are becoming common strategies adopted by Multi National Companies (MNCs). The Meta Group (2004) predicts that the global offshore outsourcing market, which is currently worth in excess of £5.6 billion, will grow by nearly 20% each year until 2008. China and India are two top outsourced destinations for many European, Japanese and American enterprises. In Europe, the UK is the largest user of offshore services and the main target for Indian offshore outsourcing due to historical links, language benefits and the maturity of sourcing models and strict labour transfer legislation in other countries.

In a globalised supply chain where suppliers, manufacturers and distributors are in different geographical locations, the interactions and communications become more complex. For example, orders received in the demand channel will be passed to the distributors, distributors will make an order to the manufactures, the manufactures will order the necessary parts from the suppliers, the suppliers will then deliver the ordered parts to its customer. In this context, it is clear that the information channel and material flow channel are intertwined and production is pulled by market demand. Without high amount of off-shoring and outsourcing, these operations could be managed and directly controlled by the enterprise concerned. Nevertheless, an increasing amount of remote location operations have resulted in many uncontrollable operations by the enterprise concerned, such as lack of real-time logistics information, late delivery from supplier, resources shortages, and so on. The decision making process becomes more complex owing to this type of uncertainty in the supply chain.

Various production planning and control systems are widely used by enterprises to assist material and resources planning. Material Requirements Planning (MRP) is defined as a set of techniques that uses Bill of Material (BOM), inventory data and the Master Production Schedule (MPS) to calculate requirements for materials. Its time-phased logic determines what and when to manufacture or purchase from the highest to the lowest levels of the BOM via back flushing from due date (Cox and Blackstone, 1998).

MRP is the core within Enterprise Resources Planning (ERP) systems used as a production planning and control system under manufacturing environments. The inherent structure of MRP/ERP systems is deterministic, requiring high levels of data accuracy and a relatively stable market place. By their nature, such systems do not provide great scope for flexibility. Modern market requirements place great emphasis on supplier responsiveness and it is therefore necessary to rethink how the performance of MRP/ERP systems can be improved in order to meet the changing nature of customer delivery requirements.

Uncertainty is defined as unknown future events that cannot be predicted quantitatively within useful limits (Cox and Blackstone, 1998). Koh (2004) defined uncertainty as any unpredictable event that affects the production of an enterprise. An example of this uncertainty is late delivery of a purchased part from a supplier. The effect of this would be a delay in the manufacture of the parent assembly that is difficult to quantify because the loading on the shop floor when the part ultimately arrives may be very different to that planned when it was due to arrive. This may result in other planned work being displaced to make way for the delayed assembly or delay may be increased if a suitable production space cannot be found. Bull-whip effect of uncertainty in a supply chain can be easily identified due to parts interdependency of shared resources in a production network.

Many uncertainties in MRP/ERP environments are treated as 'controllable' elements, with a variety of buffering, dampening and smoothing techniques such as overtime, multi-skilling labour, subcontract and holding of safety stock for parts being used to cope. However, such techniques are often found wanting, with enterprises being forced into emergency measures to ensure on-time delivery to customer (Koh *et al.* 2000). To this end, more dynamic handling of uncertainty is required to support responsive decision making on ways to deal with uncertainty.

2. Literature Review

Murthy and Ma (1991) found that the nature of MRP/ERP has driven researchers towards the quantity and timing decisions inherent in the systems. Little research has been identified that addresses the implications of uncertainty; rather the result of this trend is that many buffering, dampening and smoothing techniques have been examined to cope with quantity and timing uncertainties. The techniques found from the literature include, for example, freezing part of the planning horizon, increasing quantities in the MPS, using safety stock, using safety lead-time, using appropriate lot-sizing rules and scheduling the systems more frequently.

Freezing part of the planning horizon was suggested by Blackburn *et al.* (1986), as a means to cope with demand uncertainty. Although fixing the schedule does introduce elements of certainty into the process, it does not tackle uncertainties occurring within the manufacturing systems after the MRP run is made.

Hedging by increasing the quantities in the MPS has been proposed to cope with demand uncertainty (Orlicky, 1975, Miller, 1979, New and Mapes, 1984). The effect of this technique is to increase finished products inventory that, while it may be advantageous to the customers, has profoundly negative costs, inventories and resources implications.

Keeping safety stock is the most common technique suggested to cope with quantity uncertainty, which has been studied by Lambrecht *et al.* (1984), Blackburn *et al.* (1986), Carlson and Yano (1986), Vargas and Dear (1991), Buzacott and Shanthikumar (1994), Molinder (1997) and Ho and Ireland (1998). However, modern approaches to the removal of inventory within 'lean' manufacturing call this technique into question.

Melnyk and Piper (1981), Blackburn *et al.* (1986), Buzacott and Shanthikumar (1994), Ho *et al.* (1995) and Molinder (1997) studied the use of safety lead-time to cope with timing uncertainty. Again the effects are reducing resources utilisation, increasing Work in Progress (WIP) and overall lead-times, thus affecting delivery capability. This technique is one of the 'masking' for the uncertainties rather than addressing them directly.

Using appropriate lot-sizing rules to dampen demand uncertainty is a technique proposed by Blackburn *et al.* (1986), Ho (1993), Ho and Ireland (1998), and Ho and Ho (1999). Application of this technique requires sufficient knowledge and information about demand patterns; forecast errors and seasonality; and ordering, holding and shortage costs within the enterprises. Although this information could be available, the proposal on using the appropriate lot-sizing rules requires continuous updating of the systems performance.

Scheduling the systems more frequently has been proposed as being able to cope with uncertainty by taking into account the updated lead-time and quantity (Steele, 1975, Schmitt, 1984). This technique at least has the merit of seeking to react to uncertainties shortly after they have occurred, but will create a very nervous manufacturing system, incorporating too frequent changes to schedules.

Managers do report seeing their work environment as uncertain (Trought, 1999). Improvement in meeting customer due date is found to be the most important organisational goal (Koh and Jones, 1999). It is an apparent truth that if all the variables causing late delivery to customer are deterministic, then 100% on-time delivery will be able to achieve If 100% on-time delivery is not achieved, it is axiomatic that it is the elements of uncertainty that cause the shortfall.

This review has shown that past research on uncertainties has been focused on finding the appropriate techniques to cope with uncertainty. As a result, a mix of MRP/ERP systems performances is reported and there is no benchmark against their levels of delivery performance and relative implications of uncertainties.

To close this research gaps, a benchmarking tool is proposed and developed in this research to enable enterprises to understand how they perform with respect to their own inherent uncertainties and also how they compare against the oppositions, measured by late delivery to customer. It must be noted that this particular research does not aim to examine the causal relationship of uncertainty; hence detailed modeling approaches are not deployed.

3. Research Methodology

3.1. Development of a Benchmarking Tool

A survey based on a questionnaire was applied as part of our research method to conduct empirical data collection at its source. The survey was administered by post. A benchmarking tool was developed during the design of the questionnaire to analyse the level of uncertainties; late delivery to customer; and the buffering, dampening and smoothing techniques used to cope with uncertainties, enabling comparison of these amongst enterprises.

Cox and Blackstone (1998) found that uncertainty is difficult to quantify. To ease collection of empirical data, bandwidths for the measurement of levels of uncertainties have been introduced. Exponential bandwidths of <2%, (2–5)%, (6–15)%, (16–30)% and >30 were used to allow coverage of a wide span of levels of uncertainties in a very few bandwidths and also to highlight the fact that further increment of any bad performance is still deemed a bad performance. For example, a 20% late delivery could be deemed a bad performance and although a record of 30% is empirically higher, its effect is equally bad in the context of categorization of the performances of the enterprises.

To enable benchmarking of levels of uncertainties and levels of late delivery between enterprises, the bands were converted into a classification scale. Table 1 shows the correspondent survey bandwidths and benchmark classification scale.

Eight uncertainties were examined, derived from Mather (1977) and Turner and Saunders (1994). These were late delivery from suppliers, engineering design changes, rework, machine breakdowns, materials shortages, labour shortages, customer order changes, and MPS changes. Late delivery to customer was examined using the same bandwidths and scale.

To examine the buffering, dampening and smoothing techniques used by industry, respondents were asked to identify which they used from a list of thirty-one. To enable benchmarking of the techniques used between enterprises, the enterprises were classified into four categories, namely *Good, Poor, Laggard and Improver*. These classifications allow direct mapping of which techniques being used in each category and their effectiveness. Table 2 shows the definitions of each category. Any responses not falling within these categories were excluded from further analysis to ensure concentration on the extremes of performance.

Table 1. Survey bandwiths and benchmark classification scale

Survey bandwidths	Benchmark classification scale
< 2%	1
(2–5)%	2
(6–15)%	3
(16–30)%	4
> 30%	5

Table 2. Definitions of *Good, Poor, Laggard and Improver*

Category of enterprises	Late delivery to customer	Levels of uncertainties
Good	≤5%	≤5%
Poor	≥16%	≥16%
Laggard	≥16%	≤5%
Improver	≤5%	≥16%

Good category is defined as enterprises with low level of late delivery to customer and low level of uncertainty. Low, herein, refers to ≤5%, suggesting the benchmark classification scale of 1 and 2. *Poor* category is defined as enterprises with high level of late delivery to customer and high level of uncertainty. High, herein, refers to ≥16%, suggesting the benchmark classification scale of 4 and 5.

Due to the effect of buffering and dampening techniques used by the enterprises, the resulting delivery performance may be improved or otherwise. This calls for a third category, *laggard*, which is defined as enterprises with high level of late delivery to customer (benchmark classification scale of 4 and 5) and low level of uncertainty (benchmark classification scale of 1 and 2). The fourth category is *improver*, which is defined as enterprises with low level of late delivery to customer (benchmark classification scale of 1 and 2) and high level of uncertainty (benchmark classification scale of 4 and 5).

Good and *improver* clearly depict the best performing categories, which show the most effective and efficient use of buffering and dampening techniques. The effect from the former category is not as significant as those from the latter category due to the low level of late delivery to customer achieved under a highly uncertain environment in the latter environment. *Poor* category clearly shows its bad delivery performance. Under its highly uncertain environmemt, this is not as bad as those experienced by the *laggard*. *Laggard* is the worst performing category, which does not manage to improve the delivery performance by effective and efficient use of buffering and dampening techniques under its relatively stable environment.

3.2. Sampling Procedure and Survey Response

Various attempts were made to find information on enterprises that use MRP/ERP systems in the UK to form the survey sample. These include MRP/ERP software vendors, MRP/ERP information web sites, MRP/ERP software implementers/partners, The Oliver Wight Companies, The Business Link Database, and The Institute of Electrical Engineers *Bite* Database.

Software vendors and implementers/partners were reluctant to provide details of their customer base. From some thirty enterprises contacted, none were prepared to provide a list of customers, with a few referring to web sites that give details only of successful implementations. As a long-standing organisation within the field, The Oliver Wight Companies were expected to have information on enterprises that use such systems, but they would only provide their 'Class A' examples, representing users in the excellent category. If any survey were directed only to the 'successful' enterprises, it would have produced unacceptable bias. Neither of the databases examined contain any relevant data.

In the absence of any definitive data regarding MRP/ERP users, the survey was directed to general manufacturing enterprises in the UK, with a first filter of identifying MRP/ERP users. With a non-representative sample population and an expected low response rate, a large sample was needed to ensure a sufficient number of MRP/ERP users would be found.

The questionnaire was sent to a sample of 2500 manufacturing enterprises in the UK, addressed to named individuals identified from the Department of Trade and Industry (DTI) Business Link Hertfordshire Database.

The overall response rate was 17.36% with over a quarter of the responses (126) being current and useful users of MRP/ERP systems, which were included in the analysis. Due to the non-representative sample involved and without any follow-up activity this response rate was considered acceptable. Thirty-one respondents (excluded from the 126 sample) were excluded for detailed analysis as they were in the early stages of MRP/ERP implementations.

4. Results, Analysis and Discussions

4.1. Development of a Measure for Relative Implications of Uncertainties

It is inevitable when measuring late delivery to customer that it is only possible to measure what was achieved. Enterprises use a variety of techniques to cope with uncertainties in order to satisfy their customers and are therefore actively overcoming uncertainties, which then go unrecorded. The level of late delivery to customer is therefore a dependent variable, with the disadvantage that dependencies are largely unknown and unrecorded.

Similarly, uncertainties themselves are invariably only partially recorded. For example, a stock-out is only significant if the part is required, rather than when the stock-out occurs. The implication this has on delivery performance is therefore dependent upon when an order for the part is placed. Therefore the stock-out could have no implications or it could cause a delay. Only the latter would ever be recorded and therefore the true incidence of the event is unlikely to be known.

It is not possible to achieve an absolute measure of the implications of uncertainty on late delivery to customer at this stage as both the causes and effects are being masked. In this research, the approach taken has been to derive a benchmarking measure of performances allowing comparisons of magnitude to be made in relative rather than absolute terms.

Due to the use of different buffering and dampening techniques and their effectiveness and efficiencies, the level of individual uncertainties (X) do not necessarily has to be directly proportional to the level of late delivery to customer (Y). To assess the relative implications of uncertainties on late delivery, *K-factor* is developed.

4.2. Examining the Relationship Between *X* and *Y*

Although the levels of individual uncertainties (*X*) do not necessarily directly proportionate to levels of late delivery to customer (*Y*), *X* is likely to cause *Y*. In this study, the *X* variable has also been manipulated via translation to the benchmark classification scale. In these contexts, linear regression analysis is suitable to examine the relationship between *X* and *Y*.

After the conversion of the results collected from the respondents from the survey bandwidths to the benchmark classification scale for both *X* and *Y*, linear regression analysis was carried out.

Figure 1 shows the results of the linear regression analysis between *X* and *Y*. Considering the calculated value of correlation factor (r = 0.8037), which is equivalent to the r value from a correlation test, it can be inferred that this regression model suggests that there is a goodness-of-fit on the positive correlation between *X* and *Y*. The r and r^2 values signify the ability to predict using the regression line, compared with not using it. This regression model shows a reasonable prediction capability (r value closes to 1).

The results from the regression analysis do not indicate the statistical significance of the correlation between *X* and *Y*. The goodness-of-fit and its predictability are sufficient in this study to open the possibility of using *X* and *Y* to derive a benchmark to compare performance of enterprises and to highlight areas for further examination of causality.

4.3. Derivation of the *K-Factor*

The regression model as shown in Figure 1 gives a C intercept value of –0.0591. Since this value is so close to zero, it is negligible. The regression analysis was then repeated, adjusting to zero intercept, with the results shown in Figure 2.

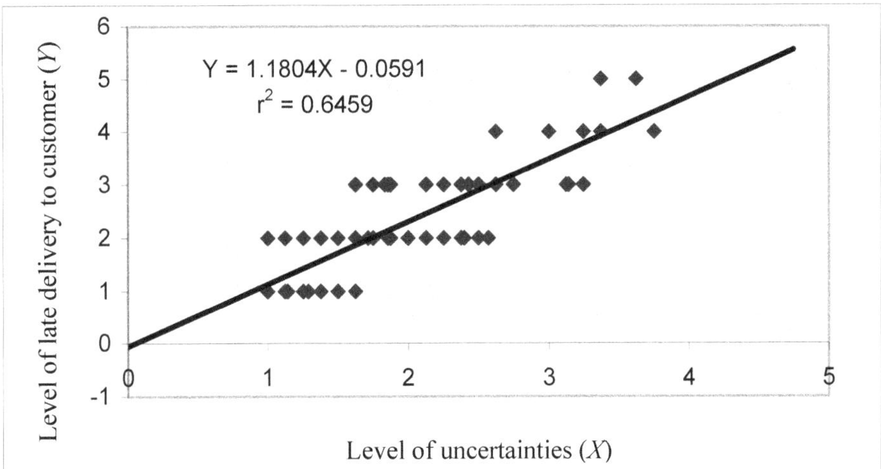

Figure 1. Linear regression model between level of uncertainties (*X*) and level of late delivery to customer (*Y*)

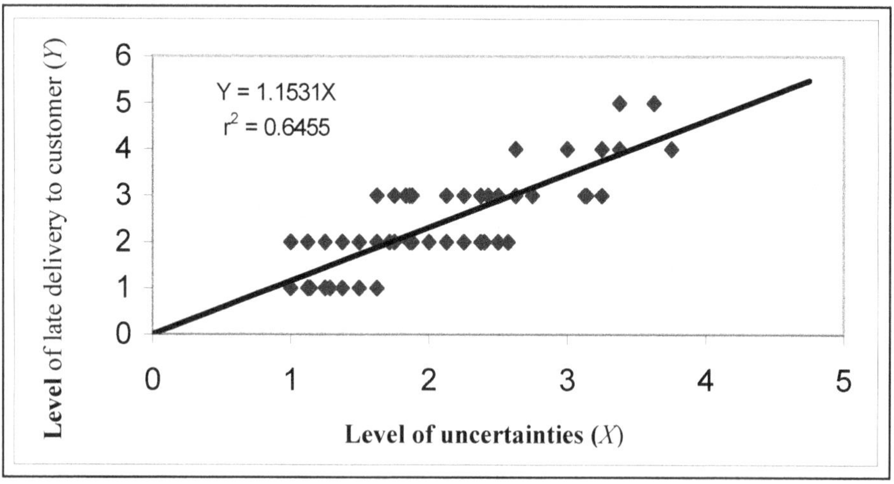

Figure 2. Adjusted linear regression model between level of uncertainties (*X*) and level of late delivery to customer (*Y*)

Knowing that correlation exists between *X* and *Y* allowed us to state that $X \propto Y$. The introduction of the *K-factor* to quantify the strength of the relationship resulted in $Y = KX$. Our sample produced a *K-factor* of 1.15. Thus our benchmark for measuring the relative implications of uncertainties for MRP/ERP enterprises on late delivery is:

$$\boldsymbol{K} = Y/X \qquad \ldots [1]$$

Where,
K-factor = the relative implications of level of uncertainties on level of late delivery to customer, where $K = 1.15$
Y = level of late delivery to customer
X = level of uncertainties

Analysis of Variance or t-test was not pursued after the regression analysis because the data itself is subjective and even if further statistical significance of the correlation could be established, the p value itself does not provide a benchmark figure for practitioners. Hence, *K-factor* is introduced to assist the benchmarking in this research.

4.4. Calculation of Survey *K-Factor*

K-factor measures the relative implications of uncertainties on late delivery to customer. The higher is the *K-factor*, the greater is the relative implications of the uncertainties measured on late delivery to customer, or vice-versa. For a *K-factor* of one, the level of uncertainty (*X*) is equal to the level of late delivery to customer (*Y*). For a *K-factor* less than one means a relatively higher level of uncertainty (*X*) than late delivery to customer (*Y*). A *K-factor* greater than one signifies a relatively lower level of uncertainty (*X*) than the level of late delivery to customer (*Y*). In other words, a

K-value = 1 indicates a balanced approach (not necessarily the best performance or outcome), a K-value < 1 indicates a good performance or outcome from the use of buffering and dampening techniques, whilst a K-value > 1 indicates the worst performance or outcome from the use of buffering and dampening techniques. These will be explained using the following results.

The survey mean of late delivery to customer (Y) has been calculated as 2.23 i.e. (6–15)%, which denotes a median performance level. This Y value was used as the baseline to calculate the relative implications of each uncertainty. Table 3 shows the survey X, Y and *K-factors*.

The overall uncertainty has a mean value of 1.94, which is in the range of (2–5)%. The corresponding *K-factor* is 1.15 (reading from Figure 2) implying that the enterprises on average have relatively lower level of uncertainty than the level of late delivery to customer. Relating this result to their use of buffering and dampening techniques, it could be inferred that in general the enterprises have not managed to cope with the uncertainties effectively and efficiently using buffering and dampening techniques, thus resulting in a relatively higher level of late delivery to customer. This finding implied that there are combinations of uncertainties, which in aggregate result in higher late delivery. This can be supported by the results which show all uncertainties identified have a higher late delivery individually with the exception of late delivery from suppliers with a *K-factor* of 0.91. This finding highlighted that poor delivery performance from suppliers has a disproportionately small implications on late delivery to customer relatively.

The relative implications of changes to MPS and customer order changes were found to be 1.15 and 1.14 respectively. These *K-factors* indicate that both these uncertainties impact directly and detrimentally on late delivery to customer. To reduce or remove these uncertainties may not necessarily result in an improved on-time delivery to customer.

Effective and efficient use of overtime, safety stock for parts, subcontract and supplier partnerships, may have contributed to the apparently small implications of late delivery from suppliers on late delivery to customer. Material/subassembly shortages,

Table 3. Survey X, Y and *K-factors*

Uncertainties	Mean (X)	*K-factor*
Changes to MPS	1.96	1.14
Customer order changes	1.94	1.15
Material/subassembly shortages	2.17	1.03
Labour shortages	1.71	1.31
Late delivery from suppliers	2.46	0.91
Engineering design changes	2.05	1.09
Machine breakdowns	1.47	1.52
Rework	1.79	1.24
Overall ($Y = 2.23$)	1.94	1.15

engineering design changes and labour shortages have relatively greater implications on late delivery to customer. This implies that the latter require closer attention than the former. Although much research on Supply Chain Management (SCM) can be identified to cope with late delivery from suppliers, it is the uncertainties inherent within the manufacturing environments that found to have relatively larger implications on late delivery.

Machine breakdowns were identified to have the greatest relative implications on late delivery. Hence, machine breakdowns need to be monitored closely even if the frequency of its occurrence tends to be low. Likewise, rework was also identified having relatively significant implications on late delivery.

4.5. Buffering, Dampening and Smoothing Techniques

It was identified that MRP/ERP users adopted a variety of buffering, dampening and smoothing techniques to cope with uncertainties. Figure 3 shows the frequency of the techniques used in percentages ranked in descending order.

Overtime and multi-skilling labour are the most frequently used techniques. The use of overtime normally resulted from failure of or lack of adherence to the manufacturing schedules generated from MRP/ERP. Such failures could be due to material/subassembly shortages, labour shortages, late delivery from suppliers, machine breakdowns, rework, engineering design changes, customer order changes or changes to MPS or any combination of these.

Overtime is shown to be a versatile technique to cope with a range of uncertainties. Multi-skilling labour could provide a degree of flexibility when faced with labour shortages, which have relatively high implications on late delivery. Holding safety stock for parts and supplier partnership/vendor rating are also frequently used. The former plays an important role in cushioning against material/subassembly shortages, labour shortages, rework, machine breakdowns and late delivery from suppliers, ensuring finished products can be delivered on time. For tackling late delivery from suppliers, supplier partnership/vendor rating could be used to minimise the probability of supply shortages. Other frequently used techniques include holding safety stock for finished product, performance review, subcontract, preventive maintenance, reschedule and training for planner/scheduler. Thereafter, frequency of the techniques used drop off rapidly in popularity.

A mapping analysis was carried out to identify the most frequently used techniques within each category of enterprises. This highlights the effectiveness and efficiencies of those techniques. In this study, an inclusion rule of greater than 70% usage was applied to define the most frequently use. Tables 4, 5, 6 and 7 show the most frequently used techniques against the types of uncertainties in each category of enterprises. Learning from the enterprises in the *improver* and *good* categories is certainly worthwhile, but knowing the techniques which do not work (extracting from the *poor* and *laggard* categories) is equally important.

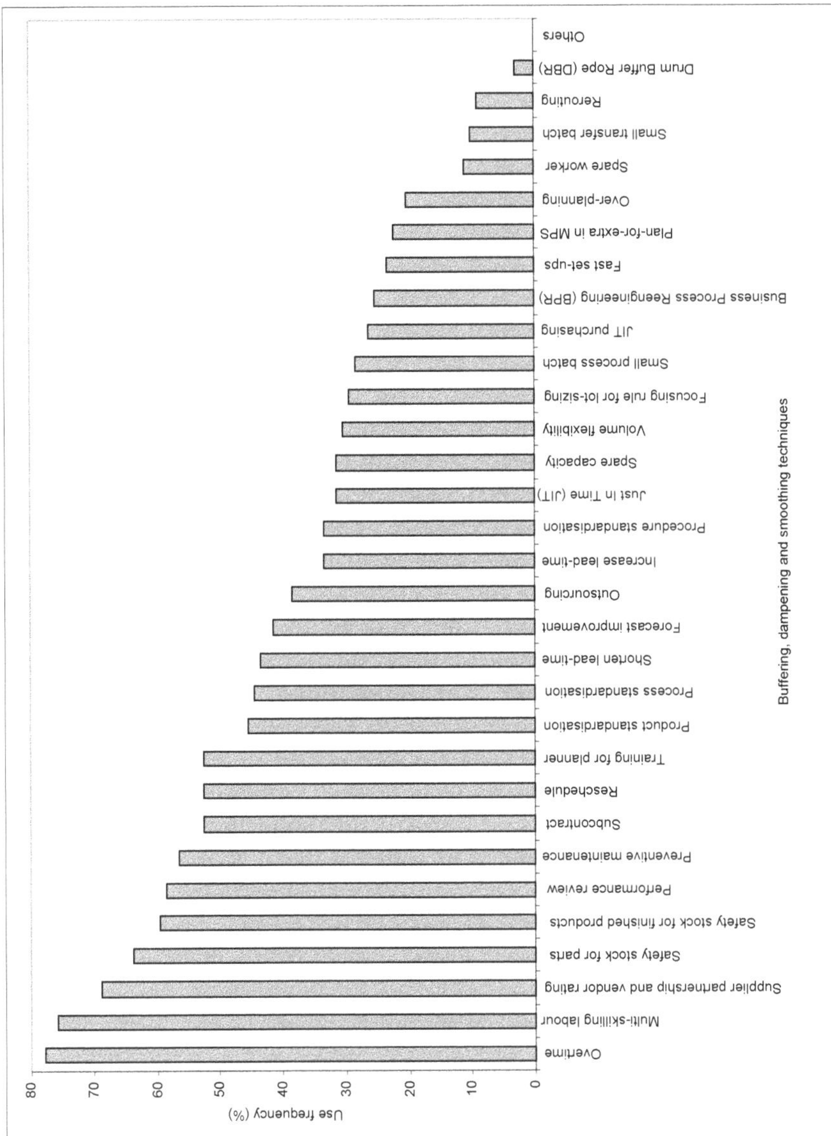

Figure 3. Frequency of buffering, dampening and smoothing techniques used

Enterprises in the *good* category appear to use fewer techniques but they use them well as compared to the other categories. Enterprises in the *good* category only adopt overtime and multi-skilling labour to deal with uncertainties. Multi-skilling labour is the most robust technique because it is used to cope with all types of uncertainty. Overtime, however, is a robust technique in general, but it has only been adopted by the enterprises in the *good* category to tackle material/subassembly shortages, rework, changes to MPS, customer order changes and engineering design changes. According to the results in this study, for enterprises with a relatively low level of these uncertainties, it is advisable to use these techniques to cope with them.

Table 4. Techniques used by enterprises in the *Good* category

Good enterprises	Material/subassembly shortages	Labour shortages	Machine breakdowns	Rework	Changes to MPS	Customer order changes	Late delivery from suppliers	Engineering design changes
Overtime	✓			✓	✓	✓		✓
Multi-skilling Labour	✓	✓	✓	✓	✓	✓	✓	✓

Enterprises in the *poor* category use a whole range of buffering and dampening techniques to cope with uncertainty. Here, overtime is found to be the most robust technique. Multi-skilling labour is used to cope with all uncertainties except material/subassembly shortages. It has also been identified that labour shortages and machine breakdowns seem to be the most dominant uncertainties within the enterprises in the *poor* category. As can be seen in Table 5, these uncertainties are coped with using many different techniques, but the most interesting finding is that subcontract has not been applied. This may be due to the lack of trust within the third party resources provider market. It is important to note that the enterprises in the *poor* category experience the highest level of uncertainty and the highest level of late delivery to customer, hence it could be safely regarded that the buffering and dampening techniques applied by this category are not the good examples to be followed.

Not too dissimilar to the *poor* category, enterprises in the *laggard* also use a wide range of buffering and dampening techniques to cope with uncertainty. The survey did not seek to identify specific uncertainties for each category of enterprises, but the results give some guidance to perceptions of each as to the uncertainties that they faced. Labour shortages are not an issue within the enterprises in the *laggard* category. In this case, the most robust technique is multi-skilling labour. It is also interesting to note that JIT is used to deal with late delivery from suppliers. Given that enterprises in the *laggard* category experience the highest level of late delivery to customer even under the environment of the lowest level of uncertainty, this suggests that their JIT performance is exceedingly poor and JIT is not a guarantee solution for supplier delivery problem. To summarise, the techniques used by enterprises in the *laggard* category are not the best practices to be learnt.

Table 5. Techniques used by enterprises in the *Poor* category

Poor enterprises	Material/subassembly shortages	Labour shortages	Machine breakdowns	Rework	Changes to MPS	Customer order changes	Late delivery from suppliers	Engineering design changes
Overtime	✓	✓	✓	✓	✓	✓	✓	✓
Subcontract				✓				✓
Reschedule		✓	✓		✓			
Finished safety stock		✓	✓			✓		
Part safety stock		✓	✓					
Increase lead-time		✓	✓		✓			
Spare capacity			✓					
Spare labour		✓						
Hedging		✓						
Over planning		✓			✓			
SCM			✓					
Preventive maintenance			✓					
Multi-skilling labour		✓	✓	✓	✓	✓	✓	✓
Forecast improvements		✓	✓					
Product standardisation		✓						
Process standardisation		✓	✓			✓		
Procedure standardisation			✓					
Volume flexibility		✓	✓			✓		

Enterprises in the *improver* category use a mixture of buffering and dampening techniques but not as many as those in *the poor* and *laggard* categories. Multi-skilling labour and overtime are the most robust buffering and dampening techniques used by enterprises in the i*mprover* category to cope with all uncertainties except material/subassembly shortages. Knowing that this category experiences the lowest level of late delivery to customer with the highest level of uncertainty, these techniques certainly have an influence on such performance.

Table 6. Techniques used by enterprises in the *Laggard* category

Laggard enterprises	Material/subassembly shortages	Labour shortages	Machine breakdowns	Rework	Changes to MPS	Customer order changes	Late delivery from suppliers	Engineering design changes
Overtime	✓			✓	✓	✓		✓
Subcontract	✓			✓			✓	
JIT							✓	
Finished safety stock			✓	✓	✓			
Part safety stock	✓		✓	✓	✓	✓		
Preventive maintenance			✓					
Multi-skilling labour	✓		✓	✓	✓	✓	✓	✓
Product standardisation			✓					
Performance review								✓
Shorten lead-time			✓					
Fast set-ups			✓					

To deal with material/subassembly shortages, enterprises in the *improver* category use preventive maintenance and planner training. It has been confirmed from the overall analysis that multi-skilling and overtime are very robust themselves, i.e. are flexible to deal with different types of uncertainty; hence it is not surprising to see their robustness here within the *improver* category. It is very interesting to find that the preventive maintenance technique has not been used to cope with machine breakdowns, which is what the technique intends to serve. In turn, preventive maintenance is applied by enterprises in the *improver* category, leading to increased uptime and availability of resources, and hence whenever material/subassembly shortages occur, they have the flexibility to respond to any new order. Planner training could certainly assist in tackling material/subassembly shortages by simply ensuring skillful re-planning to re-align resources with orders in the supply chain.

It can be seen that reschedule and Supply Chain Management (SCM) are solely used to cope with rework, and SCM is not used to tackle late delivery from suppliers. The result suggests that SCM is not always necessarily applicable to deal with universal late delivery from suppliers problem. SCM is a broad concept and its inclusion of various techniques within it makes it robust to deal other uncertainties.

Table 7. Techniques used by enterprises in the *Improver* category

Improver enterprises	Material/subassembly shortages	Labour shortages	Machine breakdowns	Rework	Changes to MPS	Customer order changes	Late delivery from suppliers	Engineering design changes
Overtime		✓	✓	✓	✓	✓	✓	✓
Subcontract		✓	✓	✓		✓	✓	
Reschedule				✓				
SCM				✓				
Preventive maintenance	✓							
Multi-skilling labour		✓	✓	✓	✓	✓	✓	✓
Planner training	✓							

Although SCM is not found to be as robust as it could be, the key finding in this respect is the indication of quality problem from supplier within the enterprises in the *improver* category, not delivery.

Although the uncertainties experienced and the techniques that appeared to be effective and efficient for each category of enterprises were identified, no direct statistical correlation could be drawn to show the most effective technique for specific uncertainty due to the multiple uncertainties scenario. However, this research provides a starting knowledge base to enable MRP/ERP enterprises to identify their implications of uncertainties and the potential buffering and dampening techniques to be adopted. In the past, the absence of this knowledge made enterprises apply the techniques indiscriminately and therefore not knowing whether application of those techniques will result in maximum benefits.

5. Conclusions, Implications and Future Research

The contribution of this research focused on the benchmarking tool, *K-factor measure*, comparing the uncertainty experienced by different types of companies, classification of enterprises, and the types of buffering and dampening techniques used to tackle the uncertainty. It was envisaged that in the decision making process, future companies could refer to the buffering and dampening techniques found to be effective and efficient under specific category of enterprises experiencing similar types and levels of uncertainty. In this research, the 'dynamism' referred to *learning* from environments which experienced similar and multiple types and levels of uncertainty.

This research has shown that MRP/ERP underperformance is due to uncertainties in the manufacturing environments. A benchmarking tool has been developed to enable enterprises to understand how they perform with respect to their own inherent uncertainties and also how they compare against the oppositions. Two performance measures were used within the benchmarking, the first being the *K-factor* and the second being the classification of enterprises. Both were measured against the levels of uncertainties and levels of late delivery to customer.

The *K-factor* and classification of enterprises provided a step forward to dynamically handle uncertainty through an empirical analysis of the position of an enterprise on implications of uncertainty on delivery performance with evaluation of suitable and unsuitable buffering, dampening and smoothing techniques. Such evaluation supported a better decision making on the adoption of suitable buffering, dampening and smoothing techniques on specific uncertainty.

Due to different buffering, dampening and smoothing techniques used and its effectiveness and efficiencies within enterprises, levels of individual uncertainties do not necessarily directly proportionate to levels of late delivery to customer. To enable the relative implications of uncertainties on late delivery to be assessed, *K-factor* has been developed. Using *K-factor* it was found that machine breakdowns have the greatest relative implications on late delivery, while late delivery from supplier has the lowest relative implications.

The enterprises were classified into four categories, namely *Good, Poor, Laggard and Improver* to enable comparison of the techniques used to cope with uncertainties. It has been identified from the survey that overtime and multi-skilling labour are most frequently and widely used by all enterprises categories, while a mixture of techniques was used by *Poor, Laggard and Improver* enterprises categories. Other system dynamic models suggested that while there may be short-term benefit to overtime, in the long run, the outcome could be increasing rework due to errors and a high turnover rate due to burnout. This would be part of the dynamics in the enterprises and deserve future research attention.

It is essential to understand the relative implications of uncertainties on late delivery so that benefits from the techniques used can be maximised. However, as the *K-factor* measures the relative implications after a variety of techniques have been applied to cope with uncertainties, the implications of uncertainties before any influence from the techniques used have not been identified. The absence of this knowledge made enterprises apply the techniques indiscriminately and therefore not knowing whether application of those techniques will result in maximum benefits. Future research could be carried out in light of filtering such noises.

References

Blackburn, J.D., Kropp, D.H.K. and Millen, R.A., 1986, "A comparison of strategies to dampen nervousness in MRP systems", Management Science, Vol 32, No 4, pp. 413–429.

Buzacott, J.A. and Shanthikumar, J.G., 1994, "Safety stock versus safety lead-time in MRP controlled production systems", Management Science, Vol 40, No 12.

Carlson, R.C. and Yano, C.A., 1986, "Safety stocks in MRP systems with emergency set-ups for components", Management Science, Vol 32, No 4, pp. 403–412.
Cox, J.F. and Blackstone J.H., 1998, APICS Dictionary Ninth Edition, APICS-The Educational Society for Resource Management, VA, USA.
Ho, C.J. and Ho, C.J.K., 1999, "Evaluating the effectiveness of using lot-sizing rules to cope with MRP system nervousness", Production Planning and Control, Vol 10, No 2, pp. 150–161.
Ho, C.J. and Ireland, T.C., 1998, "Correlating MRP system nervousness with forecast errors", International Journal of Production Research, Vol 36, pp. 2285–2299.
Ho, C.J., 1993, "Evaluating lot-sizing performance in multi-level MRP systems: A comparative analysis of multiple performance measures", International Journal of Operations and Production Management, Vol 13, pp. 52–79.
Ho, C.J., Law, W.K. and Rampal, R., 1995, "Uncertainty dampening methods for reducing MRP system nervousness", International Journal of Production Research, Vol 33, pp. 483–496.
Koh, S.C.L. and Jones, M.H., 1999, "Manufacturing uncertainty and consequent dilemmas: causes and effects", Proceedings of the 15th International Conference on Production Research, University of Limerick, Ireland, Vol 1, pp. 855–858.
Koh, S.C.L., 2004, MRP-controlled batch-manufacturing environment under uncertainty. Journal of The Operational Research Society, vol. 55, no. 3, pp. 219–232.
Koh, S.C.L., Jones, M.H., Saad, S.M., Arunachalam, S. and Gunasekaran, A., 2000, "Measuring uncertainties in MRP environments", Journal of Logistics Information Management, Vol 13, No 3, pp. 177–183.
Lambrecht, M.R., Muckstadt, J.A. and Luyten, R., 1984, "Protective stocks in multi-stage production systems", International Journal of Production Research, Vol 22, No 6, pp. 1001–1025.
Mather, J.G., 1977, "Reschedule the reschedules you just rescheduled – Way of life for MRP?", Production and Inventory Management Journal, Vol 18, pp. 60–79.
Melnyk, S.A. and Piper, C.J., 1981, "Implementation of material requirements planning: Safety lead-times", International Journal of Operations and Production Management, Vol 2, No 1, pp. 52–61.
Meta Group, 2004, Offshore outsourcing market to grow 20% annually through 2008, says Meta Group, press release, 6 October. http://www.metagroup.com
Miller, J.G., 1979, Hedging the master schedule - Disaggregation Problems in Manufacturing and Service Organisations, Martinus Nijhoff, Mass., USA.
Molinder, A., 1997, "Joint optimisation of lot sizes, safety stocks and safety lead-times in an MRP system", International Journal of Production Research, Vol 35, No 4, pp. 983–994.
Murthy, D.N.P. and Ma, L., 1991, "MRP with uncertainty: A review and some extensions", International Journal of Production Economics, Vol 25, pp. 51–64.
New, C. and Mapes, J., 1984, "MRP with high uncertainty yield losses", Journal of Operations Management, Vol 4, No 4, pp. 1–18.
Orlicky, J.A., 1975, Material Requirements Planning, McGraw-Hill, New York, USA.
Schmitt, T.G., 1984, "Resolving uncertainty in manufacturing systems", Journal of Operations Management, Vol 4, No 4, pp. 331–345.
Steele, D.C., 1975, "The nervous MRP systems: How to do battle", Production and Inventory Management Journal, Vol 16, No 4, pp. 83–88.
Trought, B., 1999, "The managerial dilemma: Uncertainty or complexity", Proceedings of the 15th International Conference on Production Research, University of Limerick, Ireland, Vol 2, pp. 1351–1354.

Turner, E. and Saunders, D., 1994, "Surveying the manufacturing planning and control environment", Proceedings of the 4th International Conference on Factory 2000, York, UK, pp. 613–619.

Vargas, G.A. and Dear, R.G., 1991, "Managing uncertainty in multi-level manufacturing systems", Integrated Manufacturing Systems, Vol 2, No 4.

Chapter 4

Using System Dynamics and Multiple Objective Optimization to Support Policy Analysis for Complex Systems

Jim Duggan
Department of Information Technology,
National University of Ireland, Galway.
Jim.Duggan@nuigalway.ie

1. Introduction

This chapter proposes the synthesis of two analytical approaches to support decision making in complex systems. The first, system dynamics, is a proven methodology for modeling and learning in complex systems. The second, multiple objective optimization, is an established approach for exploring the search space of a decision problem, in order to generate optimal solutions to problems with more than one objective. The chapter presents:

- An overview of decision making in complex systems;
- A summary of the tradition approach to single objective optimization used in system dynamics modelling;
- An explanation of multiple objective optimization;
- An illustrative case study[1], based on the Beer Game, which integrates multiple objective optimization with system dynamics, and provides a practical way for decision makers to fully explore policy alternatives in a complex system micro-world.

[1] The software for this case study can be obtained by emailing the author.

2. Complex Systems: Dynamic Complexity and Conflicting Objectives

2.1. Complex Systems

For decision makers in modern organizations, making sense of a complex world is a difficult and challenging task. The pace of change, complexities of societal relationships, and increased and often conflicting demands from customers, markets and suppliers mean that decisions – in order to be effective – require increased attention and the support of improved analytical techniques. Part of the difficulty in arriving at effective decisions is due to our tendency, in the face of overwhelming complexity, to "fall back on rote procedures, habits, rules of thumb, and simple mental models to make decisions" (Sterman 2000).

However, because of what Simon (1982) defines as our "bounded rationality", which is our limitation in conceptualizing and solving complex problems, it is vital that decision makers continue to search for ways, methods and techniques to enhance our learning and our ability to make better, more informed, decisions. A key challenge is to produce insight and promote creativity (Goodwin and Wright 2003), and find ways to allow decision makers think more deeply about problems. Better decision making approaches can enable the design of policies that are sustainable and viable, rather than creating poorly thought out policies that lead to an escalating cycle of additional unintended problems.

Fortunately, there are a number of practical and established methodologies which can assist decision makers in analyzing complex systems. Two of these are: system dynamics and multiple objective optimization, and this chapter will demonstrate how these different views on complex problem solving can be integrated in order to give decision makers a new way of exploring difficult and challenging problems.

2.2. Dynamic Complexity

A key theme of system dynamics (Forrester 1961) is that policy interventions are diluted, and often fail, because decision makers are not fully aware of the feedback structures – the *dynamic complexity* - that are present in modern complex systems. These feedbacks can be reinforcing (positive) or balancing (negative), and it is the interaction of feedback systems that give rise to system behaviors. Furthermore, when decision makers are unaware of these underlying feedbacks, their policies can give rise to a number of undesirable side effects, and in many cases the solution is even more problematic than the original problem. Within the system dynamics literature (Senge 1990, Wolstenholme 2003), archetypal feedback structures have been identified, and these are a valuable aid in diagnosing the current and future behavior of a given complex system.

For example, the *limits to growth* archetype (Figure 1) defines a common dynamic that can occur when a system grows rapidly, fuelled by a reinforcing feedback process. However – and often as a source of puzzlement to the participants in the system – the growth begins to slow, comes to a halt and may even reverse itself and begin an accelerating collapse (Senge 1990). The slowing is due to a limit being reached, for example, a resource constraint, and this slowing can have the adverse effect of tipping the reinforcing loop in the opposite direction, and so the growing action is transformed into a shrinking action, and a virtuous cycle is replaced with a vicious cycle.

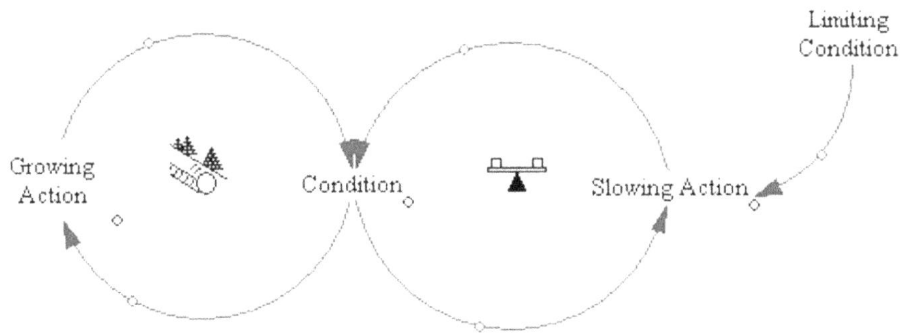

Figure 1. Sample feedback structure of limits to growth archetype (Senge 1990)

For example, consider a company that releases an innovative product where word of mouth fuels customer demand (growing action), but the company's limited production limit means that lead times for the product increase (slowing action), causing cancellation of orders, which in turn lead to a poor reputation, which then leads to a decline in demand. Senge (1990) argues that the key management principle that should be implemented is not to "push on the reinforcing (growth) process," but instead "remove (or weaken) the source of limitation." In this scenario it would mean that as demand for a product grows, companies would increase their capacity to cope with this demand, and so lead times would remain constant and competitive.

Using system archetypes to model these scenarios can greatly enhance our understanding of how complex systems behave over time. Furthermore, system dynamics is also firmly grounded in the mathematical theory of non-linear differential equations, and so robust, practical and insightful models can be created, which can then be run as virtual worlds and comprehensive learning environments. These enable decision makers to "test drive" their policies, and deepen their understanding of the real world.

2.3. Trade-Offs and Conflicting Objectives

While the focus of system dynamics centers on dynamic complexity, there is another approach to supporting decision making in complex systems where the primary motivation is to find the optimal solution to problems that involve more than one objective. Multiple objective optimization (MOO) (Deb 2001, Coello Coello 2000) is a method that generates solutions to problems that have more than one objective, and where these objectives are often conflicting. Applications of the MOO are found in many design problems in engineering, for example:

- In civil engineering, is it possible to maximize the reliability of a support structure while minimizing the amount of material needed to make the structure?
- In electronic engineering, is it feasible to maximize the speed of a processor while minimizing its footprint (i.e. design something that is fast and takes up less space)?
- In software engineering, is it achievable to minimize the time taken to produce a product while simultaneously minimizing defects?

These classes of problems are challenging, because both objectives are equally important, and because they conflict with one another. For example, in the case of software engineering, there is a tricky trade off between quality and time to market. If, for example, at a high-level company board meeting a decision must be made on a new product's release date, these two objectives may well be discussed in great detail. The quality manager may advocate taking extra time to ensure that most of the bugs are detected, in order that the customer will does not have any negative experience with the system. The marketing manager, on the other hand, argues that if the product is not shipped immediately, a crucial window of opportunity will be missed, and the consequence of that delay is that the customer will go elsewhere.

The choice here is stark: either a perfectly working system with no immediate customer, or a hurriedly released system that captures market share. The benefits of gaining market share is that the company leapfrogs their competitors and secures a foothold in the marketplace, especially if that is an emerging market with high possibilities for sustained growth. The dangers of rushing to the marketplace are that while initial gains are made, poor quality will lead to customer dissatisfaction, which in turn will lower the company's reputation and long-term viability.

In this scenario, what the decision makers need is sound analytical support, through a combination of simulation and optimization, that enable them to fully explore these challenging trade-offs, and thus improve the chances of arriving at the best decision. Before outlining this synthesis of approaches in more detail, we will review the literature to date on integrating system dynamics with single objective optimization.

3. System Dynamics and Single Objective Optimization

The complimentary roles of system dynamics and optimisation are analyzed by Coyle (1996), who writes that "it would be highly desirable to have some automated way of performing parameter variations", and explains that, theoretically, the number of possible combinations and conceivable values of parameters is "colossal" and possibly "infinite." Within this context, some form of guided search is needed whereby good approximations to optimal solutions can be found, and Coyle employs a hill climbing heuristic algorithm to search for and find optimal solutions.

The analogy of a hill or mountain range provides a useful way of thinking about optimisation. As a starting point consider two parameters that need to be optimised. For example, in a stock management structure problem (which is the basis of the case study later in the chapter), these parameters could be the supply line adjustment time and the stock adjustment time. Before optimisation proceeds, an additional variable must be added to the model, one that represents the payoff, which captures the value that must be either minimized or maximized. The idea is that an optimisation algorithm will run the model many times, compare the payoffs for different parameter values, and record the combination of parameters that produces the best payoff.

Figure 2 maps two parameters along with their payoff values to produce what Coyle terms the "response surface", which can be likened to a mountain range, with peaks and dips in the landscape. In this diagram, two possible solutions are: (p2a, p1a) and (p2b, p1b). Each of these have an associated payoff, and in this case we can see that payoff at (p2b, p1b) is at a "higher altitude" than the payoff for (p2a, p1a), and so, if our goal is to maximise the payoff, (p2b, p1b) is the optimal solution.

The challenge for an optimisation algorithm is to find the best possible combination of parameters that maximises (or minimises) a payoff function. To date, in addition to the work of Coyle, there have been a range of research papers that have explored the use of optimisation in system dynamics. Dangerfield and Roberts (1996) present an overview of strategy and tactics in system dynamics optimisation. Here they make a distinction between the two different uses of optimisation in model development. One is to use optimisation techniques in determining the best fit for historical data, while the other use is "policy optimisation to improve system performance."

Keloharju and Wolstenholme (1989) present "the use of optimisation as a tool for policy analysis and design in systems dynamics models", and demonstrate their approach using the "Project Model", which was developed earlier by Richardson and Pugh (1981). According to Kelharju and Wolstenholme, one of the major benefits of using optimisation with system dynamics models is that "the saving in computational effort required by the analyst in producing policy insights in enormous." They also

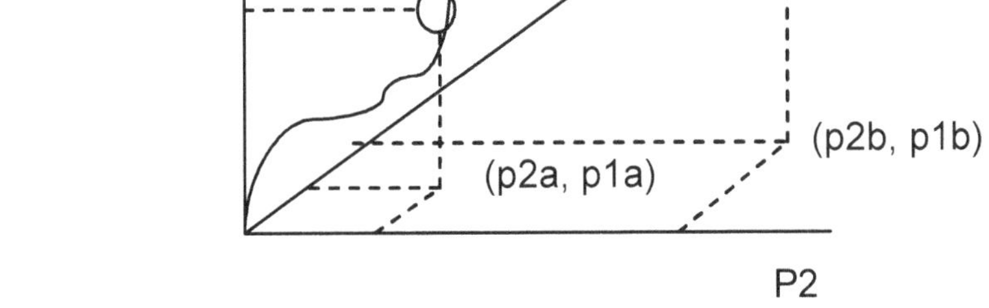

Figure 2. A possible response surface for an optimisation problem with two parameters

suggest that one of the most important issues is the "careful selection of parameters and their allowable ranges," and that, significantly, "all values within ranges must be feasible and practical from a managerial point of view."

This concern is supported by Coyle (1996), who cautions that "it is not usually a good idea to throw all parameters into the optimiser and take the results on trust" as to do this would be to "abandon thought and rely on computation." He acknowledges that the power of optimisation is "to provide a much more powerful guided search of the parameter space that could possibly be achieved by ordinary experimentation." However, he also recommends that simple experimentation is a "fundamental precursor to optimisation" and that "one cannot devise an intelligent objective function without having first 'played' with the model," and that "an unintelligent objective function is worse than useless."

More recently, studies have focused on the use of genetic algorithms for policy optimization (Chen and Jeng 2004, Grossman (2002)). And these describe the weakness of tradition gradient algorithms that stop once they reach a local optimum, whereas genetic algorithms, if well implemented, are unlikely to mistake a local optima for its global equivalent.

Regardless of the algorithm used, the process for optimising a system dynamics model for policy optimisation involves a number of initial steps.

- The stock and flow model must be completed, with all equations formulated robustly.
- The set of parameters that reflect the "levers" upon which a decision maker can adjust and vary the system performance must be selected.
- For each parameter, a minimum and maximum value must be specified.
- A model variable for the overall payoff must be identified. This variable represents the value that needs to be optimized.

Once these steps have been followed, the optimization algorithm can be run. In all cases, this will involve an iterative cycle between the optimizer and the simulation model, where the optimizer will send a set of parameter values to the simulation model (see Figure 3). The model is then be executed with these parameters, and the result, or payoff of the model, is sent back to the optimizer, which can compare this to previous permutations of the parameters.

Coyle (1996) terms this "optimization through repeated simulation", and this process will repeat until a defined stopping criteria is reached. For genetic algorithms – which is the basis of the approach demonstrated in this chapter – the stopping criteria is usually when one hundred generations have evolved the fittest solutions, through a succession of operators known as selection, crossover and mutation.

While clearly optimization is a well-established and valuable technique that is widely used in system dynamics (most available commercial tools for system dynamics contain an optimizer module), the drawback of these current approaches is that they can only optimize one variable at a time, and so do not provide explicit support for a common type of problem that arises in modern decision analysis, namely, problems with multiple and conflicting objectives.

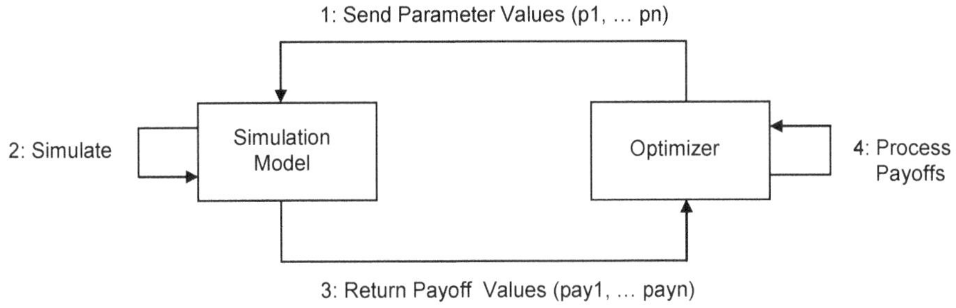

Figure 3. Optimization process in system dynamics

4. Multiple Objective Optimization

MOO comprises a two-stage approach to solving problems with many objectives (Deb 2001).

- First, find multiple trade-off optimal solutions with a wide range of values for the objectives.
- Second, choose one of the obtained solutions using higher level information.

The relationships between these stages is shown in Figure 4, based on Deb (2001).

This diagram captures the essence of a "hard/soft" synthesis of approaches employed as part of this procedure. Step one involves the formulation of a multiple objective optimization problem in terms of enumerating the objective functions, and capturing the constraints that apply. The MOO optimizer generates a set of multiple trade-off solutions. These solutions are optimal, but they represent a spectrum of optimality from which a decision maker must select a single solution. Out of this solution set, the most appropriate must be identified, and this is the goal of the second step. Here, the decision maker selects their preferred solution using what Deb terms "higher level information", which is that knowledge that cannot be easily codified, and is based on qualitative factors such as intuition, judgment and past experience.

The optimal solutions generated through the first step are known as the non-dominated set of solutions. A solution dominates another solution if it is better on all optimization objectives, and this simple rule is used to compare solutions and ensure that the fittest solutions survive. In order to explore the process by which solutions are ranked, consider the following solution set where six parameters are being optimized, as shown in Table 1 and Figure 5. This is the output of an optimization run, where the parameter values are the inputs, and these give rise to values for the two payoffs, which in this case are the retailer's and wholesaler's costs. The goal in this case is to minimize both the retailer and wholesaler costs.

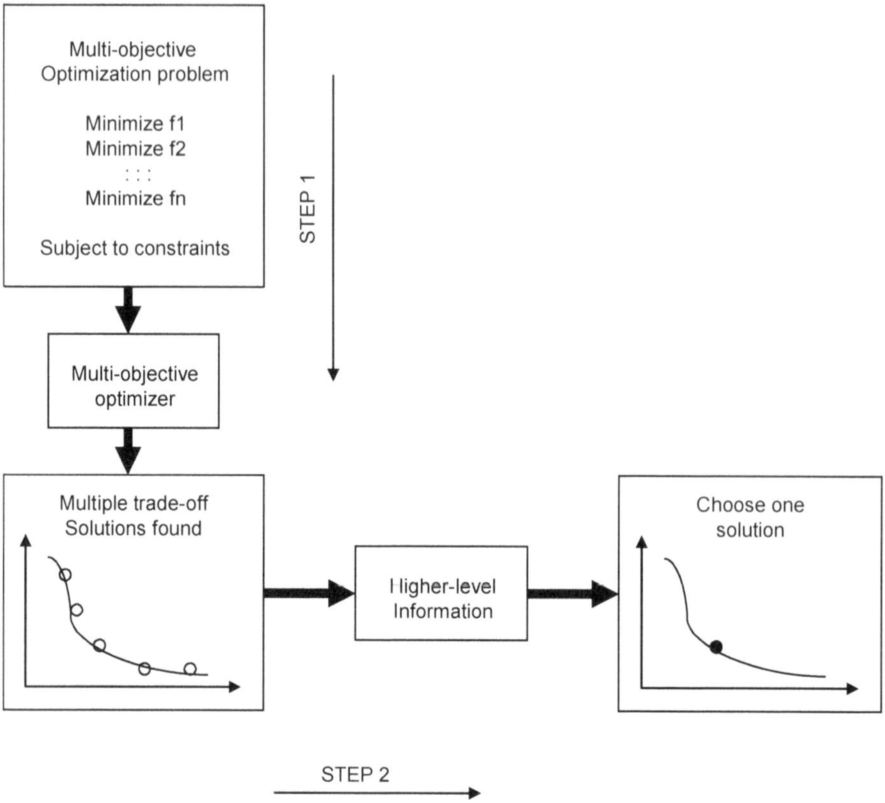

Figure 4. The two-stage problem solving approach of MOO. (Deb 2001)

Table 1. A sample set of solutions

Solution Number	Parameter Values	Retailer Cost	Wholesaler Cost	Rank	Is Dominated	P(S)
A	[2][18.02][47.19][6.96][10.33][25.12]	26.424	23,077	5	FALSE	0.29
B	[24.18][2.99][30.05][31.49][8.71][45.79]	53,303	65,321	3	TRUE	0.18
C	[6.69][42.81][33.69][17.74][28.51][44.51]	63,108	42,358	3	TRUE	0.18
D	[24.26][31.54][16.12][41.13][2.05][16.2]	64,712	3,351	5	FALSE	0.29
E	[13.07][12.22][10.46][29.53][39.57][37.07]	71,090	53,202	1	TRUE	0.06

Based on this set of possible solutions, a rank order is produced, and this ordering determines those solutions that are non-dominated. In order to calculate this ranking, solutions must be compared in a pair-wise manner. From this, it can be calculated that:

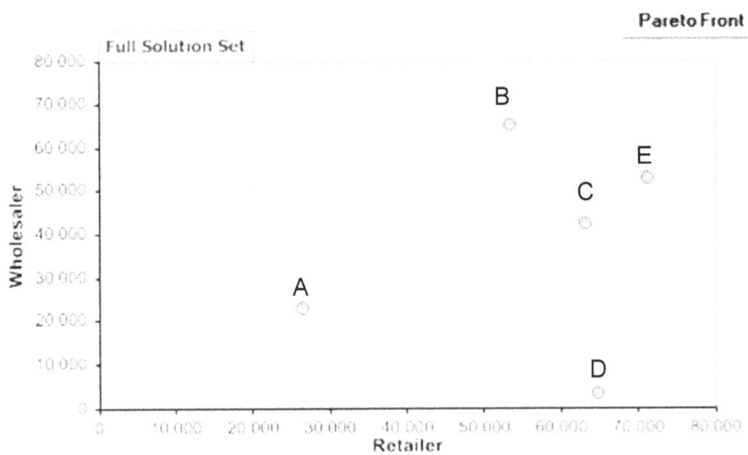

Figure 5. A sample solution space

- A dominates B, C and E, as it is better on both counts against these less-fit solutions.
- D dominates E, again because it is better on both counts.
- B dominates E, since it has a lower retailer cost, and a lower wholesaler cost.
- D dominates E, because it has a lower cost on both fronts.
- Note that even though D, which is a non-dominated solution, does not dominate either B or C, both B and C are dominated because they are inferior to A.

Therefore, A and D are not dominated by any solution, and are given the highest rank, which is 5 (the total number of solutions). These two solutions are equally optimal, because solution A wins out on minimizing the retailer's cost, and solution B is the best in terms of minimizing the wholesaler's cost. Solutions B and C both dominate E, so they are given the next order of ranking (3), and E is the lowest ranked solution, with a ranking score of 1.

Each solution is then assigned a fitness value, which determines the probability that it will be selected for future generations, and this probability P(S) is the ratio of their rank to the sum of all ranks. Table 1 shows that the two fittest solutions have an equal probability (0.29) of being selected for future generations, while the lowest ranked solution only has a selection probability of (0.06).

This idea of dominance is key to understanding MOO, because this is the ranking procedure used by the genetic algorithms as they search the solution space for the optimal solution. Genetic algorithms (GAs), are computational algorithms that are inspired by Darwin's ideas of evolution, where the fittest solutions survive and their characteristics are carried from one generation to the next. For optimizing system dynamics models, each GA represents an individual solution as an array of elements, where each position in the array corresponds to a possible parameter value.

The algorithm starts with a random population of solutions (normally 100), and these evolve by assessing their strength (e.g. running a simulation), selecting the strongest of these, performing crossover between the fittest, and also using a process known as mutation to ensure that variety is maintained in the overall population. The typical GA algorithm is of the form:

Step	Description
1. Generate Random Population - P	*A random set of solutions are generated (normally 100), based on generating random numbers from the [upper-lower] range of each parameter that needs to be optimized.*
2. While(Counter <= Number Of Generations)	*Keep going until all generations have evolved. Normally 100 generations are produced in a genetic algorithm.*
Calculate Payoffs for P	*Run the simulation model once for every solution, and return the payoff values*
Rank all solutions	*Based on these payoff values, rank each solution, where the non-dominated solutions are given the highest ranking.*
Calculate Selection Probabilities	*Based on the rankings, estimate the probability of selection for each solution*
Generate Mating Pool	*Randomly select – using roulette wheel selection – 100 of the fittest solutions.*
Crossover mating pool	*Perform crossover on a percentage of the mating pool.*
Mutate Mating Pool	*Perform mutation on a percentage of the mating pool.*
Let P = MP	*Replace the old population with the new mating pool*
Increment Counter	*Move on to the next generation*

In order for an optimal solution to emerge, a number of additional GA operators are used: *crossover* and *mutation.* But before they are described, it is worth mentioning the mechanism of roulette wheel selection, which is often employed to select solutions for the mating pool (see Figure 6).

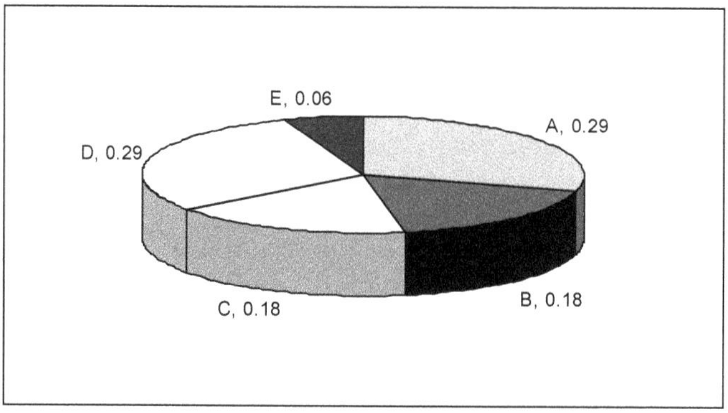

Figure 6. A roulette wheel representation of each solution's (Table 1) probability of selection

For roulette wheel selection, a random number is generated between [0..1], and solutions are given a range proportional to their strength. For example, solution A could be in the range [0.. 0.29], and if the random number generated was within this range, solution A would be selected.

Crossover (Figure 7) involves taking two solutions from the mating pool, and combining elements of those solutions to produce two new solutions. The procedure for performing this is:

- Identify a random crossover point for both parents (P1, P2), and mark the two solutions at this point.
- Join the first half of the first solution with the second half of the second solution to produce the first child solution (C1).
- Join the first half of the second solution with the second half of the first solution to produce the second child solution (C2).
- Remove original solutions (P1, P2), and replace these with the newly defined solutions (C1, C2).

While crossover works well in generating new solutions that combine the strengths of their parent solutions, there is a danger that the offspring solutions will converge too rapidly, thus leaving the overall optimization process stuck on a local optima.

Mutation (see Figure 8) addresses this problem of premature convergence, as it generates a new solution by randomly changing one of its component parts, namely, the value of one of the simulation parameters. This often results in the generation a solution that lies on a radically different point in the solution space. The procedure for mutation is:

- For each generation, randomly select a small number of solutions from the mating pool (according to a predefined probability).
- Randomly select an element of that solution.
- Generate a new value for this element, taking into account the constraints for each parameter in terms of highest and lowest possible values.

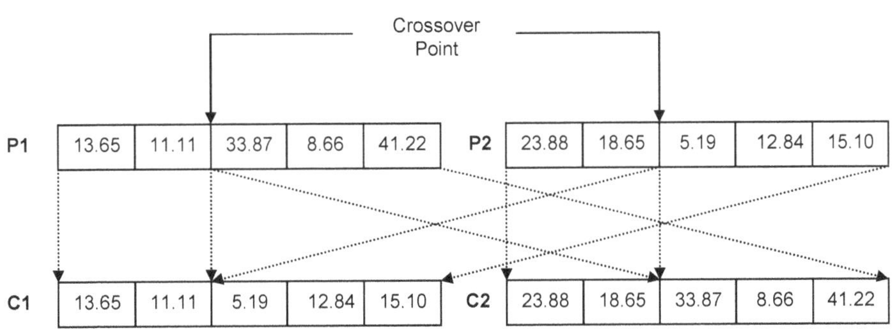

Figure 7. Crossover of two solutions

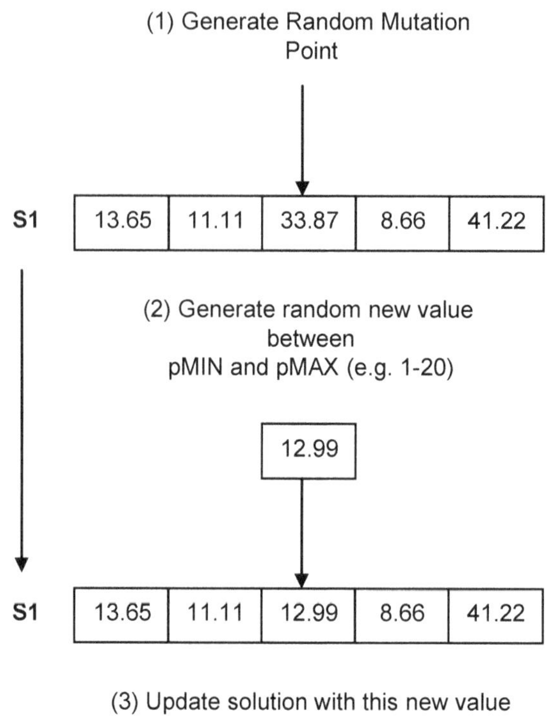

Figure 8. Mutation of a solution

In summary, as the GA evolves through each generation, the overall solution set becomes "fitter" and converges to an optimum. This optimum is known as the Pareto-optimal front, and contains the final set of non-dominated solutions from which the decision maker can select the most ideal, based on their unique judgment and experience. An example of how this works is now presented.

5. The Two-Agent Beer Game

In order to demonstrate how system dynamics and multiple objective optimization can be integrated, a purpose-built simulation and optimization demonstrator is presented. The underlying premise of the software is to take the well-known beer game model (Sterman 1989), simplify it to two agents (the normal game involves four agents), and then allow the user to perform multiple objective optimization and explore the solution space. For the purposes of the demonstrator, the following assumptions are made:

- A two echelon supply chain is the basis of the model (Figure 9). This model consists of two actors: a retailer and a wholesaler. Each actor has a similar control structure, which is based on the standard *stock management structure.* The essence of this structure is a decision heuristic

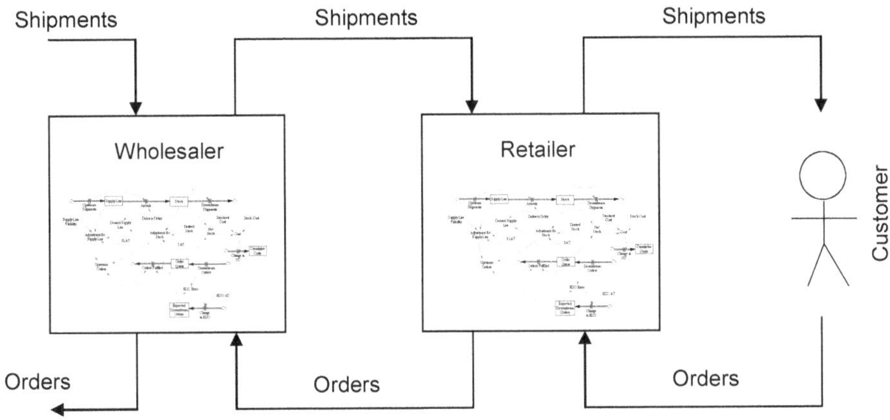

Figure 9. A two echelon supply chain

which strives to maintain each stock at a desired level – in effect, its goal is to maintain a dynamic equilibrium between inflows (arriving goods) and outflows (goods being sent to customers). It does this by replacing the expected losses, and also ordering sufficient material to move the stock towards the desired level.

- Two conflicting objectives are selected. These are: Retailer Cost and Wholesaler Cost. This is a slight change from the original game, where the overall cost is to minimize the total cost of a supply line (by aggregating the individual costs of each participant). The premise for this model is that the decision maker needs to trade off costs between the two agents, and observe the range of possible optimal solutions.

5.1. The Stock and Flow Model

The stock and flow structure for each agent is shown in Figure 10, and is the standard representation for an agent in the beer game.

Each agent receives downstream orders (1) from the downstream agent. For example, in the case of a retailer the orders would be placed by customers, and for a wholesaler agent, the orders would be placed by the retailer. These orders accumulate on an order queue (2), and orders are fulfilled depending on whether enough stock is available to fulfill the order (3).

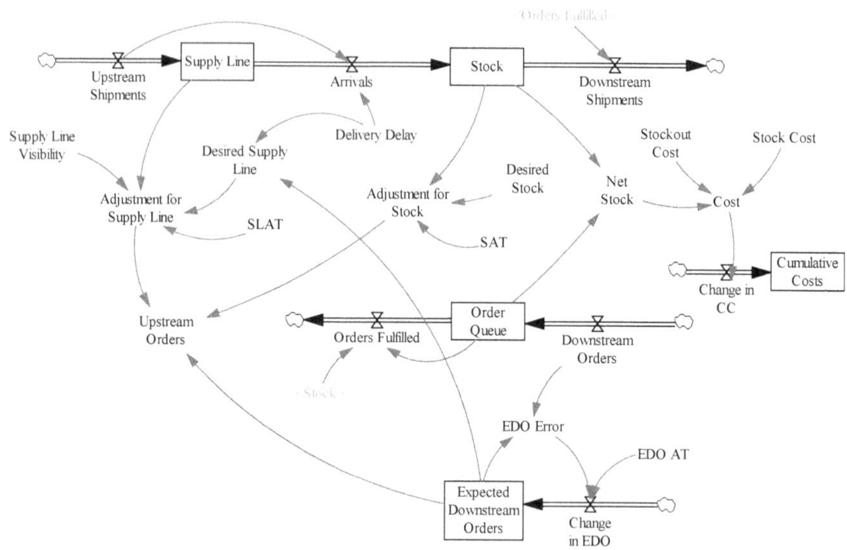

Figure 10. Generic Structure for an Agent in the Beer Game

Downstream Orders = Customer Demand	(1)
Order Queue = INTEGRAL(Downstream Orders – Orders Fulfilled, 100)	(2)
Orders Fulfilled = MIN(Order Queue, Stock)	(3)

The agent's stock (4) is depleted by downstream shipments (5) and incremented by arrivals (6). These arrivals are as a result of previously placed orders, which have worked their way through the supply line (7), following a pipleline delay distribution with a fixed value (9). The supply line is filled by upstream shipments (8) that are determined by the upstream agent.

Stock = INTEGRAL (Arrivals – Downstream Shipments)	(4)
Downstream Shipments = Orders Fulfilled	(5)
Arrivals = DELAY FIXED (Upstream Shipments, Delivery Delay, 100)	(6)
Supply Line = INTEGRAL(Upstream Shipments – Arrivals, 300)	(7)
Upstream Shipments -Determined by upstream agent using (5)	(8)
Delivery Delay = 3	(9)

The key decision rule for each agent is the calculation of upstream orders (10). This is a summation of the supply line adjustment (11), expected downstream orders (12), and adjustment for stock (13), and its value must always be positive. The supply line adjustment can be either on or off, depending on the value for supply line visibility (14). Therefore the demonstrator can model the earlier mentioned concept of bounded rationality, by switching off the supply line information cue by setting equation (14) to zero.

Upstream Orders = MAX(0, Adjustment for Supply Line + Expected Downstream Orders + Adjustment for Stock)	(10)
Supply Line Adjustment = Supply Line Visibility * ((Desired Supply Line – Supply Line) /SLAT)	(11)
Expected Downstream Orders = INTEGRAL(Change In EDO, 100)	(12)
Adjustment for Stock = (Desired Stock – Stock) / SAT	(13)
Supply Line Visibility = [1 = Use Supply Line, 0 = Ignore Supply Line]	(14)

Sterman (2000) explores the question "why do we ignore the supply line?" He discusses a range of examples, ranging from young people experimenting with alcohol, where there is a delay before the alcohol is fully absorbed and impacts on their behavior, to people clicking furiously on an unresponsive web page, unaware that all their previous clicks are being stacked up. He concludes that our decision making is tuned to the "here and now" and that our experiences tend "to reinforce the view that cause and effect are closely related in time and space".

A feature of these types of problems is a deterioration in our decision making performance due to the fact that our intuitive understanding of cause and effect breaks down. For example, in the Beer Game example, there is a significant delay between the time an agent orders and the time that order arrives, and this delay causes people to forget what they have previously ordered. In a sense, because we are not "programmed" to recognize the impact of time delays on cause and effect, this behavior must be learned, and hence the importance of methodologies such as system dynamics as a means of learning about complex systems.

A number of additional variables are used to in the calculation of upstream orders. These include, desired supply line (15), which is the product of expected downstream orders and the delivery delay, and reflects the steady state target for the number of orders that should be in the supply line, give the expected demand. The supply line adjustment time (16) is a parameter that determines how quickly the difference between the desired and actual supply line levels is bridged. Desired stock (17) represents the agent's goal, and the stock adjustment time (18) is a parameter that determines how quickly this goal is reached, and in effect is a measure of how aggressive the agent is in reaching its goal.

Desired Supply Line = Expected Downstream Orders * Delivery Delay	(15)
Supply Line Adjustment Time (SLAT) = 1	(16)
Desired Stock = 400	(17)
Stock Adjustment Time (SAT) = 1	(18)

The remainder of the model comprises the additional variables needed to calculate the expected downstream orders, including EDO Error (19), change in EDO (20), and EDO adjustment time (21) – which represents the information delay in estimating future downstream orders. Finally, the net stock (22) is calculated, and used in determining the cost (23), the change in cumulative costs (24) and the cumulative costs (25). The variables stock cost (26) and stockout cost (27) capture the relative inventory holding and stock penalty costs.

EDO Error = Downstream Orders – Expected Downstream Orders	(19)
Change in EDO = EDO Error / EDO AT	(20)
EDO AT = 3	(21)
Net Stock = Stock – Order Queue	(22)
Cost = IF THEN ELSE (Net Stock < 0, ABS(Net Stock * Stockout Cost), Net Stock * Stock Cost)	(23)
Change in CC = Cost	(24)
Cumulative Costs = INTEGRAL(Change in CC, 0)	(25)
Stock Cost = 0.5	(26)
Stockout Cost = 2.0	(27)

5.2. Reflections on the Model

Before exploring the role of simulation and optimization for the two-agent beer game, it is instructive to reflect on key aspects of the model. While each agent uses the same heuristic, there are important variations in the use of the heuristic that can lead to a wide range of different behaviors.

- First, there is the choice of whether or not an agent takes the supply line into account. In effect, this models whether the agent has reflected on the dynamics of delayed cause and effect, or whether the agent has a fundamental misperception of time delays and their effect on the system. The variable that determines the level of feedback awareness of each agent is Supply Line Visibility (14). If this value is one, the agent is aware of the supply line, if it is zero, the agent suffers from misperceptions of feedback.
- Second, there are a number of parameters in the model which relate to the adjustment times an agent uses in order to formulate a decision. These adjustment times are: supply line adjustment time (16), stock adjustment time (18) and expected downstream orders adjustment time (21). The first two of

these are a measure of how quickly an agent adjusts to a goal, and the third is the time constant employed for exponential smoothing in order to calculate the expected value for downstream orders. Low values indicate rapid adjustment, higher values model the scenario of gradual adjustment, and more conservative decision making. The choice of values for these adjustment times can have significant impact on the overall performance of the model. Furthermore, the task of finding the optimum performance of the model is dependent on a search of the solution space for varying values of each parameter.

- Third, agent interactions are an important feature of the model, and there are two classes of interactions. The first is where a downstream agent sends an order signal to an upstream agent. In effect, this means that the value determined by the downstream agent in equation (10) is copied to the inflow of the upstream agent's order queue, which is described by equation (1). The second interaction models the shipment of goods from an upstream agent to a downstream agent. For this scenario, equation (5) of the upstream agent determines the value for equation (8) in the downstream agent. Finally, because there are only two agents in the model, it is assumed that an unlimited supply of goods is available to the agent furthest upstream in the supply chain, and in this scenario, upstream shipments (8) is determined by the value for upstream orders (10).
- Fourth, payoffs are a key feature of the model, because they determine how well an agent performs. The measure used to assess agent performance is total cumulative cost (25), where the cost is a function of the agent's inventory. If the agent performs poorly and allows their net inventory to slip below zero, they incur a penalty cost which is four times higher than the normal carrying cost. Therefore, for each agent they must strive to employ a good stock control heuristic which keeps their actual stock levels close to the target values.

5.3. Running the Model

The software system is a learning environment for supply chain dynamics, where the goal is to find the best combination of parameters that will minimize each agent's total costs. The system can run in two modes:

- Simulation mode, where the user can run repeated simulations by varying the parameter values, and observe the system performance (see Figure 11).
- Optimization mode, where the user specifies the valid range for parameters, gives initial values for configuring the genetic algorithm, runs the algorithm to generate the Pareto optimal front, and browses through the solution space, using their qualitative knowledge and judgment – their *higher level information* - to select the best solution.

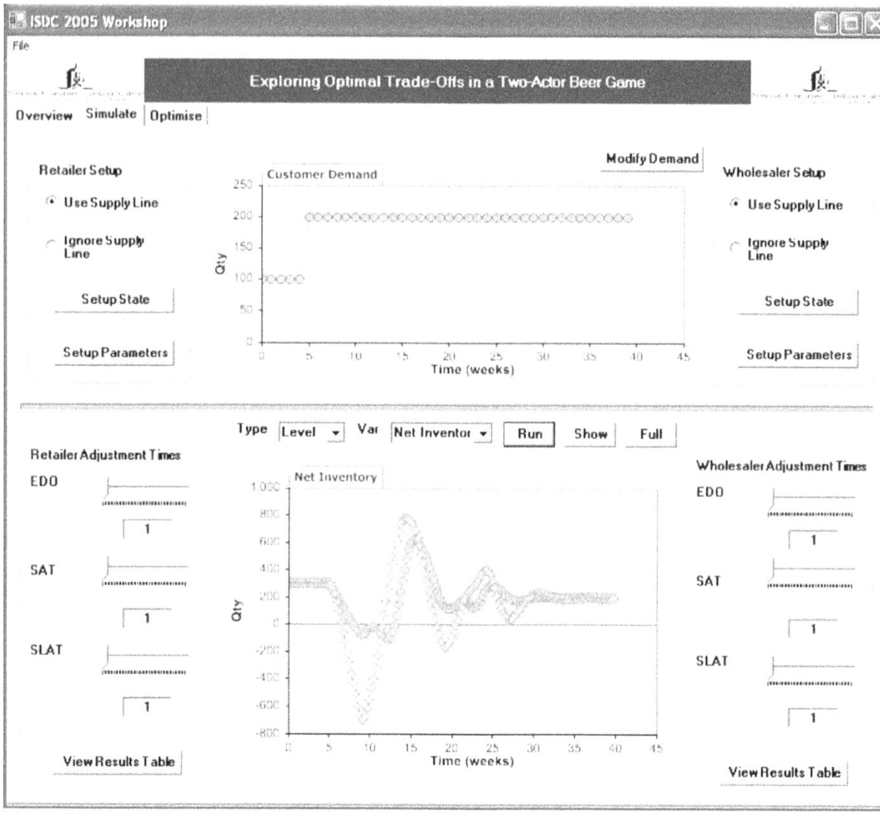

Figure 11. The Simulation Console

In order to run the simulation, the user specifies the appropriate demand function that models customer's orders. The default value, and the one commonly used to demonstrate the Forrester effect of oscillation and amplification, is a step function where demand doubles shortly into the simulation run. The default values for each variable ensure that the model initially operates in equilibrium, and the effect of the change in demand disturbs this equilibrium and forces the agents to seek out the new equilibrium by applying their decision rules. In the screen shot shown, equilibrium is achieved after thirty time units or so. This is because each agent is using the supply line as part of their decision heuristic, if the supply line was "switched off", the inventories would continue to undershoot and overshoot their target value: this is the classic behavior that is so often replicated in the beer game.

Furthermore, for the initial simulation, all the parameters are set initialized at one. These are arbitrary selections, and as the user runs consecutive simulations, they can experiment with different parameter values and observe the overall impact on the system performance with a view to minimizing both retailer and wholesaler costs.

The key benefit of running the simulation is that it allows the user to experiment by changing parameter values and observing the result immediately. In this way, they get a good intuition for how the system behaves, and the impact different parameters have on system performance. For example, they might well start off with the question: "if I vary the retailers SAT and keep all other parameters unchanged, how will that effect system performance?" By observing the impact of this, they can achieve a greater awareness and understanding about the system, and how sensitive it is to even small changes in parameter values.

However, when a user has a good understanding of the overall system structure, and has gaining a good insight into how the choice of parameters, they will want an answer to the question: "what combination of parameters will optimize the retailer and wholesaler costs?" The answer to this is not repeated experimentation. This complex computational is best solved by an optimization algorithm.

Before the multiple objective optimizer can run, a number of specific values need to be decided on, including:

- The range of values for all six parameters. For the purposes of this experiment, the minimum value for each parameter is one, the maximum value fifty.
- Whether or not the supply line will be used as part of the agent's heuristics. For this experiment, the supply line is user by both agents, although this can be easily changed for subsequent optimization runs.
- The initial values for the genetic algorithm, including population size (100), number of generations (100), crossover probability (80%), and mutation rate (15%).

With these constraints for the parameters and the configuration information for the genetic algorithm, the multiple objective optimizer is run. Because these optimizations are stochastic, the exact same results will not be replicated every time, but for one run of this experiment, the entire solution set (10,000 solutions) is shown (Figure 12).

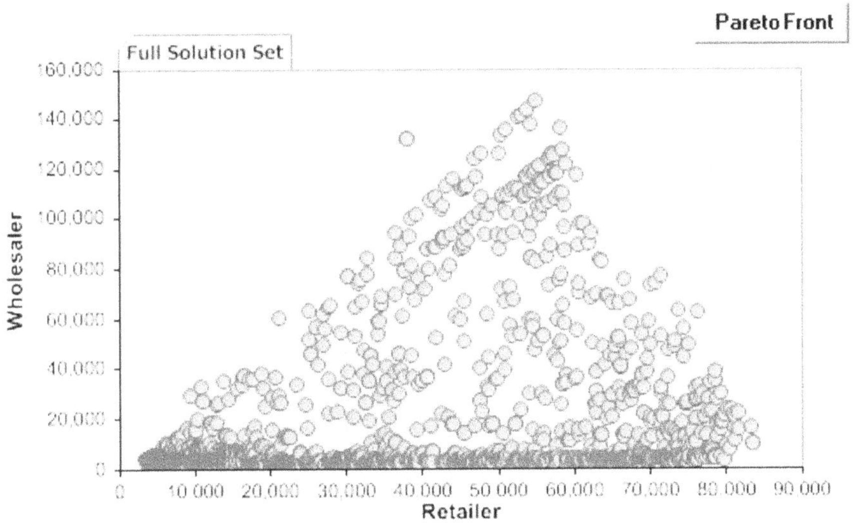

Figure 12. The entire solution set for an optimization run

The overall picture is one where, after successive generations, the fittest solutions evolve, and these are found closest to the origin (0,0). Each solution represents a particular combination of parameter values that produced the two payoffs, namely wholesaler cost and retailer cost. This diagram also clearly shows that a number of solutions are extremely poor, for example, the one at the apex of the triangular shape which costs the wholesaler over 140,000 and results in a retailer cost of approximately 55,000. However, as discussed earlier, poor solutions will be easily dominated and because of this receive a much lower rank, and so have a far lower probability of being selected for future generations.

In the final generation, only a certain number out of the population size will be non-dominated solutions, and this is the final solution set presented to the user. In this particular experiment, fourteen solutions are non-dominated, and this comprises the Pareto optimal front. The front is visualized in Figure 13, while the data points and corresponding parameter values are displayed in Table 2.

At this point the multiple objective optimizer has completed its task, which corresponds to step one of the two-stage process described earlier. It has generated and presented a valid set of optimal solutions to the decision maker, who must now employ their judgment and experience, and select the most appropriate optimal solution.

Using the demonstrator, the decision maker can iterate between the Pareto front and the specific simulation results (for example, view total costs, net inventory, phase plots and other variables for each optimal solution), and so observe the overall performance of the model (see Figure 14, which shows the net inventory behavior for both wholesaler and retailer generated by solution 80).

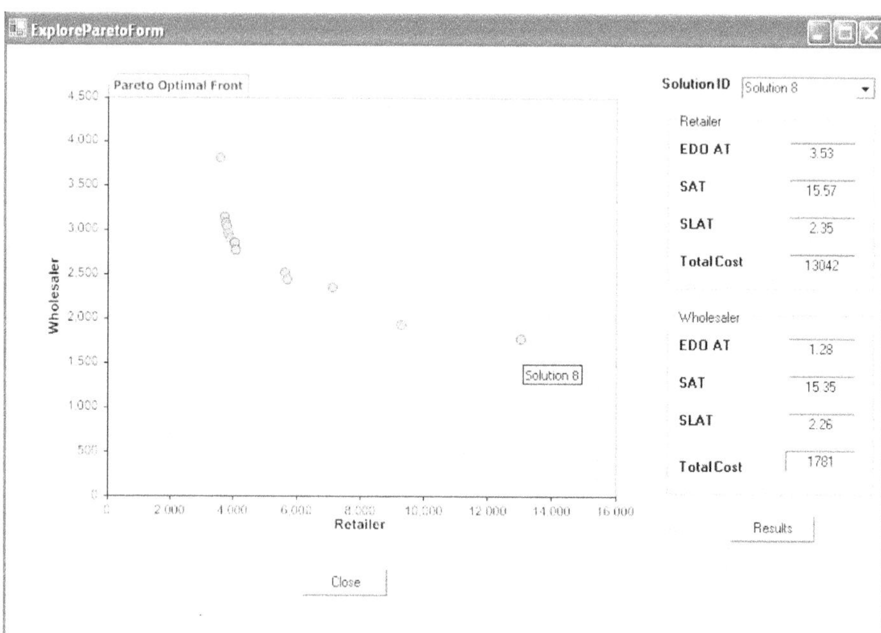

Figure 13. Visualization of the Pareto optimal front

Table 2. The Pareto Optimal Front

	RETAILER				WHOLESALER			
Solution	EDO AT	SAT	SLAT	Cost	EDO AT	SAT	SLAT	Cost
80	1.62	10.9	2.35	3588	1.28	4.68	2.26	3814
7	1.62	10.9	2.35	3736	1.28	10.96	2.26	3144
4	1.62	10.9	2.35	3774	1.28	12.87	2.26	3079
11	1.62	12.72	2.35	3784	1.28	10.96	2.26	3044
91	1.62	12.72	2.35	3826	1.28	12.87	2.26	2961
99	1.62	12.72	2.35	3885	1.28	15.35	2.26	2915
10	1.62	15.57	2.35	4036	1.28	12.87	2.26	2851
19	1.62	15.57	2.35	4073	1.28	15.35	2.26	2781
5	1.62	15.57	6.79	5624	1.28	12.87	2.26	2530
25	1.62	15.57	6.79	5674	1.28	15.35	2.26	2444
49	1.62	31.92	3.73	7093	1.28	15.35	2.26	2351
16	3.02	12.49	2.35	9296	1.28	15.35	2.26	1934
8	3.53	15.57	2.35	13042	1.28	15.35	2.26	1781

The question now remains: which solution should be selected?

There is no right answer to this question, because – objectively speaking – all solutions are equally optimal, and therefore the final selection depends on the viewpoint and perspective of the decision maker, and also is a function of what extra qualitative knowledge, judgment and experience they can make use of. We could take three scenarios, based on the two extreme points of the Pareto optimal front and the middle point (see Table 3), and speculate on the kind of issues that could swing the balance towards one solution over another.

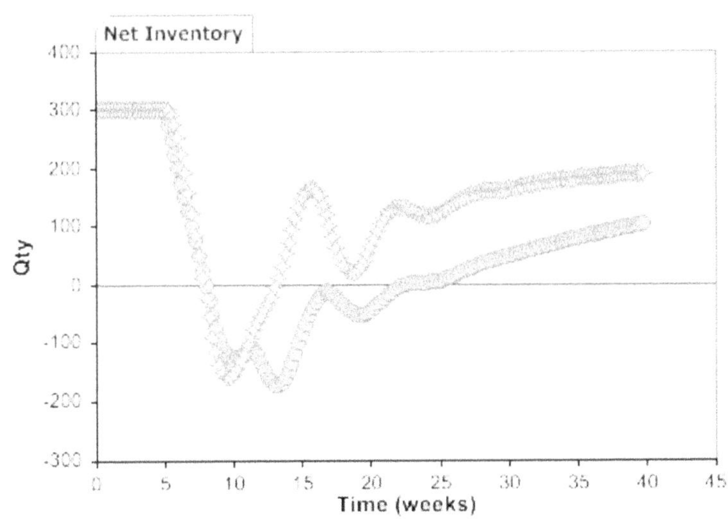

Figure 14. Simulation of inventory for solution 80

Table 3. A subset of the Pareto optimal front

	RETAILER				WHOLESALER			
Solution	EDO AT	SAT	SLAT	Cost	EDO AT	SAT	SLAT	Cost
80	1.62	10.9	2.35	3588	1.28	4.68	2.26	3814
10	1.62	15.57	2.35	4036	1.28	12.87	2.26	2851
8	3.53	15.57	2.35	13042	1.28	15.35	2.26	1781

For example, one decision maker could advocate solution 8, on the basis that the wholesaler costs is minimized, and that their view is that the retailer can more easily absorb the extra cost. Another person could argue strongly that there is no way the retailer could be expected to incur such a high cost, and that solution 80, which does the best job at balancing the costs should be selected. A third decision maker may well advocate solution 10, because the retailer costs are significantly reduced compared with solution 8.

Whatever the merit of each argument, the key point is that decision makers are participating in an informed discussion, where the given quantitative options are transparent and observable. The discussion can then focus on the relative merits of individual arguments, based on the additional qualitative information that is provided by individual decision makers.

6. Conclusion

A key challenge for management science is to continue to search for ways, methods and techniques to enhance learning in complex systems, in order to improve decision maker's ability to make better, more informed, decisions. By presenting a synthesis of system dynamics and multiple objective optimization, this chapter has demonstrated how this hybrid approach can be used to help decision makers apply a quantitative and qualitative approach in order to decide between conflicting trade-offs. The quantitative element involves building a system dynamic feedback model, formulating the optimization problem, and generating a set of non-dominated optimal solutions. This is followed by an iterative qualitative step, where decision makers can bring higher level information to bear, in order to arrive at an agreed, and viable, overall optimal solution.

References

Dangerfield, B. and C. Roberts. 1996. "An Overview of Strategy and Tactics in System Dynamics Optimisation." *Journal of the Operational Research Society,* 47, pp 405–423.

Chen, Yao-Tsung, and Bingchiang Jeng. (2004). "Policy Design to Fitting Desired Behaviour Pattern for System Dynamics Models." *Proceedings of the 22nd International Conference of the System Dynamics Society,* Oxford, England.

Coello Coello, C.A., Van Veldhuizen, D.A., and Lamont, G.B. 2002. Evolutionary Algorithms for Solving Multi-Objective Problems. Kluwer Academic Publishers, New York, NY 10013.

Coyle, R.G. 1996. *System Dynamics: A Practical Approach.* Chapman and Hall, London, UK.

Deb, K. 2001. *Multi-Objective Optimisation Using Evolutionary Algorithms.* John Wiley and Sons, Baffins Lane, Chichester, UK.

Forrester, J.W. 1961. *Industrial Dynamics.* Productivity Press.

Goodwin, P. and G. Wright. 2003. *Decision Analysis for Management Judgment.* Wiley.

Grossman, B. 2002. "Policy Optimization in Dynamic Models with Genetic Algorithms" *Proceedings of the 20th International Conference of the System Dynamics Society,* Palermo, Italy.

Keloharju, R. and E.F. Wolstenholme. 1989. "A Case Study in System Dynamics Optimisation." *J. Ops. Res. Soc.,* Vol. 40, No.3., pp 221–230.

Richardson, G.P and A.L Pugh III. 1981. *Introduction to Systems Dynamics Modeling with DYNAMO.* MIT Press.

Simon, H.A. 1982. *Models of Bounded Rationality.* The MIT Press, Cambridge MA, 1982.

Senge, P.M. 1990. *The Fifth Discipline: The Art & Practice of The Learning Organization.* Doubleday, New York.

Sterman, J.D. 1989. "Modeling Managerial Behaviour: Misperceptions of Feedback in a Dynamic Decision Making Environment." *Management Science.* Vol. 35. No. 3, March 1989.

Sterman, J.D. 2000. *Business Dynamics. Systems Thinking and Modeling for a Complex World.* McGraw Hill Higher Education.

Wolstenholme, E.F. 2003. "Towards the definition and use of a core set of archetypal structures in system dynamics." *System Dynamics Review.* No. 1. (Spring 2003). pp 7–26.

Chapter 5

Creation of Hydroelectric System Scheduling by Simulation

Jesús M. Latorre, Santiago Cerisola, Andrés Ramos
Instituto de Investigación Tecnológica
Universidad Pontificia Comillas
Andres.Ramos@iit.icai.upcomillas.es

Rafael Bellido, Alejandro Perea
Iberdrola Generación

1. Introduction

In electric energy systems, hydro scheduling plays an important role for several reasons [Wood 1996]:

- Hydro plants constitute an energy source with very low variable costs. Operating costs of hydro plants are due to operation and maintenance and, in many models, they are neglected with respect to thermal units' variable costs. This is the reason for employing this technology as much as possible.
- Hydro plants provide a greater regulation capability than other generating technologies, because they can quickly change their power output. Consequently, they are suitable to guarantee the system stability against contingencies.
- Electricity is difficult to store, even more when considering the amount needed in electric energy systems. However, hydro reservoirs and pumping storage units give the possibility of accumulating energy, as a volume of water. Clearly, management of water accumulation suffers from the energy losses due to the efficiency of pumping storage units, but benefits from avoiding spillages and preventing floods. Altogether, hydro management is usually oriented to guarantee long-term electricity supply, even in recent electricity markets.

In the Spanish electric system, hydro production ranges from 15 % to 20 % of the total output, depending on meteorological conditions. Therefore, it is very important to have a tool to determine water management.

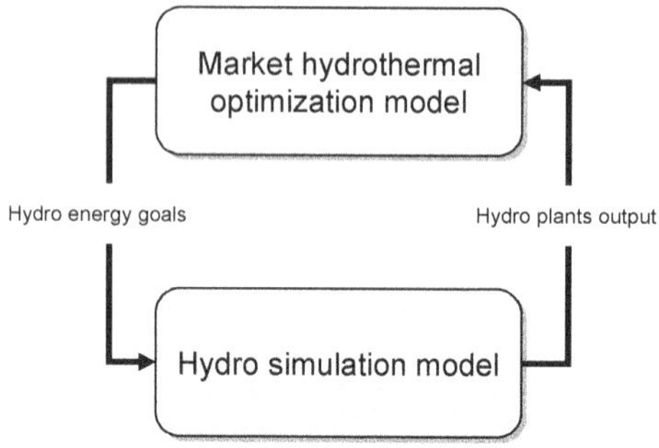

Figure 1. Hydro scheduling models

In particular, the simulation model we are presenting takes into account the detailed topology of each basin and the stochasticity in hydro inflows. This hydro simulation model is directly related to a higher-level long term market hydrothermal model, based on optimization, see figure 1. The optimization model conducts the simulation model through a collection of hydro production goals that should be done in different weeks at the largest hydro reservoirs, left arrow of figure 1. Later, the simulation model checks the feasibility of these objectives, may test the simplifications made by the optimization model, determines the energy output of hydro plants, the reserve evolution of the reservoirs and, therefore, a much more detailed operation (right arrow of figure 1). The simulation model deals with plausible scenarios and generates statistics of the hydro operation. Its main applications are:

- Comparison of several different reservoir management strategies.
- Anticipation of the impact of hydro plant unavailabilities for preventing and diminishing the influence of floods.
- Increment of hydro production by reducing the spillage.

This double hierarchical relation among different planning models to determine the detailed hydro plants operation has also been found in [Turgeon 1998].

Simulation is the suitable technique when the main objective is to analyze complex management strategies. Simulation models are classified according to the next three features [Law 2000]:

- Models are called static when time is not considered. On the other hand, dynamic models regard system evolution along the time.
- Models are deterministic when input data are deterministic. Otherwise, they are stochastic or probabilistic.
- Simulation models are discrete when system state changes occur at selected moments of the complete simulation scope. Otherwise, if system state changes continuously, simulation models are termed continuous.

Taking this into account, the simulation model proposed is a dynamic model, whose input data are series of historical inflows in certain basins spots. Historical or synthetic series (obtained by forecasting methods) can be used for simulation. For this reason, it is also a stochastic model. Finally, system state changes take place once a day. These events can be changes of inflows, scheduled outages, etc. Consequently, the model is discrete with one day as time step. This is a reasonable time step because the usual model scope is one year and no hourly information is needed for planning purposes.

This chapter is organized as follows. In section 2, we discuss and review some similar references found in the literature. In section 3, we present the data representation adapted to Object Oriented Programming. In section 4, we show the proposed simulation method. In section 5, we present a realistic case study. Finally, in section 6 we extract some conclusions of the chapter.

2. Literature Review

The Object Oriented Programming paradigm [Wirfs-Brock 1990] results very attractive for simulation because it allows the independent simulation of each system element. An example can be found in [Allen 2001], where a simulation model applied to electric energy systems is carried out in an object-oriented environment. The model allows the study about frequency control in those systems.

The idea behind object-oriented programming is that a computer program is composed of a collection of individual data and code units, or objects, that act on each other, as opposed to a traditional view in which a program is a list of procedures to the computer, that operate on separate data. In the hydro simulation model of this chapter, those objects are the basin elements (reservoirs, plants, channels, inflows and river junctions). Each object is capable of receiving data, processing this data and its own state, and sending information to other objects. In this model, the most basic information corresponds to the exchanged water flow between elements. One of the main reasons to adopt this paradigm stems from the fact that the resulting simulation model will be useful for a general basin. With this purpose, an object-oriented approach allows a more direct analysis and understanding of complex hydro basin operation.

Simulation of hydrothermal systems has been used in the past for several purposes. One is reliability analysis of electric power systems. An example of this is [Roman 1994], where a complete hydrothermal system is simulated. The merit order among all the reservoirs to supply the demand is determined as a function of their reserve level. Simulated natural hydro inflows and transmission network are considered. The goal is to determine the service reliability in thermal, hydro or hydrothermal systems. In [Sankarakrishnan 1995] and [Van Hecke 1998] the electric systems simulated do not consider the hydro topology. This simulation model takes a time step of one hour, and considers the electric network, whose node demand is defined as a function of the expected user type (industrial, residential, commercial, etc.). Variance reduction techniques are applied and the results are reliability measures. In [Sankarakrishnan 1995] sequential simulation is compared with non-chronological methods based on the load-duration curve.

In [Cuadra 1998] it is proposed a simulation scheme for hydrothermal systems where medium and long term goals (maintenance, yearly hydro scheduling) are established. The system is simulated with stochastic demand and hydro inflows. For each day an optimization problem is solved to achieve the goals obtained from long-term models.

The model presented in this chapter is based on the object-oriented paradigm and defines five objects that are able to represent any element of a hydro basin. The model incorporates a simulation algorithm in three phases that decides the production of the hydro plants following several strategies about reservoir management. These management strategies and a description of the five objects are presented next.

3. Data Representation

A natural way to represent the hydro basin topology is by means of a graph of nodes, each one symbolizing a basin element. Those nodes represent reservoirs, plants, inflow spots and river junctions. Nodes are connected among them by arcs representing water streams (rivers, channels, etc.). Each node is independently managed, although that may require information about the state of other upstream basin elements. As a result of this data structure, object oriented programming is a very suitable approach to solve the problem of simulating a hydro basin.

Analyzing real hydro basin patterns, we have concluded that five objects are enough to represent adequately every possible case. These object types are described in the next section. Additionally, different reserve management strategies can be pursued in a reservoir element. In section 3.2 we discuss these strategies.

3.1. Object Types

In this section, we introduce the node types we have identified. These nodes represent reservoirs, channels, plants, inflows and river junctions, which are described from now on.

1) Reservoirs

The objects representing reservoirs have one or more incoming water streams and only one outgoing. They may have a minimum outflow release, regarding to irrigation, sporting activities or other environmental issues.

Besides, they may have rule volume curves guiding their management. Examples are minimum and maximum rule curves, which, in the second case, avoid spillways risk.

Reservoirs are the key elements where water management is done. The chosen strategy decides its outflow (see section 3.2). This management takes into account minimum and maximum guiding curves, absolute minimum and maximum volume levels and pre-calculated production tables. Pre-calculated tables are used for hyper-annual management reservoirs and determine the optimal reservoir outflow as a function of the week of the year, the level of the involved reservoir and other important reservoirs of the same basin and of the incoming hydro inflows. These tables are computed with other long-term tools that use stochastic optimization as commented in the introduction of the chapter, see figure 1.

2) Channels

These elements carry water between other basin elements, like real water streams do. They do not perform any water management. They just transport water from their origin to their end. However, they impose an upper limit to the transported flow, which is the reason to consider them.

3) Plants

Kinetic energy from water flow that goes through the turbine is transformed into electricity in the plant. In electric energy systems, hydro plants are important elements to consider. However, in hydrothermal simulation they do not manage the water, which it is decided by the reservoir located upstream. From this point of view, they are managed in the same fashion as channels: they impose an upper limit to the transported flow.

As a simulation result, electric output is a function of the water flow through the plant. This conversion is done by a coefficient of efficiency, whose value is a linear[1] function of the water head. Water head is the height between the reservoir level and the maximum between the drain level and the level of the downstream element. In hydro plants, once the water flow has been decided, daily production is divided between peak and off-peak hours, trying to allocate as much as energy possible in peak hours where expensive thermal units are producing.

In addition, some plants may have pumping storage units, which may take water from downstream elements and storage it in upstream elements (generally, both elements will be reservoirs). It is important to emphasize that, in the model presented in this chapter, pumping is not carried out with an economic criterion but with the purpose of avoiding spillage.

Preventive maintenance of hydro units is also taken into account. This maintenance influences the available resources for producing electricity.

4) Inflows

These objects introduce water into the system. They represent streamflow records where water flow is measured, coming from precipitation, snowmelt or tributaries. These elements have no other upstream elements. The outflow corresponds to the day been simulated in the series. This series may come from historic measures or from synthetic series obtained from forecasting models based on time series analysis.

5) River junctions

This object groups other basin elements where several rivers meet. An upper bound limits the simultaneous flow of all the junction elements. An example of this object appears when two reservoirs from two rivers drain to the same hydro plant, as in figure 2. As both reservoirs share the penstock, this element has to be managed independently.

[1] This dependency is nonlinear but for short time intervals, as a day, where there are only small variations in the reservoir head it can be supposed to be linear.

Figure 2. Example of river junction

Management of this river junction is organized in two steps. Firstly, default and individual operation of all the elements is obtained. Secondly, when this operation exceeds the maximum flow, outflows are reduced following a priority order. This priority order may follow different criteria. Electric output maximization is the one considered for the numerical results presented later.

3.2. Reservoir Management Strategies

Reservoir management is the main purpose of the simulation; the rest of the process is an automatic consequence of this. Different strategies represent all possible real alternatives to manage reservoirs with diverse characteristics. These strategies are discussed in the following paragraphs.

1) Pre-calculated decision table

This management is used for large reservoirs that control the general basin operation. Typically, those reservoirs are located at the basin head. This table is determined by a long-term optimization model and gives the optimal reservoir outflow as a function of the following input data:

- Week of the year of the simulated day.
- Hydrological index of the hydro basin state. It can represent the inflow at a certain important and distinctive spot or the total basin inflow. Another possibility is to consider the ratio of total inflow divided by the historical mean.
- Volume of the reservoir being simulated. The greater the volume, the greater the possible outflow and vice versa.
- Volume of other reservoirs of the same basin, if that exist. These data provide additional information about the basin state and allows managing the main reservoirs together.

A multidimensional interpolation among the corner values read from the table is made to obtain the result.

2) Production of the incoming inflow

This strategy is specially indicated for small reservoirs. Given that they do not have much manageable volume, they must behave as run of the river plants and turbine the incoming inflow.

3) Minimum operational level

In this strategy, the objective is to produce as much flow as possible. Of course, when the reservoir level is below this operational level no production is done. When the volume is above this operational level, the maximum available flow must be turbined to produce electricity. This strategy can be suitable for medium-sized reservoirs when little energy is available at the basin.

4) Maximum operational level

With this strategy, the volume is guided to the maximum operational level curve. This is a curve that prevents flooding or spillage at the reservoirs. However, in case of extreme heavy rain it can be dangerous to have the reservoir at this level. This strategy can be suitable for medium-sized reservoirs when enough energy is available at the basin.

4. Simulation Method

Simulating a hydro basin allows to observe its evolution for different possible hydro inflows. Operation of hydro units follows the management goals of the reservoirs. These goals are given by the management strategy decided for each reservoir, as presented in the previous section. However, other factors can force changes in previous decisions, for example:

- Spillage has to be avoided. If a reservoir spills because it reaches its maximum level, at the same time it is losing the opportunity to produce at another moment where an expensive thermal unit will be used. Those costs would be avoided if a better reservoir management were done.
- Minimum stream flow has to be assured at certain sections of the river for irrigation, sporting activities, water supply to cities or any other ecological reason.

To achieve this purpose we propose a three-phase simulation method, consisting of these phases:

1. Decide an initial reservoir management.
2. Modify the previous management to avoid spillage or to supply irrigation and ecological needs.
3. Determine the hydro units output with previous outflows.

Once obtained the results for several series of inflows, the detailed results for each series and mean and quantiles values are shown.

4.1. First Phase

In the first phase, we simulate the operation of the basin elements to determine an initial reservoir management. To simulate each element it is necessary to know how much water is available, so simulation is performed from upstream to downstream elements.

Each element is managed independently, that is the reason why we have chosen Object Oriented Programming. In this phase, for each element whose inflow is known (i.e., whose upstream elements have already been simulated) the outflow is calculated. For reservoirs, outflow depends on the reservoir volume v, on the incoming flow f and on the chosen strategy. Each element is considered individually:

- *Reservoirs* decide outflow as a function of the chosen strategy and it is determined as follows:
  - By their pre-calculated decision table.
  - Spending the inflow, as if they were a channel

$$o = f \tag{0.1}$$

  - Guiding the volume to the minimum operational curve v^{min}

$$o = f + \frac{v - v^{min}}{0.0864} \tag{0.2}$$

  - Guiding the volume to a percentage k^{max} of the maximum operational curve v^{max}

$$o = f + \frac{v - k^{max} v^{max}}{0.0864} \tag{0.3}$$

- *Channels*: as mentioned previously, channels only carry water and put the inflow as the outcome

$$o = f \tag{0.4}$$

- *Plants*: similarly to channels, in this phase plants only move the inflow to the output

$$o = f \tag{0.5}$$

- *Inflows*: the outflow of this element is the series data corresponding to the simulated day.
- *River junction*: river junctions are groups of elements that have a common maximum flow. As a first step in their process, each junction element is managed individually, without considering its belonging to a river junction. Then, the maximum total flow is checked. If this constraint is exceeded it will be necessary to reduce the flow upstream using some criterion.

Besides determining the flow for each element, some additional values are computed that will be used in the second phase to correct spillage and lack of minimum ecological flow. These values are: additional volume v^a that the element can provide, additional volume v^k that the element can keep, additional volume rv^a that the element requires their upstream elements to provide, and volume rv^k that the element requires their upstream elements to keep. These magnitudes take only a value different from zero for the reservoirs, where this computation is made as a function of the pre-calculated output flow for the reservoir o, the maximum outflow o^{max}, the minimum outflow o^{min} and the spillage spl. Moreover, accumulated volumes v^k_{acc} and v^a_{acc} that can be retained and provided, respectively, from that upstream element are additionally computed. Then the following controls are made:

- If *spillage* is taking place, increase reservoir outflow to avoid it. If reservoir is at maximum flow and its spillage is spl', then upstream elements must retain water

$$rv^k = \min\left(v^k_{acc}, 0.0864 spl'\right) \tag{0.6}$$

- If no enough ecological *flow* is produced, increase the reservoir outflow as much as possible to achieve it. Even if with this increased flow $\overline{o}$ the ecological flow o^{min} is not reached, upstream elements must provide additional flow

$$rv^a = \min\left(v^a_{acc}, 0.0864\left(o^{min} - \overline{o}\right)\right) \tag{0.7}$$

- If the reservoir is not in any of the *above* cases, it may contribute keeping water or producing more water to avoid problems with downstream elements. In this case

$$\begin{aligned} rv^a &= 0 \\ rv^k &= 0 \end{aligned} \tag{0.8}$$

The additional volumes the reservoir can keep or provide are then calculated. If the final reservoir volume v^f is above a value $\overline{v}$, the volume to keep is set to zero, to avoid spillage in the future

$$\begin{aligned} v^a &= v^f \\ v^k &= \overline{v} - v^f \end{aligned} \tag{0.9}$$

This computation is made individually for each element of the river basin and the remaining magnitudes take value zero.

$$\begin{aligned} v^a &= 0, \quad rv^a = 0 \\ v^k &= 0, \quad rv^k = 0 \end{aligned} \tag{0.10}$$

In all the elements, the volume to be additionally provided or kept in this element and their upstream elements is updated as follows

$$v_{acc}^{a} = v_{acc}^{a} + v^{a} - rv^{a}$$
$$v_{acc}^{k} = v_{acc}^{k} + v^{k} - rv^{k} \tag{0.11}$$

With these auxiliary variables, we store information for modifying the initial values in the second phase to avoid spillage (using the volumes that can be kept) and supply the minimum ecological flow (using the volumes that can be additionally provided).

4.2. Second Phase

This phase is performed from the downstream to upstream elements. Now, the requests for additional flow or keeping water are divided among upstream elements. This split is made proportionally to the capacity of each element for keeping or providing water, information that has been stored in the first phase as capacities (v^a and v^k) and cumulative capacities of upstream elements (v_{acc}^a and v_{acc}^k).
In each element that has requests from downstream elements, one of these two cases is possible:

- Request for additional flow from downstream elements rv_{acc}^{a}. Then, element outflow is augmented as follows

$$o = o + \min\left(v^{a},\, rv_{acc}^{a} \frac{v^{a}}{v_{acc}^{a}} \right) / 0.0864 \tag{0.12}$$

- Request for keeping water from downstream elements rv_{acc}^{k}. Then, element outflow is reduced as follows

$$o = o - \min\left(v^{k},\, rv_{acc}^{k} \frac{v^{k}}{v_{acc}^{k}} \right) / 0.0864 \tag{0.13}$$

Although in theoretical cases one element can be requested to keep water for some downstream elements and to produce more water for some other, these pathological cases are not frequent in reality. An example can be an extremely dry year where minimum ecological flows can not be provided, but at the same time there is an extreme flood in some reservoir that was keeping water. However, this situation is very unlikely to happen in reality.

4.3. Third Phase

In this third phase, the starting point is the set of outflows determined in the two previous phases. With this information, the units output can be determined.

As it has been previously said, the coefficient of efficiency that converts water flow in electric energy depends linearly on the water head. This head is the difference in altitude between the upstream element of the plant and the downstream element of the plant or its drainage pipe. This head is computed as the average value between the situations at the end and the beginning of the day, because this is the time step of the simulation model. Once obtained both heights, upstream a^u and downstream a^d, the coefficient of efficiency ec is computed as

$$ec = k^{ec0} + k^{ec1}\left(a^u - a^d\right) \tag{0.14}$$

where k^{ec0} y k^{ec1} are coefficients that depend on the plant characteristics.

Besides, in this phase the electric output is divided between peak and off-peak hours. (h^p and h^{op} respectively). The objective is to produce as much energy as possible in peak hours, where more expensive thermal units will be used. If output water flow o is the maximum allowed, then peak hours flow o^p and off-peak hours flow o^{op} will be also at maximum

$$\begin{aligned} o^p &= o \\ o^{op} &= o \end{aligned} \tag{0.15}$$

If daily flow is lower than the maximum, then it is produced in peak hours with priority

$$\begin{aligned} o^p &= \min\left(o^{max}, o\frac{h^p + h^{op}}{h^p}\right) \\ o^{op} &= \max\left(0, o\frac{h^p + h^{op}}{h^{op}} - o^p\frac{h^p}{h^{op}}\right) \end{aligned} \tag{0.16}$$

Finally, using flows and coefficients of efficiency it is possible to compute the total electric production p and that of peak p^p and off-peak hours p^{op}

$$\begin{aligned} p &= 0.0864 \cdot ec \cdot o \\ p^p &= 0.0036 \cdot ec \cdot o^p h^p \\ p^{op} &= 0.0036 \cdot ec \cdot o^{op} h^{op} \end{aligned} \tag{0.17}$$

where the coefficient 0.0864 converts from m^3/s to hm^3/day and the coefficient 0.0036 converts from m^3/s to hm^3/h.

5. Results

In this section, we are going to present the simulation results of real cases taken from the Spanish hydroelectric basin. The model has been run with two different time scopes: a first case with a time scope of the last eight months and a second case with a time scope of one year. There have been used 27 different historical series.

Figure 3 corresponds to the first case, eight-month scope and therefore the first four months are taken from reality. In this figure, volume curves for a large reservoir can be seen. This reservoir has no minimum and maximum operational rule curves. Therefore, the reservoir volume can evolve without any constraint. The upper side of the figure represents the quantiles of the volume curves of the hydro inflow series depicted in the lower side.

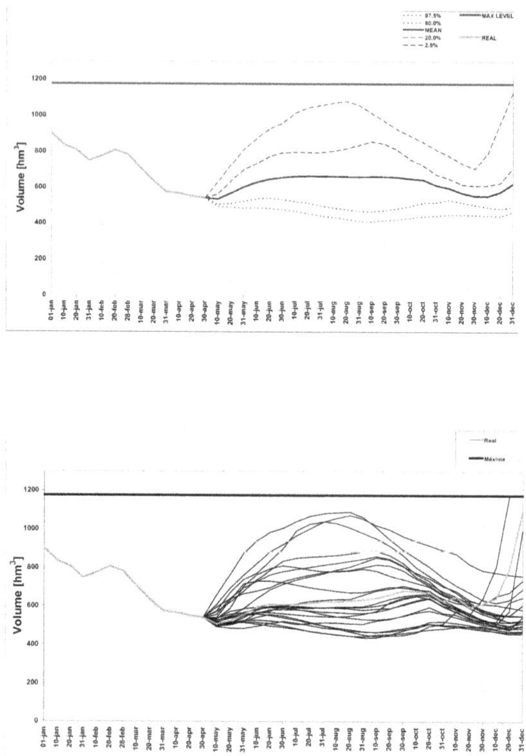

Figure 3. Volume curves for a large reservoir

Figure 4 corresponds to the second case, one-year scope. In this figure, volume curves for a small reservoir can be seen. This reservoir has minimum and maximum operational rule curves guiding the volume evolution along the time. The strategy of this reservoir consists on going to the minimum operational curve. However, the reservoir begins the simulation period with a very low reserve level (below the minimum operational level) and as it can be seen most of the time is below that curve. Only beginning March the mean volume (over all the historical series) touches this curve. This situation is the consequence of guiding the reservoir volume to the minimum level. If in a certain day, the volume is at minimum level and there are no inflows in the following days and downstream elements have to supply irrigation needs, then the reservoir volume decreases below the minimum operational level.

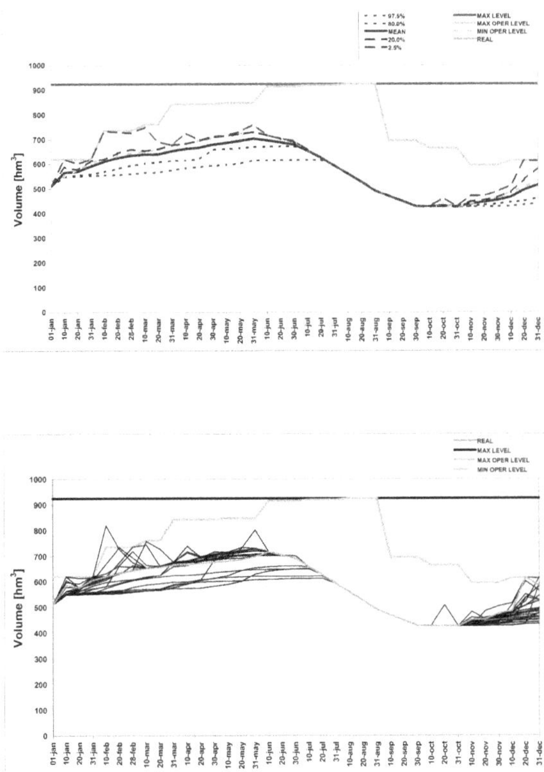

Figure 4. Volume curves for a small reservoir

The results of this three-phase simulation model have been compared with those obtained by an intraday direct optimization. This simulation model achieves a 95% of the total optimal hydro production at a much lower computational cost.

6. Conclusions

In this chapter, we have proposed a simulation method for hydro basins, based on Object Oriented Programming paradigm. Hydro reservoir management is obtained with a three-phase algorithm, whose main purpose is to adapt the given strategies to the fulfillment of the minimum outflows and avoiding superfluous spillages. Besides, this simulation method has been implemented in a computer program whose results have been shown. A realistic case for a Spanish hydro basin has also been presented.

References

Allen, E. et al., 2001, "Interactive Object-Oriented Simulation of Interconnected Power Systems Using SIMULINK", *IEEE Transactions on Education*, Vol. 44, No. 1, pp. 87–95.

De Cuadra, G., 1998, *Operation Models for Optimization and Stochastic Simulatiuon of Electric Energy Systems* (original in Spanish), PhD Thesis, Universidad Pontificia Comillas.

Law, A.M. and Kelton, W.D., 2000, *Simulation modelling and analysis*, 3rd edition, McGraw-Hill.

Roman, J., & Allan, R.N., 1994, "Reliability assessment of hydro-thermal composite systems by means of stochastic simulation techniques", *Reliability Engineering & System Safety*, Vol. 46, No. 1, p. 33–47.

Sankarakrishnan, A. and Billinton, R., 1995, "Sequential Monte Carlo simulation for composite power system reliability analysis with time varying loads", *IEEE/PES Winter Meeting*.

Turgeon A. and Charbonneau R., 1998, "An aggregation-disaggregation approach to long-term reservoir management", *Water Resources Research* Vol. 34, No. 12, p. 3585–3594.

Van Hecke, J. et al., 1998, "Sequential probabilistic methods for power system operation and planning", CIGRE TF38.03.13, *Elektra* No. 179.

Wirfs-Brock, R., Wilkerson, B., Wiener, L., 1990, *Designing Object-Oriented Software* Prentice Hall.

Wood, A.J. and Wollenberg, B.F., 1996, *Power generation, operation, and control*, 2nd edition John Wiley & Sons.

Chapter 6

Complex Decision Making using Non-Parametric Data Envelopment Analysis

Dr. Michael Alexander
Department of Information Systems
Wirtschaftsuniversität Wien
malexand@wu-wien.ac.at

1. Data Envelopment Analysis

Many problems in decision science involve a complex set of choices. The study of choices tends to be multivariate in nature, which is sometimes difficult to model in parametric form. Complex decision analysis, depending on analysis methodology, may require assumptions on distributions or functional forms of e.g. a production function. The study of choices could furthermore involve problems with a large number of inputs and outputs whereby some methodologies can be sensitive to the frequency distributions of the respective inputs. Data Envelopment Analysis (DEA) is a nonparametric mathematical programming approach which allows for comparisons of decision options with similar objectives, canonically termed decision making units (DMUs) in the following. This chapter discusses Data Envelopment Analysis in the context of decision science and is based on [Alexander 1997 and 2006]. It first examines the method's characteristics relative to decision analysis problems, followed by the discussion of the formation of frontiers – hyperplanes as part of the comparison process. Consecutively, the main base analysis models are introduced. The sections Returns to Scale and Target choice present advanced model extensions which for the former are helpful in estimating otherwise unknown output returns to scale properties of a problem and for the latter to ex post establish option input variables so to make a choice efficient. Data Envelopment Analysis can be applied to many complex decision problems, such as in evaluating the relative efficiency of various decision alternatives, optimize – maximize for a certain output decision variable, establish target values of given decision input variables, identify peer decision alternatives, besides others. Its main benefit is its ability to handle the evaluation of alternatives with input and output vectors consisting of elements on different units and scales, which are common to complex decision problems. No assumptions on the form of the decision functions, common to parametric approaches, have to be made. Also, Data Envelopment Analysis is not susceptible to parametric modeling challenges in parameter estimation, such as for input variables showing conlinearity, or the presence of autocorrelation of an error term with an output variable.

Data Envelopment Analysis is an approach to decision analysis, which uses an empirical production function describing the set of DMUs under consideration [Dyson et al. 1996 and Tim Anderson 1995]. It determines the relative efficiency of units within the DMU set. Efficiency is determined as relative to a virtual (constructed) frontier curve, which describes the best possible achievable output for a given level of input. Being a frontier method, it measures each option as the distance to a hyperplane, constructed from the best attainable outcome or minimum input vector of a set of peer options. Statistical techniques, in turn, frequently measure the deviation from a medium reference curve, which is where their sensitivity on input data frequencies stems from. Furthermore, Data Envelopment Analysis allows for use of different units of input and output measure, such as time, cost, units of labor etc. A DMU is Pareto efficient, if no other DMU or combination of DMUs can improve one output, without lowering another output level or the increase of an input level [Thanassoulis and Dyson 1992, p. 80].

Data Envelopment Analysis is an extreme point method. In the context of the study of choices, it constructs a virtual (composite) decision option out of a set of alternatives. The relative efficiency of a given option vis-à-vis the composite option of a solution is the – e.g. radial – distance of the option under evaluation to a hyperplane formed of all alternatives. It converts a fractional to a linear problem and uses the Linear Programming (LP) Method for numerical computation – one LP model has to be solved per decision option. It has first been described by [Charnes et al. 1978] with further refinements for optimal input-output targets by [Thanassoulis and Dyson 1992] and unit raking by [Andersen and Petersen 1993], amongst others.

The following diagram in Figure 1 shows a simplified graphical representation of a Data Envelopment Analysis problem with two inputs x_1, x_2 and one output. The convex envelope is the efficiency frontier, spawning a hull over the feasible region.

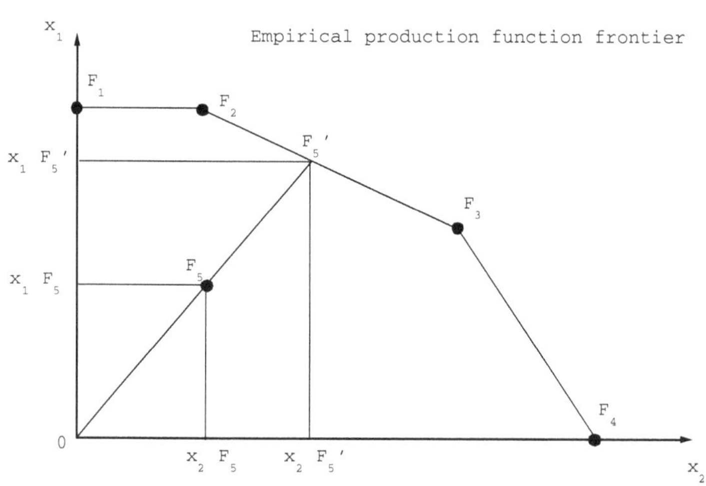

Figure 1. Graphical representation of a DEA problem

x_1 and x_2 are two inputs, and $F_{1..5}$ denote the $DMUs_{1..5}$. F_{1-4} are all relatively efficient, being on the efficiency frontier. F_5, in turn, is deemed inefficient. The graphical representation of the measure of X-inefficiency is given by the quotients of the rays:

$$\theta_{F5} = \frac{\overline{0F_5}}{\overline{0F_5'}}$$

The DMU's F_2 and F_3 are the peer units of F_5; their input times a scaling-down factor determined by linear programming produce the same output as F_5. Their unit weights, $\lambda_{2,3}$ are described by the respective ray quotients:

$$\overline{F_3F_5'}/\overline{F_3F_2} \qquad \overline{F_2F_5'}/\overline{F_3F_2}$$

The principal idea, on which Data Envelopment Analysis is based, is to measure relative efficiency for a set of n DMUs not by using an average composite with fixed weights for inputs x_{ij}, but by allowing input weights to be variable for each x_{ij}. Hence, DMUs can be classified relatively efficient through strength in a particular factor. This assumption corresponds to reality with firms and individuals frequently showing causal relations to specific factors they attribute their success to.

Generally, an m+s dimensional supported hyperplane is formed [Ali and Seiford 1993, p. 127]. The generic equation for a hyperplane in R_{m+s} with the normal variables $\{\_\mu_{1..s}, \_-\upsilon_{1..m}\}$ is:

$$\sum_{r=1}^{s} \mu_r y_r - \sum_{i=1}^{m} v_i x_i + w = 0$$

The conditions for fulfilling the "supported" criterion are that a) all points having to be located on or below the envelope surface and b) at least one point has to be an element of the surface. The following figure is a graphical example representation of a VRS (variable returns to scale) surface.

Data Envelopment Analysis, as introduced by Charnes et al. (1979), is derived from the mathematical programming problem of maximizing the virtual (composite) output to input quotient. This is the output oriented form of the problem. Alternatively, an input oriented Data Envelopment Analysis problem minimizes the inputs, while holding the output level constant. Most Data Envelopment Analysis applications in decision analysis use the input oriented variant of the method.

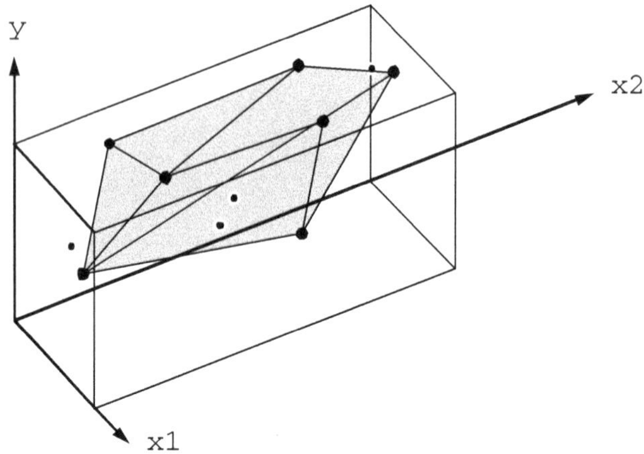

Figure 2. VRS Envelopment Surface Example
Source: [Ali and Seiford (1993), p. 122]

$$\max h_0(u, v) = \frac{\sum_r u_r y_{r0}}{\sum_i v_i x_{i0}}$$

s.t.

$$\frac{\sum_r u_r y_{rj}}{\sum_i v_i x_{ij}} \leq 1 \qquad \text{for } j = 0,\ldots,n$$

$$u_r, v_i \geq 0$$

The nonlinear programming problem above is transformed into a linear programming model shown in the CCR and BCC sections [see Charnes, A. and W.W. Cooper 1962]. Like other linear programming problems, Data Envelopment Analysis models can be formulated in one of two forms, which are reciprocal-dual against each other. [Anderson et al. 1991 pp. 236–243] and [Meschkowski 1972 pp. 883–893] show a detailed comparison of the properties of a primal vs. the dual problem form. The most important ones are:

1. The values of the primal and dual optimal solutions are equal.
2. A primal has m constraints and n variables, the dual n constraints and m variables.

3. A primal maximization problem is a dual minimization and vice versa.
4. The optimal value of a dual variable is the shadow price for the primal resource.

Based on computational considerations for large problems, the form which has the lower number of constraints should be chosen. In Data Envelopment Analysis, the primal problem is also called multiplier problem, the dual problem is called an envelopment problem.

Furthermore, Data Envelopment Analysis assumes that if DMU j is capable of producing a given input-output combination, another DMU can do the same. Restrictions here need modified Data Envelopment Analysis models by adding linear programming constraints.

The method carries disadvantages as well: the main one being that in theory, in a system of m inputs and s outputs m x s efficient units are possible, supporting the hyperplane. Consequently, the strongest propositions Data Envelopment Analysis produces are those on inefficient units. In a corollary to the prior proposition, the dimensionality of a decision problem should be chosen so that the number of decision options exceeds the number of input and output variables. In an empirical formula it should be exceed the product of inputs and outputs. Below that level, the method's discriminatory abilities are reduced due to a potentially resulting in a high number of efficient options. Another disadvantage is that currently, no statistical hypothesis testing methods are available. There are also no provisions for handling disturbances-random error, such as measurement error.

1.1. Data Envelopment Analysis Base Models

The following are the main models suitable for complex decision analysis. In addition, there are many further Data Envelopment Analysis variants (characterized by different envelopment surface construction) and extensions. Important ones are: a) multiplicative problems in which inputs and outputs in the non-linear problems are related through multiplication, b) stochastic DEA in which $E(x_{ij})$, $E(y_{rj})$ are modeled through $E(x_{ij})$-x_{ij} and corresponding output constructs [Charnes et al. 1958], variance-covariance matrixes for all variables and probabilities for constraint satisfaction leading to non-linear programming problems, c) DEA with bounds on the multiplier ranges (used in the empirical analyses), e) models which allow for categorical variables, f) the inclusion of non-discretionary exogenous variables directly into the model, g) dynamic models, which distinguish between period specific factors, such as direct costs and input factors which contribute to output in multiple periods, and h) target choice, which is shown in the following as a typical decision problem that can be solved with Data Envelopment Analysis .

A good summary of the various Data Envelopment Analysis models can be found in [Binder 1995] and [Seiford and Thrall 1990]. The following sections discuss the two prime models for decision analysis, the Charnes, Cooper and Rhodes (CCR) and the Banker, Charnes and Cooper (BCC) model. The former applies to problems in which marginal additional inputs lead proportionally to increased outputs (constant returns to scale), whereby the latter applies to problems where input increases result in either over proportionally increased or decreased output (variable returns to scale). The estimation of the returns to scale itself form a class of decision problems which is discussed consecutively.

1.1.1. The CCR Model

The basic Charnes, Cooper and Rhodes (CCR) model [Charnes et al. 1978] is the first, and most widely used Data Envelopment Analysis model which assumes constant returns to scale. As described above, it can be solved either in its primal or dual form. A separate linear programming solution for each DMU_j has to be calculated. The solution to the primal are optimal variable weights μ_r*, υ_i*, to the dual it is the efficiency measure θ. A DMU is relatively efficient with $\theta_j = 1$, and inefficient with $\varepsilon \geq \theta j < 1$ (input oriented). The CCR efficiency measure is precisely a product of technical and scale efficiency.

Primal (Multiplier) Problem

$$\max h_0 = \sum_{r=1}^{s} \mu_r y_0$$

$$\sum_{i=1}^{m} \upsilon_i x_i = 1$$

$$\text{s.t.} \quad \sum_{r=1}^{s} \mu_r y_{rj} - \sum_{i=1}^{m} \upsilon_i x_i \leq 0 \qquad j = 1, \ldots, n$$

$$-\mu_r \leq -\varepsilon \qquad r = 1, \ldots, s$$

$$-\upsilon_i \leq -\varepsilon \qquad i = 1, \ldots, m$$

Denotations equal those previously used with ε being a non-Archimedean infinitesimal. $h_0 = \theta_0$ is the efficiency of DMU_0.

Dual (Envelopment) Problem

$$\min \theta_0 - \varepsilon \sum_{r=1}^{s} \overline{s_r} - \varepsilon \sum_{i=1}^{m} s_i$$

$$\text{s.t.} \quad \sum_{j=1}^{n} y_{rj}\lambda_j - \overline{s_r} = y_{r0} \qquad r = 1, ..., s$$

$$\sum_{j=1}^{n} x_{ij}\lambda_j - \theta_0 x_{i0} + s_i = 0 \qquad i = 1, ..., m$$

$$\lambda_j, \overline{s_r}, s_i \geq 0$$

The actual relative difference of an inefficient unit to the envelope is given by the distance of their output and input vectors to their projected points on the surface. This vector is denoted by $\overline{s}$ for the output and s_i for the input slack. The projection formula for the optimal values is:

$$Y_E = Y_0 + \overline{s}^* \qquad X_E = \theta_0^* X_0 - s^*$$

1.1.2. The BCC Model

The Banker, Charnes and Cooper model [Banker et al. 1984] is an extension of the CCR, providing for variable returns to scale. The following convexity constraint is added to the model for decreasing, constant and increasing returns:

$$\sum_{j=1}^{n} \lambda_j = 1$$

For non-increasing returns (as shown in Figure 3):

$$\sum_{j=1}^{n} \lambda_j \leq 1$$

is added. The variable u_0 denotes the corresponding returns to scale indicator in the primal. The BCC efficiency measure represents true technical efficiency, compared to the CCR model.

Primal BCC Problem

$$\max \sum_{r=1}^{s} \mu_r y_{r0} - u_0$$

$$\text{s.t.} \quad \sum_{i=1}^{m} \upsilon_i x_{i0} = 1$$

$$\sum_{r=1}^{s} \mu_r y_{rj} - \sum_{i=1}^{m} \upsilon_i x_{ij} - u_0 \le 0 \qquad j = 1, \ldots, n$$

$$-\mu_r \le -\varepsilon \qquad r = 1, \ldots, s$$

$$-\upsilon_i \le -\varepsilon \qquad i = 1, \ldots, m$$

Dual BCC Problem

$$\min \theta_0 - \varepsilon \sum_{r=1}^{s} \overline{s_r} - \varepsilon \sum_{i=1}^{m} s_i$$

$$\text{s.t.} \quad \sum_{j=1}^{n} y_{rj}\lambda_j - \overline{s_r} = y_{r0} \qquad r = 1, \ldots, s$$

$$\sum_{j=1}^{n} x_{ij}\lambda_j - \theta_0 x_{i0} + s_i = 0 \qquad i = 1, \ldots, m$$

$$\sum_{j=1}^{n} \lambda_j = 1$$

$$\lambda_j, \overline{s_r}, s_i \ge 0$$

The dual again uses DMU weights and the slack variables.

2. Returns to Scale

The decision problem of estimating the presence of either decreasing, constant or increasing returns to scale has been brought forward by [Banker and Thrall 1992]. This Data Envelopment Analysis-based method is regarded as very reliable, with many DEA applications being based on it. [Färe Grosskopf and Lovell 1985] (FGL) provide an alternate, considerably simpler approach with updates being [Färe and Grosskopf 1994] and [Färe 1997]. [Sheldon and Hägler 1993, p. 115], for example, use this alternative method in a study on Swiss Banking efficiency. The approach is a procedure which a) determines the BCC variable returns to scale efficiency and the BCC non-increasing RTS of each DMU; b) the scores are compared with the set of those DMUs whose efficiency falls compared to VRS belonging to either the CRS, or the decreasing RTS group. The others exhibit increasing returns to scale; c) the CRS theta is computed for all DMUs; if it is lower than the efficiency for the {CRS, DRS} group from a) then the point where a DMU operates shows DRS, otherwise CRS.

X in Figure 3 denotes the input and Y the output vector: variable (VRS) and constant returns to scale (CRS). VRS generally wraps closer to the options set than CRS, leading to higher efficiency measures. Sheldon and Hägler found their results to differ substantially, compared to a ray scale elasticity measure. Other papers are finding inconsistencies between the two approaches as well[1]. The discussion which RTS approach to prefer in Data Envelopment Analysis is ongoing at the time of the writing of this thesis. Tests comparing the FGL, Banker and Thrall measurements to DMUs on average cost plots using data sets from the following analyses, led to the selection of the latter method. It generally provided a better fit with intuitive point scale elasticity expectations and the plots.

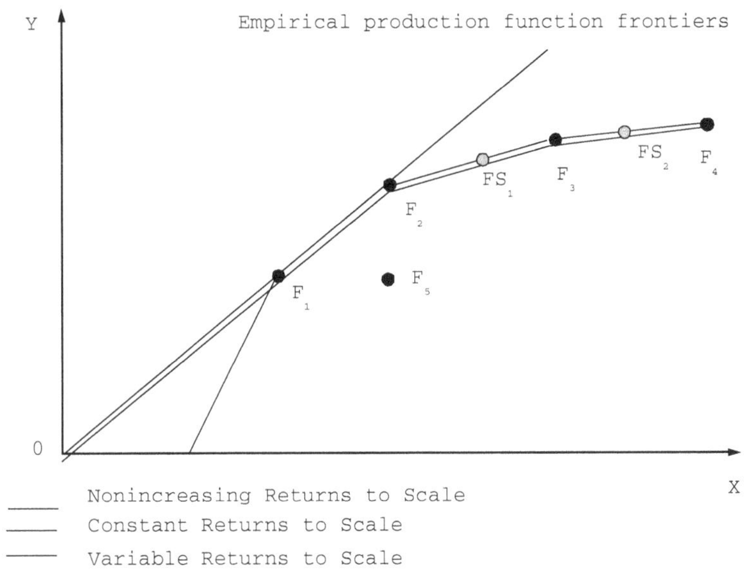

Figure 3. Constant and Variable Returns to Scale

[1] A notable exception being [Zhu 1996] who in a longitudinal efficiency analysis finds the results from the two approaches to conform.

The [Banker and Thrall 1992] approach extends the concept of a most productive scale size (mpss) to Data Envelopment Analysis models, which can have multiple optimal solutions. It therefore allows to deduct the economies of scale property in cases where multiple options are located on the efficiency frontier. Different to the FGL approach which just gives indication of the three RTS categories, it calculates a quantitative range of scale elasticities. These bounds are necessary since the supporting hyperplane may not be unique for efficient options. They show, that if there are multiple production possibilities with mpss units for an input-output vector, all mpss DMUs come to lie on the same line segment. Their approach rests on identifying the intersection of the constant RTS plane and a hyperplane running through a technical efficient point. [Førsund and Hjalmarsson 1996] provide a good summary on how the Banker and Thrall scale properties measure is derived. First, the dual variable u_0 in the dual, is necessary to quantify the distance, when moving from the current DMU to the hyperplane in input direction.

$$\sum_{r=1}^{s} \mu_r y_{rj} - \sum_{i=1}^{m} \theta_j \upsilon_i x_{ij} + u_j = 0 \qquad j = 1, \ldots, n$$

Secondly, a general measure on scale elasticity is given by:

$$\rho_0 = \frac{\delta\beta(y, x)}{\delta\mu}$$

β is an output expansion factor, μ an input expansion factor. Finding the maximum scale elasticity value in the input oriented case means finding the maximum value of u_0--the shadow pride of the convexity constraint--from the BCC primal form. Third, the objective function from BCC primal is substituted with:

$$max \ u_j$$

and the following constraint added:

$$\sum_{r=1}^{s} \mu_r y_{rj} + u_j = 1$$

[Førsund and Hjalmarsson 1996, p. 9] provide a formulation of scale elasticity derived from the prior measure of scale elasticity as:

$$\varepsilon(v_j, \mu_j, \theta_j) = \frac{\sum_{i=1}^{m} \theta_j \upsilon_i x_{ij}}{\sum_{r=1}^{s} \mu_r y_{rj}} = \frac{\theta_j}{\theta_j - u_j}$$

The solution to the modified LP problem of the BCC primal with the substituted objective function and the constraint added using prior scale elasticity measure gives the following resulting RTS measure ρ_{max}:

$$\varepsilon_{max}(y_j, x_j) = \rho_0^+ = \frac{1}{1 - u_j^{max}}$$

Geometrically, using Figure 3, the intersection from the prior constraint corresponds to the tangential through e.g. the points $FS_{1,2}$.:

Similarly, for the lower bound, the following objective function is used along with the same constraint.

$$max\ (-u_j) \qquad \varepsilon_{min}(y_j, x_j) = \rho_0^- = \frac{1}{1 - u_j^{min}}$$

Corresponding to a minimization/maximization of β / α of a simple linear production frontier $\alpha\, y = \beta\, x + \alpha\, y_0 - \beta\, x_0$ in (X_0, Y_0), the intermediates RTSs are:

$$\rho_0^- \geq \rho_0^+ > 1$$

$$\rho_0^- \geq 1 > \rho_0^+$$

$$1 > \rho_0^- \geq \rho_0^+$$

Note that this method allows deducting RTS figures for all units - not only reference mpss units that are required to be radial-technical-efficient. The reason is that the range of point scale elasticity measures covers all optimal solutions if there is not a unique one [Banker and Thrall 1992] and reiterated by [Banker, Bardhan and Cooper 1996].

3. Target Choice

Target choice decision models allow to determine input and output target levels which render relatively inefficient decision option efficient. They support decision problems in which the right input factor to e.g. invest resources should be found. The determination of these optimal target levels is based on the projection of the inefficient DMUs towards the frontier.

Given that Data Envelopment Analysis does not distinguish between different regions under a frontier hyperplane, that is $\forall\ \theta_j < 1\ \exists\ \infty\ \theta_j < 1$ hence, the preferences for improvement are crucial.

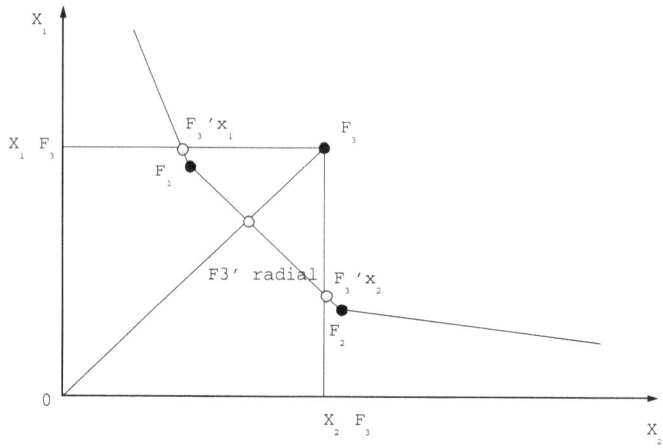

Figure 4. Target Input Choice

Figure 4 shows three possible projections of F_3 on the efficiency frontier for an input minimization case. The most common approach is radial expansion, which is depicted by the ray:

$$\overline{F_3F_3'}$$

going through (0,0). In the two input case shown, improvement priority can be on $X_{1,2}$ alternatively, shown by lines which lie orthogonally to the category axes. It is important to note that these extensions require separate minimization/maximization runs for each problem and can not be calculated from a CCR or BCC result, since changes in one input/output affect the whole input/output vector.

[Thanassoulis and Dyson 1992] have extended the CCR/BCC model for several possible preference cases: One is a radial type uniform input lowering/output expanding target choice strategy shown in the following. Here, the solution is characterized by having a minimum in the sum of slacks weighted by the inverse average value on each input or output variable before the radial efficiency is maximized. Alternatively, target choice could set a priority for each output variable expansion or input variable reduction, with an "infinite" size solution set. Their following weights based general preference structure model attaches the weights w_r+ w_i– to the outputs/inputs which are modeled as coefficients z_r, p_i. The former represents the target output increase, the latter the input decrease. I, R are input-output index sets I={1,...,m} R={1,...,s}. After specifying the relative priority of all i and s, the model gives the targets (x_{ij0}", y_{rj0}").

$$
\begin{aligned}
\max \quad & \sum_{r \in R_0} w_r^+ z_r - \sum_{i \in I_0} w_i^- p_i + \varepsilon\left(\sum_{i \in \bar{I}_0} d_i^- + \sum_{r \in \bar{R}_0} d_r^+\right) \\
\text{s.t.} \quad & z_r y_{rj_0} - \sum_{j=1}^{n} \beta_i y_{rj} = 0 \qquad r \in R_0 \\
& p_i x_{ij_0} - \sum_{j=1}^{n} \beta_i y_{rj} = 0 \qquad i \in I_0 \\
& \sum_{j=1}^{n} \beta_j x_{rj} - d_r^+ = y_{rj_0} \qquad r \in \bar{R}_0 \\
& \sum_{j=1}^{n} \beta_j x_{ij} + d_i^- = x_{ij_0} \qquad i \in \bar{I}_0 \\
& z_r \geq 1 \qquad \forall r \in R_0 \\
& p_i \leq 1 \qquad \forall r \in I_0 \\
& \beta_j \geq 0 \qquad \forall j \\
& p_i, z_r \qquad \text{free} \forall i \in I_0 \quad \text{and } r \in R_0 \\
& d_i^-, d_r^+ \geq 0 \qquad \forall i \in \bar{I}_0 \text{ and } r \in \bar{R}_0
\end{aligned}
$$

A further extension for the above model is the specification of desired relative gains (such as 20%), for a particular decision factor. Possible further extensions of Data Envelopment Analysis target choice include models which allow for exogenously fixed absolute inputs, as suggested by [Luptacik 1997].

4. Summary and Conclusion

The discussed nonparametric frontier approach method Data Envelopment Analysis provides an effective means to analyze complex decision problems: it suits problems with a large number of input and output variables, heterogeneous data types within the model and differing scales while at the same time not being sensitive to distribution, autocorrelation and colinearity issues found in parametric models. The main disadvantage pointed out was the limited discriminatory ability of the approach when the dimensionality of inputs, outputs and decision options due not fall within the outlined bounds. The chapter outlines the formation of the efficiency frontier, the two base models Charnes, Cooper and Rhodes (CCR) and the Banker, Charnes and Cooper (BCC) model. In addition, extended Data Envelopment Analysis models on scale determination and (decision option) target choice determination are discussed. Data Envelopment Analysis is shown as a decision science method that carries few required model parameter choices while providing and effective means to analyze otherwise hard to model decision problems.

References

Alexander, M., 1997, *Payment Systems Efficiency and Risk*, Doctoral Thesis, University of Vienna.

Alexander, M., 2006, Load-Balancing for Loosely-coupled Inhomogeneous Nodes using Data Envelopment Analysis. in *Proceedings of The Fourth International Symposium on Parallel and Distributed Processing and Applications* (ISPA'2006). 4–6 December, 2006 Sorrento, Italy. Springer LNCS, 2006.

Ali, A. I. and Seiford, L. M., 1993, The mathematical programming approach to efficiency analysis, in *The measurement of productive efficiency: techniques and applications*, edited by Harold Fried and C.A. K. Lovell and S. S. Schmidt, Oxford University Press (Oxford).

Anderson, D. R., Sweeney, D. J., and Williams, T. A., 1991, Management Science, West Publishing (Los Angeles).

Andersen, P. and Petersen, N. C. 1993, A procedure for ranking efficient units in data envelopment analysis, *Management Science* 39, 1261–1264.

Anderson, T. A, 2006, A data envelopment analysis (DEA) home page, http://www.emp.pdx.edu/dea/homedea.html

Banker, R. and Thrall, R., 1992, Estimation of returns to scale using DEA, *European Journal of Operations Research* 62, 74–84.

Banker, R., Barhan, I., and W.W.Cooper, 1996, A note on returns to scale in DEA, *European Journal of Operations Research* 88, 583–585.

Binder, S. (1995), *Effizienzanalyse von Bankfilialen mittels Data Envelopment Analysis*, Masters Thesis, University of Vienna.

Charnes, A., Cooper, W. W., and Rhodes, E., 1978, Measuring the efficiency of decision making units, *European Journal of Operations Research* 2, 429–441.

Dyson. R. G., Thanassoulis, and Boussofiane, 1996, A. A DEA tutorial, http://www.deazone.com/tutorial

Färe, R., 1997. Determining returns to scale, unpublished working paper. Southern Illinois University Department of Economics.

Färe, R., Grosskopf, S., and Lovell, C. K., 1985, The Measurement of Efficiency of Production, Kluwer-Nijhoff (Boston).

Färe, R. and Grosskopf, S., 1994. Estimation of returns to scale using data envelopment analysis. *European Journal of Operational Research* 79, 379–382.

Forsund, F. and Hjalmarsson, L., 1996, *Measuring returns to scale of piecewise linear multiple output technologies: theory and empirical evidence*, Göteborg University Department of Economics Memora.

Luptacik, M., 1997, Discussions, Technische Universität Wien.

Meschkowski H. (ed.), 1972, *Meyers Handbuch Mathematik*, Meyers Lexikonverlag, (Mannheim).

Seiford, L. M. and Thrall, R. M., 1990, Recent developments in DEA, *Journal of Econometrics* 46, 7–38.

Sheldon, G. and He, U., 1993, Economies of scale and scope and inefficiencies in Swiss banking, in *Banking in Switzerland*, Physica Verlag (Heidelberg, Germany).

Thanassoulis, E. and Dyson, R., 1992, Estimating preferred target input-output levels using data envelopment analysis, *European Journal of Operations Research*, 56 – 80.

Zhu, J., 1996, DEA/AR analysis of the 1988–1989 performance of the Nanjing Textiles Corporation, *Annals of Operations Research* 66, 311–335.

Chapter 7

Using Group Model Building to Inform Public Policy Making and Implementation

Aldo A. Zagonel
Systems Engineering & Analysis
Sandia National Laboratories
aazagon@sandia.gov

John Rohrbaugh
Department of Public Administration and Policy
University at Albany, State University of New York
jwr26@albany.edu

1. Introduction

Group model building (GMB) is based essentially in the system dynamics (SD) model-building method. However, deeply involving a client group in the process of model construction has benefited from theoretical and applied input from other fields, such as sociology and psychology [Vennix 1999]. In the applied research concerning group decision making being conducted at the University at Albany, this influence has been refined and expanded in a framework called decision conferencing [Rohrbaugh 2000].

While one might apply GMB to modeling any sort of dynamic problem, it seems that particular kinds of problematic situations "attract" the application of this approach. GMB interventions often will address problems that involve multiple stakeholders who contribute to the process with partial views of the system, but who are affected by the system as a whole [Huz *et al.* 1997]. They also are particularly useful in situations in which there is strong interpersonal disagreement in the client group regarding the problem and/or regarding the policies that govern system behavior. Vennix [1999] refers to the latter as messy problems, i.e., "a situation in which opinions in a management team differ considerably."

Zagonel [2002] traced a genealogy of the GMB approach used by the University at Albany research group. Two schools of thought were identified as contributing to this approach, as illustrated in Figure 1.

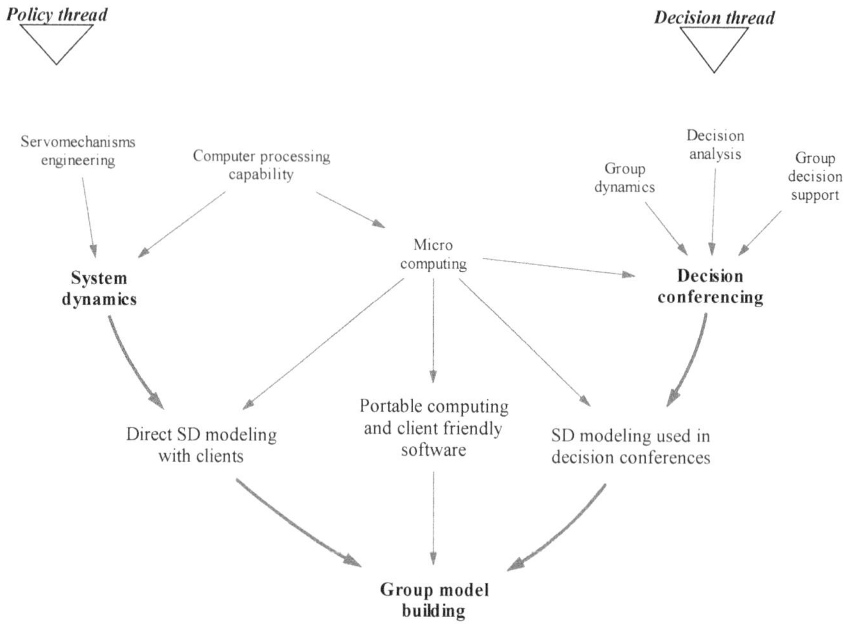

Figure 1. Tracing a genealogy of Group Model Building

System dynamics is at the root of the policy thread. The SD model-building method can be described in phases, beginning with a clear definition of the problem of interest, and ending with a conclusive statement about this problem that contains policy recommendations aimed at its solution or mitigation [Richardson & Pugh 1981]. This method is based upon an endogenous feedback view of system causes and effects. Solutions to the perceived problem are revealed through feedback thinking, the key expertise offered by system dynamicists [Sterman 1994]. The second thread, labeled as the decision thread, is formed by a confluence of schools that gave shape to the decision conferencing framework; those are group dynamics, decision analysis, and group decision support [Rohrbaugh 2000]. Scholars and practitioners who conduct decision conferences are process experts who assist groups in using appropriate techniques to achieve consensus and make decisions [Reagan *et al.* 1991].

An important characteristic of GMB is its diversity in objectives and expectations, perhaps resulting from the integration of these varied influences that give shape to it (e.g., system dynamics and group decision support). Even a superficial examination of its genealogy reveals a tension between policy *vs.* decision, and between content *vs.* process. To some extent these tensions overlap. The decision conferencing influence emphasizes the decision to be made and focuses upon the processes that lead up to this decision. Decision or process-oriented objectives in GMB may be stated as accelerating a management team's work, problem structuring and classification schemes, as generating commitment to a decision, as creating a shared vision and promoting alignment, and as creating agreement or building consensus about a policy or decision. Alternatively, policy or content-oriented objectives may be stated as improving shared understanding of the system, system improvement, and system process and outcome change. These involve changing the mental models of individuals in the group or organization, guided by reliable insights produced through the use of the modeling tools and methods.

Ideally, as Eden [1990] appropriately pointed out, astute analysis (content) and skillful facilitation (process) should be combined: "Within the context of group decision support it may be suggested that the two skills can become integrally tied together so that they are fully interdependent." In the context of GMB, this may be stated as promoting organizational learning and organizational change, or promoting insightful collaboration and cooperation among interdependent stakeholders. While this may be an ideal, Zagonel [2002] has suggested that there are tensions related to bringing these two skills together. The objectives and procedures for using the GMB process as a tool for creating a shared understanding of an interpersonal or inter-organizational problem (in the form of a "boundary-object" model) and as a tool for exploring a "micro-world" representation of reality (to address this particular problem) are not necessarily aligned. While GMB interventions are designed to achieve this ideal goal by intertwining these two threads, in reality, one or both may be sacrificed.

This chapter reports the process, content and outcomes of a GMB effort in which a complex issue in the public sector was modeled and analyzed. It reports how GMB was used to inform welfare reform policy making and implementation, drawing upon the perspectives and knowledge of the client teams. It addresses the policy aspects of the research by answering an important question using the model elicited, built, simulated, tested and extensively used throughout the intervention. The conclusion contains a summary of the findings, as well as a discussion of the limitations of the analysis.

2. Overview of the Project

The history of social welfare in the United States has come full circle. In 1935 amid the Great Depression, the federal government assumed greater responsibility for the poorest members of society through the enactment of the Social Security Act and the implementation of a federally funded program called Aid to Families with Dependent Children (AFDC). Recently, there was devolution of this task to state and local governments [Thompson 1996]. This devolution occurred in the form of the Personal Responsibility and Work Opportunity Reconciliation Act (PRWORA), enacted in 1996 and gradually implemented in five years between 1997 and 2002. In brief, PRWORA ended a federal entitlement and limited benefits to a cumulative five-year period per family –after which the family loses eligibility for the federally funded Temporary Assistance to Needy Families (TANF) program, which replaced AFDC. The law also switched the funding mechanism from matching to block grants, thereby capping federal spending while providing states and localities with more managerial flexibility.

The shift in focus, from monetary assistance (through AFDC) to services designed to promote work and avoid dependency (through TANF), actually had been initiated earlier through the Family Support Act (FSA) of 1988; FSA was designed to promote welfare-to-work initiatives [Lurie 1997]. While this transitional experience contributed both to understanding the new mode of operation and to initial restructuring of services and programs, the full impact of the newest legislation was to unfold only as states and municipalities made the effort to fully implement it. Thus, enactment of PRWORA was accompanied by criticism that its measures were too harsh on the poor, or simply not thought out carefully enough [Edelman 1997].

In 1997, Governor Pataki proposed a package of state reforms that would provide for the implementation of PRWORA within the specific context of New York State. The New York reform maintained the entitlement by creating a "safety net" program, thus respecting Article 17 of the New York State Constitution, which requires the State to help the needy. As Governor Pataki's reform package was scheduled for debate in the New York State Legislature, program administrators within the New York State Department of Social Services began a partnership with the University at Albany to undertake a joint series of GMB sessions designed to focus on how county-level officials might respond to both federal and state reforms. (In New York, it is the counties that actually provide social welfare services, and those services must conform to state and federal guidelines and funding policies.)

These GMB sessions were scheduled to take place in onc small rural county, one medium-sized county, and one large county. The purpose of these efforts was three-fold: (1) to assist the participating counties in their consideration of welfare reform strategies using a SD modeling framework, (2) to provide state policy makers with an opportunity to observe the GMB process and, hence, to learn how local communities were likely to respond to state and federal initiatives, and (3) to create a management flight simulator (MFS) for welfare reform viewed from the county level. The three counties directly involved in this project were Cortland, Dutchess, and Nassau.

Drawing upon the collective knowledge of management teams,[1] the modeling team[2] produced a series of SD models. In Cortland, the focus of discussion was on TANF; in Dutchess, it was on the safety net (SN). In preparation for work with Nassau, a combined TANF and SN model was developed. In building these models, the GMB

[1] The *management teams* are the practitioners engaged in the group modeling effort, also referred to as the client groups or client teams. In this project, the participants were primarily but not exclusively high-level local managers of social welfare agencies. Other agencies involved were health, mental health, labor, and education. State personnel often observed the sessions and sometimes participated more actively (in particular, when the TANF and safety net models were joined). On occasion, a local legislator was present. Some non-governmental actors also participated (in particular, representatives of private charities). At each site, the results of the modeling work were presented to a broader group of community leaders.

[2] Richardson and Andersen [1995] referred to the *modeling team* in a group modeling session as ideally composed of individuals performing five roles: facilitator, modeler/reflector, process coach, recorder, and gatekeeper. Except for the gatekeeper, a role filled by "a person within, or related to, the client group," all of the other roles were performed by University at Albany faculty and students. In addition to these roles, this particular intervention had a model builder, was supported by a data and parameter recorder, and an interface developer. Irene Lurie –an economist who is also a specialist in social welfare policy– assisted the modeling team, taking the role of domain expert.

method was used as initially developed at the University at Albany [Richardson & Andersen 1995, Andersen & Richardson 1997, Rohrbaugh 2000, and Zagonel 2002], while incorporating into this approach specific frameworks developed by others such as Checkland [1981], Eden [1989], Hodgson [1994], Wolstenholme [1994], Bryson [1995], Vennix [1996], and Eden & Ackermann [1998].

The basic plan for this project was conceived in discussions between the senior author and Robert Johnson, a researcher at the Office of Temporary Assistance and Disability Assistance (OTADA), previously the New York State Department of Social Services. Rohrbaugh suggested to Johnson a GMB approach, based upon prior experience using SD modeling in decision conferences (detailed in Rohrbaugh [2000]). Johnson made contacts within OTADA and promoted the idea of three county-level models culminating in a MFS, and Rohrbaugh organized a University-based GMB team to complete the technical modeling portions of the project.

Johnson frequently used a team approach in his work in OTADA, and he knew that Cortland County had an active management team, led by an energetic commissioner, Jane Rogers. He approached the Cortland team and arranged for the first full-day modeling conference. They previously had been introduced to systems approaches and were enthusiastic about working on the project. The other two counties involved were chosen later, along the same lines. Thus, in all three counties, work progressed with county commissioners who were open to a system's view and willing to involve their management teams in the processes of problem definition, decision making, and change.

2.1. A First TANF Model (February-April 1997)

The intervention began with a sequence of four full-day meetings for Cortland County. The first meeting on February 11, 1997, was focused on issue elicitation (i.e., problem definition); the pair of second and third meetings occurred five weeks later for the purpose of model elicitation (i.e., model conceptualization) on March 17 and 18; the fourth meeting was held after another five weeks on April 29 for the purpose of model presentation. In addition to the meetings themselves, documented in detail in meeting reports, there was extensive in-between-meetings work to prepare elicitation (and presentation) roles and scripts,[3] plan the meetings,[4] and develop the model.

Preliminary contacts between the modeling team in Albany and the client team in Cortland involved only Rohrbaugh, Johnson, and Rogers. The exact purpose of the modeling effort was not defined ahead of time, and the decision to model the TANF program resulted from focused discussions carried out during the first formal day of work with the local management team. It is not possible to expand here the

[3] The different *roles* of the members of the modeling team are discussed in Richardson and Andersen [1995]. *Scripts* are sophisticated pieces of small group process, planned and rehearsed for accomplishing sub goals in the course of a GMB workshop. For detailed illustrations of scripts designed to achieve different purposes, see Andersen and Richardson [1997].

[4] For more information on planning the meetings (*scheduling the day*), see also Andersen and Richardson [1997].

description of what actually took place in these meetings and between meetings, but detailed procedural and substantive descriptions are found in the reports and papers of the project, inventoried and discussed in the first author's doctoral dissertation [Zagonel 2004].

The model assembled for the Cortland management team was built in layers, with additional complexity being added one stage at a time. Concept models[5] were prepared in advance by the modeling team and introduced to spur interest and engagement. The model elicitation meetings, however, primarily aimed to represent *the clients' view* of the system. Each day ended on a high note, with the expert modeler reflecting back to the participants the key feedback-rich insights the group generated during the day. This moment of the session has been termed as "modeler's feedback insights."[6] The model that actually was built captured many of these insights, refined through this process of reflection based upon SD expertise.

The Commissioner of Cortland County summarized this early part of the intervention with the following statement [Rogers *et al.* 1997]:

> The model provokes the participants to examine the impact of changes in the administration of welfare programs as having community-wide consequences. Traditionally, changes in this department's budgeting have been viewed as being an isolated problem, which allowed a "business as usual" response for providers of services... The development process involved promotes the creation of a grass roots plan... The model provides a structure for discussions for community-wide strategy sessions by identifying validated high impact points in the system. Welfare reform partners are able to create community-acceptable goals... Most importantly, the simulations vividly expose the system of welfare by portraying the community-wide opportunities for investment and risk.

There were four other full days of group meetings in Cortland County associated with model improvement, community presentations and model use. These meetings will be discussed below in the section 2.3. At this point in the project, however, the focus of attention turned to the second county.

---

[5] *Concept models* are used to introduce SD icons, to demonstrate the connection between structure and behavior, and to initiate discussion with the client team about the real system. These models are very simple and usually purposefully "wrong," and they serve the additional objective of motivating the client team to engage, take control, and begin re-formulating the model to their liking. Guidelines in building and using concept models can be found in Richardson and Andersen [1995], Andersen and Richardson [1997], and Richardson [2006].

[6] Richardson and Andersen [1995] characterize the role of the system dynamics modeler *in a group modeling intervention* as:

> ... a reflector on the group's discussion, a "contemplator" whose job is to refine and crystallize the thinking of the group... There is great value to having a person reflecting on the group's thinking and reflecting it back to them. The modeler/reflector can perceive subtleties the facilitator might miss, can identify linkages and systems insights that emerge only from reflection, and can punctuate the discussion with points of important emphasis.

2.2. The Safety Net Model (May-July 1997)

The first four meetings in Dutchess County followed the same format described above. Commissioner Robert Allers was familiar with the TANF work done in Cortland and had decided that the focus of the Dutchess meetings should be in developing a model for the safety net (SN). During the first half of the first day on May 30, attention was given to the nature of the SN sector, and the resources, policies and actions available in it. The second half of the day was used to present the existing TANF model to the client team and to show to the group the base run and two policy simulations. A few days later on June 3 and 4, the structure and parameters of the SN model were elicited. The fourth meeting for model presentation was held seven weeks later on July 29.

The SN model assembled for the Dutchess management team also was built in layers. At the same time, Cortland's TANF model was adjusted to reflect Dutchess' numbers and parameters. The model presentation meeting drew upon both models. There were two other full-day group meetings in Dutchess County, one for model improvement and the other for a community presentation of the modeling work and insights. These meetings are discussed in the next sections.

2.3. Responding to Client Needs (June-October 1997)

In this project there was one client OTADA who paid for the GMB intervention, as well as at least three more non-paying clients. As previously stated, the OTADA contract promised that the University at Albany would deliver a stand-alone management flight simulator (MFS) to support New York State welfare reform policy making viewed from the county level. To build the welfare reform model, complete reliance was placed upon the county-based management teams for model elicitation and validation. The development of these models, however, spurred considerable interest on the part of the local client teams to use the models for county-level policy-making purposes, too.

The Commissioners in both counties were interested in using the models to share their views of the welfare system and of welfare reform with their communities. The larger boundary depicted in the models made vividly clear the community-wide implications of delivering social services, revealing inter-organizational effects and dependencies. Nevertheless, they felt the models needed to be validated before they were willing to discuss them with others. While they were very pleased with the structure of the models and the behaviors of the simulations, they were concerned with making sure that the models contained parameters specific to their county.[7] Also, there were several suggestions to add detail and complexity to the models. The modeling team strived to strike a balance between these individual client needs and the overall purpose of the modeling effort.

In terms of model improvement and GMB practice, the most fruitful outcomes of these requests were the development and use of new scripts designed for group processes related to model parameterization and calibration. Andersen and Richardson

---

[7] This was found to be a recurring theme in the project. In fact, none of the counties involved were willing to engage in a *generic* model, and they accepted the models for their own use only after parameterization and calibration specific to their county.

[1997] already described scripts for supporting equation writing and parameterization, and these were used in the course of the four meetings. However, the concerns raised, as well as the modeling team's attention devoted to resolving them, led to some innovations in these procedures.

While the parameterization and calibration meeting in Cortland County on June 6 was not documented, the meeting in Dutchess County on October 16 was. The report identifies which parameters were elicited using expert judgments, how resources were computed and aggregated, and which table functions were elicited directly from the participants. Two new protocols were used in this meeting. Because of the large number of table functions used in the models (and the time constraint to elicit each and every one of them directly from the expert team), a column-by-column and row-by-row procedure was used to identify the more significant table functions in the SN model for streamlining the elicitation and triangulating logically to the other table functions in both SN and TANF models. Cross-sectional calibration (in equilibrium) then was accomplished directly with the client group, using a spreadsheet equipped with stock-and-flow visualization [Lee *et al.* 1998]. These activities were pursued both to accommodate the concerns of the county teams and to advance the goal of developing a generic welfare model at the state level.

While this work was being completed, development of a prototype for the MFS interface also was underway. To advance this objective, modeling team members Andersen and Richardson incorporated the welfare reform work into a sequence of systems thinking workshops scheduled with the Governor's Office of Employee Relations (GOER). Since many of the participants of this workshop were social welfare services managers and staff, an excellent opportunity was offered to develop and experiment with a training module based upon hands-on use of the model. The workshop began on June 9 with an introduction to systems thinking and soft systems analysis based upon a specially developed case and ended with a demonstration of the TANF model in anticipation of the participants' hands-on use of the model in follow-up meetings.

In the meantime, the Cortland management team had gained greater confidence in their model and became eager to show the work to their partners in the community. A facilitated community-wide presentation meeting was held on July 8. This presentation provided a detailed overview of the structure of the TANF model, including several simulations of both policies and scenarios, tailored by subgroups of participants and then contrasted by all. The community-wide implications of welfare reform were explored by linking model behavior to system structure. Model-based insights contrasting high and low leverage strategies were explained in terms of the stock-and-flow structure, as well as reinforcing and balancing loop-structures in the system (the essence of which is documented in Rogers *et al.* [1997]). During breaks, participants experimented hands-on with a preliminary version of the MFS interface.

The follow-up to the GOER workshop was held on September 8 and 9. The structure of the TANF model was reviewed, and participants engaged in hands-on work with the second version of the simulator in guided simulation exercises using a tailored user's manual; a thorough debriefing followed the exercises. These workshops at the state level were followed by a resource allocation conference in Cortland County on

September 29 and 30 for the purpose of county-wide fiscal decision making. The conference included 45 participants from 30 agencies and organizations in the county. The conference began with the presentation of the TANF model, as well as policy and scenario simulations. Based upon group discussion of insights derived from this model, participants agreed on a set of planning assumptions which led to the creation of three new county-wide task forces [Rohrbaugh 2000]:

- Teen life skills – to target resources for prevention services;
- Day care – to target resources for improved employment services and self-sufficiency promotion; and
- Jobs center – to target resources for a one-stop service center for job skills assessment and training, job finding, and job mentoring.

Also, conference participants consensually agreed to five proposals costing $675 thousand, which were widely understood to be the most important, immediate initiatives that could be undertaken in Cortland County:

- Job center ($150K);
- Resource center (150K) – to coordinate community efforts toward diversion;
- A program to support employed self sufficiency ($200K) – involving case managers and employers in job counseling to keep individuals on the job;
- Computer-based comprehensive assistance (150K) – to link all providers and to develop a common resource information and referral database; and
- Expansion of child care services ($25K).

A second resource allocation conference was held a year later on September 28, 1998, for Cortland County. On this occasion the final welfare reform model (described below) was presented. The Dutchess County team did not have a resource allocation conference. They did bring together their own facilitated community-wide presentation meeting, however, similar to the one in Cortland, but they waited until much later (April 28, 1998) after the TANF and SN models had been joined.

2.4. Joining the Two Models (November 1997 – January 1998)

After the model parameterization and calibration meeting in Dutchess County in October, preparation began for joining the two models. However, a host of issues needed clarification, including important questions related to model purpose and boundary. To address these issues, two meetings were assembled. The first meeting involved a small group of financial experts at the state level, for the purpose of reviewing the finance sector of the model and discussing corrections and improvements to be incorporated in the joined TANF/SN model. Some of the specific issues of concern were the separation of program and administrative costs and how to model the administrative caps imposed through federal and state regulations. This meeting took place on December 1, 1997.

The other, more important meeting had the participation of all of the major clients of this modeling effort, including the Cortland and Dutchess Commissioners. Based upon a preliminary discussion of model purpose and audience, the group made decisions regarding the client-flow structure of the model and the form of aggregation of services to form resource clusters. This second meeting took place on December 16. These meetings drew upon the TANF model and its numbers for Dutchess and Cortland counties, Dutchess' SN model, and the working version of the MFS. Very specific questions were listed, and several decisions were made in these two meetings that carried significant implications for the boundary and structure of the model. From these decisions, the entire model was rebuilt, this time with both TANF and SN sectors.

2.5. Reaching Out to Policy Makers

From beginning to end, the project actively reached out to the people who were implementing welfare reform in New York State by engaging them as the primary source of expertise to design a simplified yet comprehensive view of the welfare system. Actually, these welfare managers and executives were not just implementing a static version of reform. Rather, they were making the reform happen. This was recognized openly by Brian J. Wing, the OTADA Commissioner [Rohrbaugh & Johnson 1998]:

> With welfare reform, everyone's role is changing. For us at the state level, we have to stay out of the business of dictating how things get done at the local level. The modeling project is a good example of this kind of positive change in state/local relations and cross-agency teamwork. In the past, we probably would have told counties –in great detail and with incredible specificity– how we wanted them to implement welfare reform. Now we are making every effort to provide sophisticated, yet practical tools such as the welfare reform simulator, that local communities can use to think through policy implementation and arrive at their own solutions. I'm convinced that flexibility will be a key to successful welfare reform.

These efforts to work closely with state and local policy makers were evident not only from the meetings previously discussed, but also from a series of briefings and presentations carried out between April, 1997, and November, 1998. Of these, one of the most important meetings was the Management and Leadership for the 21st Century Training Forum on September 22–25, 1997. This meeting of the OTADA Commission and 29 county commissioners was organized around the theme of welfare reform. This four-day forum included presentations and discussions conducted by experts on the topic, as well as discussions led by the commissioners themselves. One afternoon was devoted to presenting and experimenting with the welfare model. The work was presented in shared ownership by the commissioners of Cortland and Dutchess counties and by a liaison at the state level. The modeling team described the GMB process and facilitated the use of the model. Interested commissioners were invited to take the project into their counties. The conversations evolving from this meeting led to the decision to extend the modeling work into Nassau County as the third project site.

A return to Dutchess County for a community-wide presentation using the joined TANF/SN model and a tailored version of the MFS occurred on April 28, 1998. This meeting was very similar in format to the meeting in Cortland. Allers *et al.* [1998] depict the essential simulations presented and discussed in this meeting. A new product was created especially for this occasion: a parameter booklet carefully documenting the type, values, units, sources, and history of each model parameter and table function.[8] With a firm handle on the numbers, the leadership of Dutchess County stated:

> The group modeling sessions provided an opportunity for representatives from several agencies within the community to come together in one setting to plan for welfare reform. As discussions regarding the complexity of social services programs took place, the value of computer support and modeling was amply recognized. The model and the simulations permitted detailed exploration of the interactions between different scenarios and alternative welfare reform policies. Finally, the experience to work with the University at Albany modeling team –in the process of model conceptualization, formulation and calibration, for the purpose of examining the impact of changes in the administration of welfare programs –was both challenging and rewarding.

These experiences in Cortland and Dutchess Counties were communicated in publications such as *Empire State Report* and *Government Technology*:

> We were hoping that, with the development of the model, communities could identify the high-leverage points in their system and shift their strategies accordingly," said David Avenius, OTADA deputy commissioner [Rohrbaugh & Johnson 1998].

> The flight simulator makes the relationships between government and non-government agencies such as charities visible, thereby helping to make the entire welfare system easier to understand [*ESR* 1998].

> The simulator makes it easy to pose a large number of "what if" questions about alternative strategies, changes in the economic and social environment and alternative approaches to funding services [Rohrbaugh & Johnson 1998].

2.6. Rolling Out the Welfare Reform Model (February-September 1998)

With these two success stories, the modeling team moved on as planned to work with a third county. It was not clear how this work was to be developed further in Nassau County. The Commissioner Irene Lapidez requested that the model be presented to the local management team on February 19, 1998. When they had reviewed the model and simulations, they requested to have it parameterized and calibrated to their county. This provided yet another opportunity to build on the group processes developed for model parameterization and calibration by drawing upon the experiences with Cortland and Dutchess counties, identifying the key information needed to parameterize the model, and requesting in writing those data that would be easily obtainable (without the need for expert judgments).

---

[8] This booklet also contained a record of the cross-sectional contrast between model-generated and actual numbers for key stocks and flows (in equilibrium) and spreadsheets documenting the formulas and sources for aggregation of actual spending items into model-based resource clusters.

Scripts and materials were prepared for the several elicitation exercises using expert judgments (elicitation of unknown parameters, relative strength of the table functions, and key table functions) and for reconciling cross-sectional, model-generated stock-and-flow values versus actual data. The parameterization and calibration meeting was held on June 16, almost four months later, because the increasingly sophisticated parameterization and calibration required much more time in preparation. (In Cortland, this had occurred within five weeks and in Dutchess within two-and-half months.) Out of this meeting surfaced the desire to replicate the historical behavior of the TANF caseload. The assumption was that, if the simulated caseload corresponded to the actual caseload, then this would build confidence in the model. The calibration of the model resulted in a reasonably good fit for the TANF caseload. When the modeling team returned to Nassau three months later to present the final results on September 22, there were no challenges to the model structure or to its simulated behaviors. County leadership changes that were occurring simultaneously appeared to preclude any strong and continuing project involvement.

3. The Welfare Reform Model

The final TANF/SN welfare reform model contained nearly 700 equations. It was built in 33 views. The first seven views of the model describe the patterns of client flow through the welfare system, as conceptualized by the management teams involved in the GMB sessions. Figure 2 is a summary view of this client-flow structure.

The model contains a TANF sector and a SN sector. In the former, families are differentiated as high-need or low-need families on TANF. In the latter, there are both families and individuals: post-TANF families on the safety net, individuals on the safety net, and other families (non-TANF) on the safety net.[9] The main stock-and-flow dynamics of both sectors are:

- An inflow from the mainstream economy into the TANF and SN systems, representing families and individuals who become at risk (e.g., fall into poverty, loss of employment, teen-age pregnancy, etc.);
- In the SN sector, this inflow represents direct enrollment into the SN program;
- In the TANF sector, before families are enrolled, they are accounted for in the stock of families at risk eligible for TANF;
- An outflow of clients from welfare programs into employment; this is a "pump" that moves clients from monetary assistance into paid jobs;
- An outflow back into the mainstream economy, representing the attainment of self-sufficiency; these families and individuals are no longer at risk;
- Those families and individuals served and placed into employment who are not able to attain self-sufficiency normally return directly into the programs in a process called recidivism or eventually fall back at risk (in the TANF sector).

---

[9] Post-TANF families on the safety net are those families that exhausted their federally-funded TANF eligibility (after five cumulative years on the TANF roll) and, therefore, fell into the state-funded SN program. The individuals on the safety net constitute the clientele of the existing Home Relief program, now incorporated onto the SN program. Other (non-TANF) families are those families that do not qualify for TANF for particular reasons –such as legal aliens– but who qualify for certain services provided by the SN program.

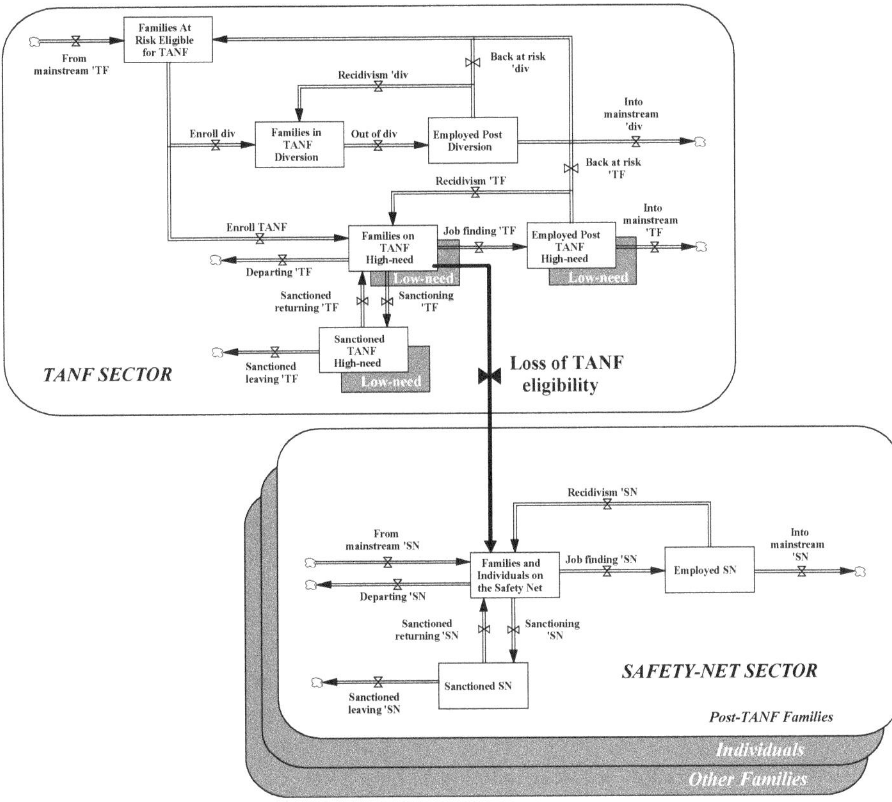

Figure 2. Summary view of the stock-and-flow structure of the Welfare Reform model

The consequence of this client-flow structure is that, at any one point in time, only a fraction of the clients served in the welfare system actually move back into the mainstream economy. Most families and individuals in this system are not permanently passing through but cycling (i.e., coming in and out of programs, finding only temporary employment or being able to maintain only low-paying jobs and eventually returning to the welfare roll). This is why, with new welfare reform policies including the implementation of time limits in the TANF program, it was expected that many initially eligible recipients eventually would run out of their allotted time and lose TANF eligibility. Thus, the connection between the two sectors is represented by the rate of loss of TANF eligibility –a one-way flow from the TANF sector to the SN sector.

4. Exploring Strategies to Improve System Performance

The PRWORA in 1996 shifted decision-making power closer to on-site administrative action, by increasing state and local flexibility and discretion in designing and running programs. States now can choose to transfer a portion of federal dollars to child care or other social services deemed important to accomplish the legislative intent of the Act; they can reduce monetary assistance in favor of other means of support; they can establish governmental and non-governmental partnerships to administer programs; and they can promote innovative means to reduce caseloads, such as an increased emphasis on child support enforcement services [Lurie 1997]. The TANF/SN models were used extensively with the client groups in a thorough search for management strategies that would improve the performance of the system at the local level. Managers were allowed to experiment with their initially preferred strategies, investing in alternative mixes of welfare services (prevention, assessment and monitoring, diversion, employment services, child support, and self-sufficiency promotion). This process of strategy selection, implementation and evaluation using the models as a laboratory proved to be an excellent means of surfacing and testing the assumptions of members of the management teams. It also provided an opportunity for a highly useful exchange of ideas among team members, as well as substantial learning about implicit mechanisms of the welfare system.

Thus, after extensive hands-on strategizing with the models based upon the solutions envisioned by the state and local policy makers themselves, can one now suggest specific, practical ways of improving system performance? In fact, as might be anticipated, the "best" strategies are heavily dependent upon what one seeks to accomplish and how one measures success in this system. There is no evident optimal strategy to improve system performance in every way. This difficulty is illustrated by drawing upon two alternative approaches to managing the system.

The first approach is a classic form of system solution, in which local social services agencies in New York State counties would invest in programs most closely associated with the social services mission and over which they have the greatest control; this strategy is labeled the "middle" investment strategy (because it is aimed at the very core of social service delivery), but it is often referred to as the "welfare-to-work" approach. It involves allocating more resources toward:[10]

- Assessment and monitoring (and diversion)
- Employment services

The second approach cannot be accomplished by local social services agencies dedicating resources to their own core programs alone, as it requires investing more in additional services associated with the missions of other actors in the community, both public and private (e.g., education, labor, private charities); this strategy is labeled the "edges" investment strategy (because it is aimed at the boundaries where county-level social services interface with a variety of other local organizations), but it can also be referred to as the "community-wide" approach. It involves allocating more resources toward:[11]

- Prevention
- Self-sufficiency promotion (or job maintenance)

---

[10] The "middle" investment strategy (or "welfare-to-work" approach) is simulated with a 25 percent increase in resources for the following services: TANF Assessment and Monitoring, SN Monitoring, Employment Services to High-need Families on TANF, Employment Services to Low-need Families on TANF, and Employment Services for SN Clients.

[11] The "edges" investment strategy (or "community-wide" approach) is simulated with a 25 percent increase in resources for the following services: Prevention, Self-sufficiency Promotion on TANF, and Job Maintenance Services on the safety net program.

These effects are illustrated in Figures 3-A and B.

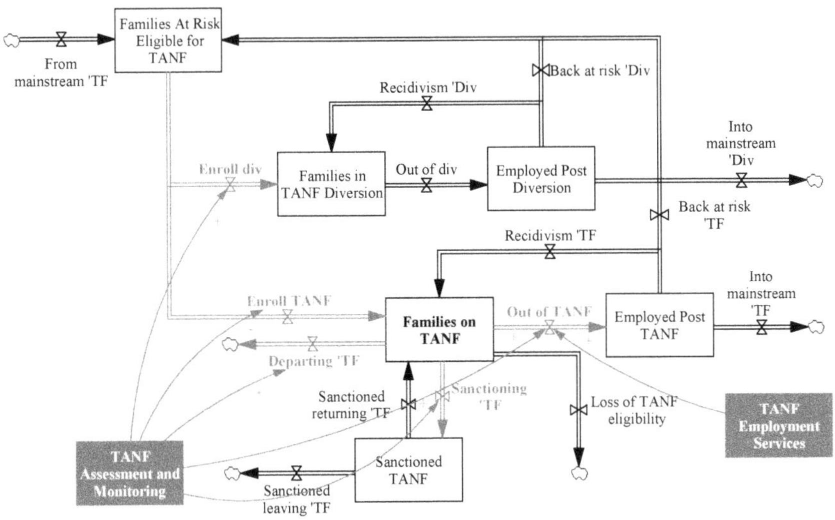

Figure 3-A. The effects of the "middle" strategy on the TANF sector

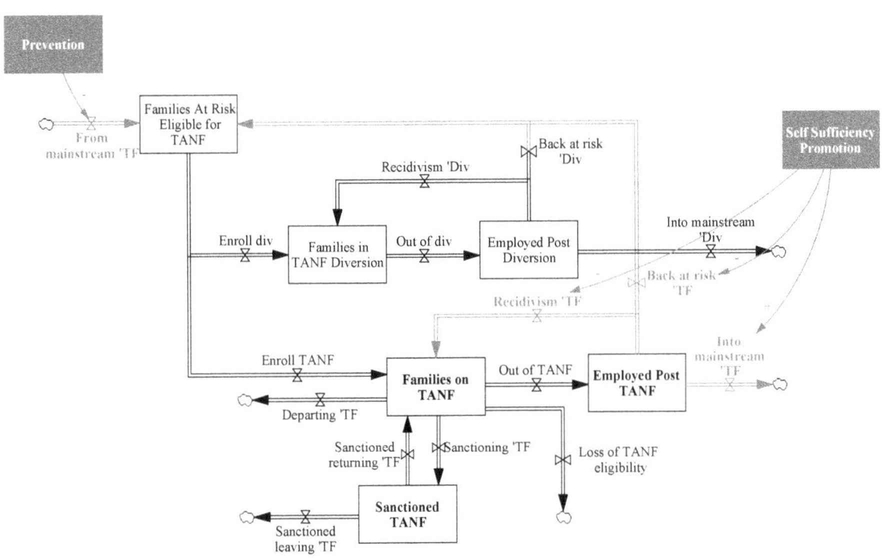

Figure 3-B. The effects of the "edges" strategy on the TANF sector

In short, the "middle" investment strategy devotes resources towards services that have an effect on the rates that pump clients from one stock to another in the middle of the system (e.g., employment services that help TANF recipients find jobs). The "edges" investment strategy devotes resources towards services that have an effect upon the rates that pump people into and out of the welfare system (e.g., job maintenance services that help those clients who find jobs to maintain those jobs, find better jobs, and move back into the mainstream economy).

4.1. Illustrating an Investment

In the TANF/SN model, any priority for the investment of a resource is expressed in terms of its "intensity." An initial resource intensity for a particular service is a function of the amount of expenditure in this service divided by the number of clients currently benefiting from it. A decision to prioritize a specific service is made by establishing a desired resource intensity that is greater than the initial resource intensity. Changing the target for the service creates a gap between desired and actual intensities. This gap determines investment and triggers additional spending in the resource until desired and actual resource intensities match. In these simulations, the budget is allowed to increase (or decrease) to adjust to the requested resources. The only delay is the time needed to reallocate resources (budget cycle) and to build (or do away with) resources.[12]

Figure 4 illustrates the policy strategy of making an investment in TANF Assessment and Monitoring (TAM). Observe how the actual resource intensity (line 2) "seeks" the desired intensity (line 1), and how the actual available resource (line 4) "seeks" the proposed (line 3). The dynamics in the behaviors after 1998 include not only this process of resource adjustment to new priorities but also changes in the client flow. The investment strategy itself begins to have its effect (after 1998), causing clients to move around in the system and changing needs (proposed resources). The shift in the client populations, from TANF to the safety net due to loss of eligibility (after 2002), causes additional changes in the client flow, and leads to further need to adjust resources. All the while, actual intensity and resource (lines 2 and 4) are "seeking" to reach desired intensity and resource levels (lines 1 and 3, respectively).

Figure 5 illustrates what is accomplished by an increased investment in TAM. Although the scales do not help to visualize the changes, the increased investment results in a larger enrollment in diversion programs (line 1) and more job-finding assistance to clients enrolled on TANF (line 2). Thus, it causes the TANF caseload (line 3) to fall slightly (between 1998 and 2002) and the number of employed families post TANF to rise (line 4). After 2002, the most significant changes are related to the loss of

---

[12] The simulations were performed under different budgeting assumptions, ranging from a fully flexible budget (based upon desired intensities of resources) to a fully rigid budget that cannot respond to desired intensities. In fact, flexibility in the budget can be set independently for each resource in question. Here, for the sake of illustration, it is assumed that resources can be adjusted freely to match desired intensities. In other words, managers have the ability to allocate resources freely and to adjust spending as necessary according to priorities. This assumption also was made for the base run.

TANF eligibility and not to the investment on TAM. As a result, the number of families employed (line 4) rises in the short term but falls toward the end of the period, due to lower numbers in the TANF system. In the following sections, the results of the "middle" versus "edges" strategies are contrasted. This exercise serves two purposes. It illustrates the advantages and disadvantages of the "welfare-to-work" and of the "community-wide" strategies, and it underscores the problem of finding an optimal strategy.

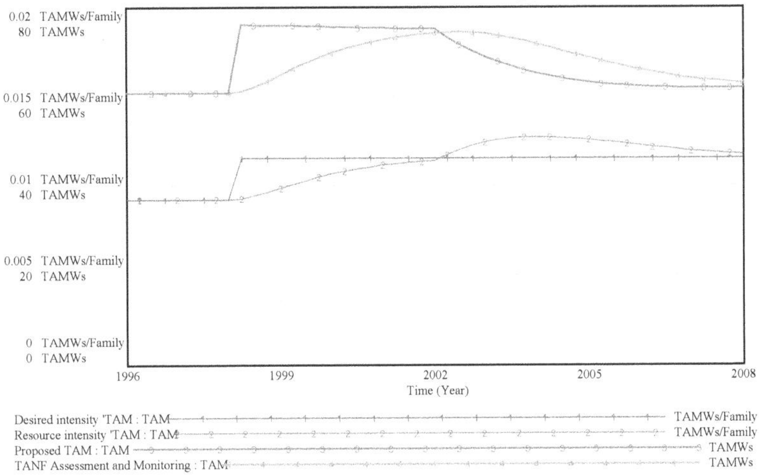

Figure 4. Investment on TANF assessment and monitoring

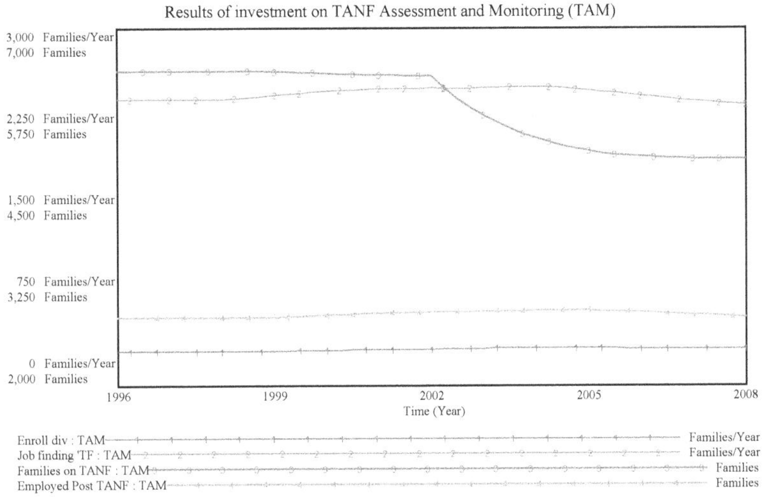

Figure 5. Results of investment on TANF assessment and monitoring

4.2. Multiple Criteria of Success

Fifteen criteria to evaluate the success of the "middle" versus "edges" strategies were examined. Four criteria are related to the size of the caseloads and populations in the welfare system:

- Number of families on TANF
- Number of families and individuals on the safety net
- Size of the TANF system (sum of all stocks in the TANF system)
- Size of the SN system (sum of all stocks in the SN system)

Six criteria are related to the rates of flow per year in the system:

- In-flow into the system from the mainstream economy
- TANF families finding jobs
- TANF families losing eligibility
- Recidivism
- Flow back at risk
- Out-flow back into the mainstream economy

Finally, five criteria are related to costs for financing the system (and who pays for those costs):

- Actual expenditures
- Local expenditures
- State and local expenditures
- Local share
- State and local share

The results of the "middle" versus "edges" strategies for all 19 criteria are summarized in Table 1. Because this GMB process was undertaken with both state and local priorities in mind, it was assumed that the best results are those that reduce their total expenditures (and shares of total expenditures), while worst results are those that increase their total expenditures (and shares of total expenditures). Table 1 indicates that neither the "middle" nor the "edges" strategy is best at all periods of time with respect to all criteria.

4.3. Analysis of the Results

The "welfare-to-work" approach that emphasizes the "middle" of the system is best at pumping clients out of the welfare roll and into jobs, as illustrated in Figure 6-A (line 2), best at lowering the size of the SN caseload, and best (but in the short term only) at reducing the size of the TANF caseload (see Figure 8-B), thus avoiding short-term loss of eligibility. This strategy also results in the lowest state and local share of costs for the short term. However, it performs the worst in the sense that it leads to "trapping" the populations of the welfare system (both TANF and SN), as illustrated for TANF in Figure 6-B, by producing the largest flows of recidivism (see Figure 7-A) and back at risk. As a result, this "middle" strategy accumulates the highest local expenditures in the short term and the highest overall expenditures in the long term (see Figure 8-A).

Table 1. Condensed results of the "middle" *vs.* "edges" strategies

Populations:	Base run:	Middle strategy:	Edges strategy:
Families on TANF	Worst	Best in short term (until 2004)	Best in long term (after 2004)
Families and individuals on the SN	Worst	Best	---
Populations on TANF system	---	Worst	Best
Populations on the SN system	---	Worst	Best

Rates of flow:	Base run:	Middle strategy:	Edges strategy:
From mainstream economy	---	---	Best
Job finding	---	Best	Worst
Loss of TANF eligibility	---	Best in short term (until 2006)	Best in long term (after 2006)
Recidivism	---	Worst	Best
Back at risk	---	Worst	Best
Into mainstream economy	Worst	---	Best

Costs:	Base run:	Middle strategy:	Edges strategy:
Actual expenditures	Best	Worst in long term (after 2006)	Worst in short term (until 2006)
Local expenditures	Best in short term (until 2002)	Worst in short term (until 2001)	Best in long term (after 2003)
State and local expenditures	Best	---	Worst
Local share	Worst	---	Best
State and local share	Best in long term (after 2002)	Best in short term (until 2002)	Worst

The "community-wide" approach that emphasizes the "edges" of the system is best at preventing families in coming out of the mainstream economy into the welfare system and best at helping families in the system transition back into the mainstream economy. Because people succeed in transitioning back into the mainstream economy, it also results in the lowest flows of recidivism and back at risk, as illustrated for recidivism in Figure 7-A (line 3). For these reasons, it deflates the welfare system and produces the best results in the sense that the size of the TANF and SN sectors (see, for example, Figure 6-A) are decreased. In the long term, it also most reduces the size of the TANF caseload (see Figure 8-B), thus avoiding loss of eligibility. Finally, it is the winner in terms of both the reduction of local expenditures and the local share of costs. However, this "edges" strategy results in the worst performance for job finding (see

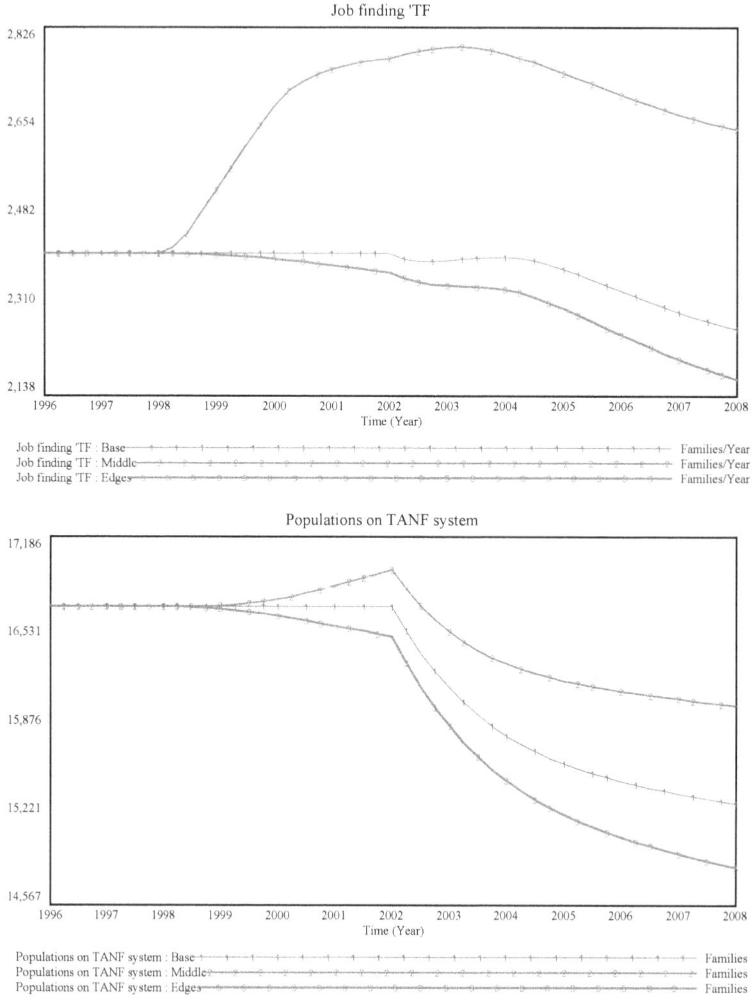

Figures 6 (A and B). Pros and cons of the "middle" strategy

Figure 6-A), for state and local expenditures, as illustrated in Figure 7-B, and for state and local share. It is also the worst strategy in the short term for overall expenditures (see Figure 8-A).

With respect to several criteria, the base run or "status quo" strategy also is shown to be best. For example, it results in the lowest overall expenditures, as illustrated in Figure 8-A (line 1), and the lowest state and local expenditures (see Figure 7-B). This is largely because no new investments are made, and for the same reason in the short term it yields the lowest local expenditures. However, the favorable results of both the "middle" and "edges" strategies make the "status quo" strategy pale by comparison in

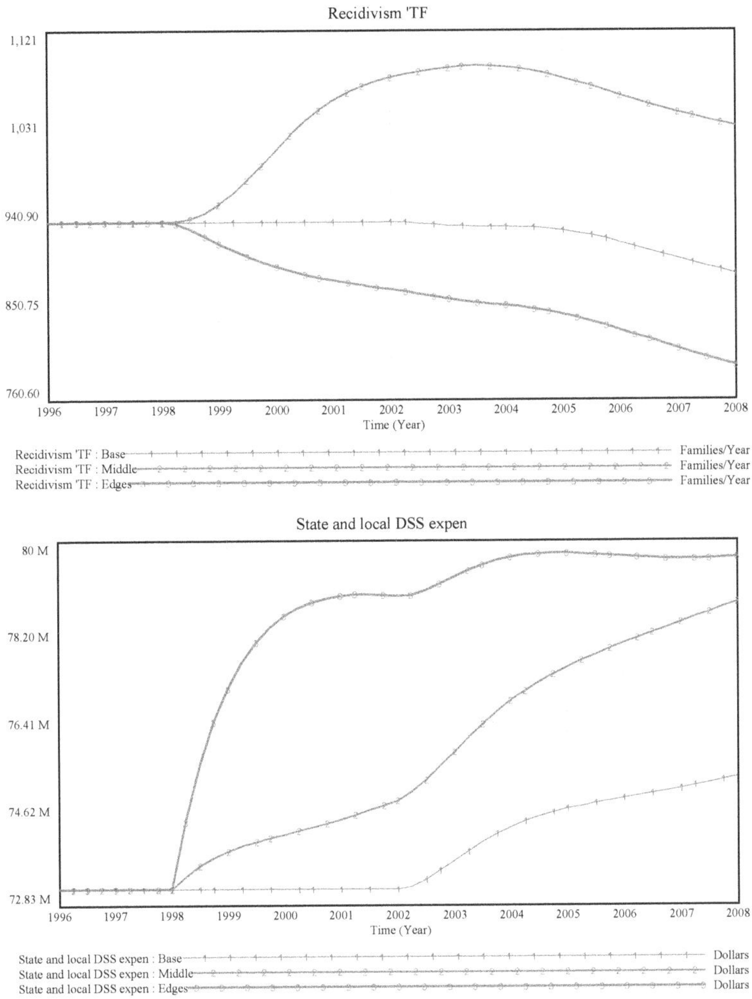

Figures 7 (A and B). Pros and cons of the "edges" strategy

their reduction in the size of the TANF and SN caseloads (as illustrated for TANF in Figure 8-B) and their increased outflow back into the mainstream economy. The local share of costs is always worst for the "status quo" strategy and the state and local share of costs is best only in the long term.

4.4. What is the Best Strategy to Improve System Performance?

No one strategy will produce the best results with respect to all of the performance criteria that are relevant in this system, in both the short and long term. Decision makers' goals and objectives, according to the positions in the system that they occupy (with respect both to the agency and the level of government—or being a private sector or nonprofit provider) are major factors in their views of their preferred strategies. There are tradeoffs between short-term and long-term results. There are differences in the importance of reductions in total expenditures or, alternatively, relative changes in the federal, state, and local shares. Individual providers are likely to ask themselves: Has the system improved with respect to the criteria affecting our workload? Has it improved with respect to the performance measures upon which we are evaluated? In the short term can we afford to wait for the long-term results expected?

The use of the model to examine what-if questions related to management policies revealed a wealth of insights with respect to issues that are absolutely necessary to resolve if system performance were to be improved. For example, key reinforcing and

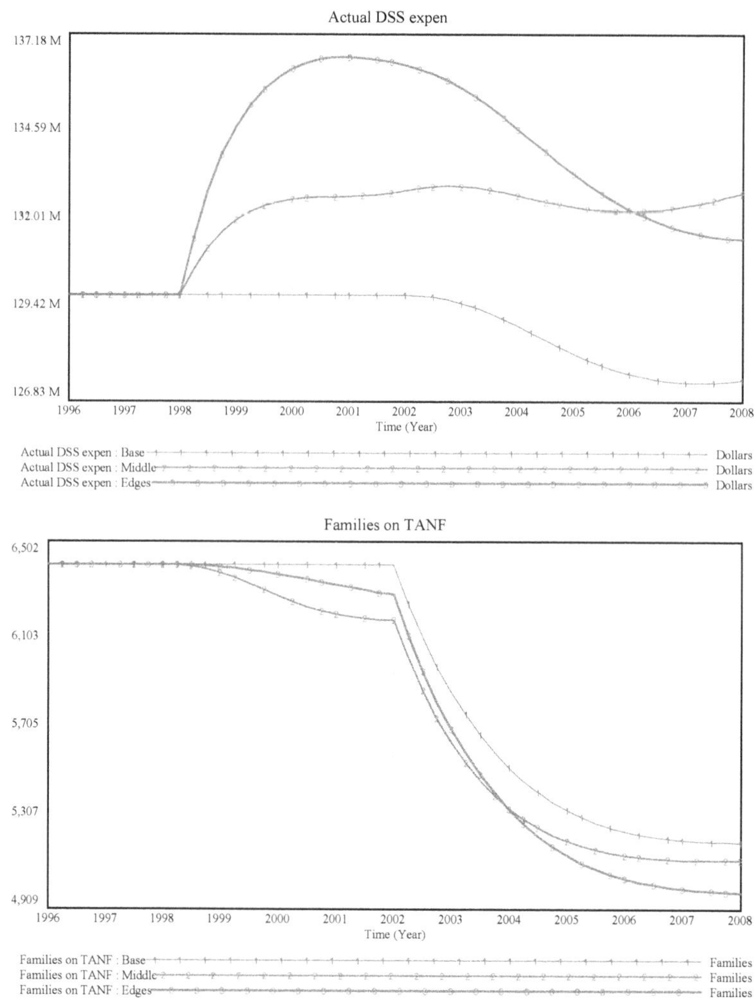

Figures 8 (A and B). Pros and cons of the "*status quo*" strategy (base run)

balancing feedback loops indicate that a powerful policy is the promotion of self-sufficiency through services that assist individuals in gaining and, especially, maintaining employment. While it is not best with respect to every performance criterion, investments in employment services deflate the welfare system and avoiding cycling. In the long term, this strategy is likely to produce many good outcomes for the welfare of the recipients themselves and for the welfare of society at large.

Most importantly, however, GMB made apparent that success or failure in welfare reform lies in the extent to which partnerships across a large and diverse number of community organizations develop, not strictly in the isolated performance of county departments of social services. Thus, what is needed to succeed is a concerted countywide response involving multiple community organizations, both public and private, in a broad leadership coalition, who make shared decisions about the allocation of county resources.

5. Conclusion

The welfare reform models were built for the purpose of helping management teams go through a difficult conversation regarding a topic filled with uncertainty. They needed some facilitation assistance to engage in this conversation, including technology to help "remember" their assumptions, integrate impacts on client flows, and deal with feedback-rich dynamic complexity. This intervention was successful in producing a consensual view of the welfare system and of welfare reform among the participants. The structure of the model and the behavioral patterns that it produced –base, policy and scenario runs– were found reasonable and plausible, even to audiences who had not engaged directly in the model-building process.

The model also proved useful for addressing what-if questions. Because the model was built based upon causal relationships elicited from the participants, it allowed a very insightful form of pattern analysis: "if we do so and so, then we expect so and so to happen..."[13] The usefulness of this modeling work is not necessarily to predict how much some quantity is increasing or decreasing or what will be the projected return on an investment but, rather, to understand the multiple and interacting relationships in a complex system and to be able to link behavior back to that structure. This case study is a good illustration of the multiple purposes usually involved in GMB efforts, as well as the difficulties related to the absence of clarity in problem definition due to a large number of diverse client and audience groups.

Although the overall purpose was broadly stated in terms of the creation of the MFS for welfare reform viewed from a county level, individual clients pursued different uses of this simulation environment. Cortland and Dutchess counties wanted to build a shared view of the welfare system and of welfare reform more than to solve a particular problem, but, once the models were built, they were eager to test their own preferred strategies for making welfare reform successful. As they built consensus and confidence in the lessons emerging from this work, they aimed to communicate and involve their partners and

---

[13] On the usefulness of GMB to address what-if questions, see Zagonel *et al.* [2004].

communities at large in the welfare reform effort. Thus, the model also was used as a focus point for decision making and resource reallocation efforts.

To accomplish this, the clients felt they needed to validate the models which, from their perspective, meant to have the models reproduce their own counties' numbers. It was only after the modeling team was able to satisfy this client need for cross-sectional and longitudinal fit that the clients' attention to the numbers could be shifted constructively to patterns of system behavior. However, there was always a lurking temptation to use the model as a forecasting tool, as revealed in this passage published in the *Empire State Report*:

> The computer model is essentially a flow diagram, with arrows and boxes showing how families would move through the welfare system and formulas calculating how factors such as an increase in the unemployment rate or additional job training programs might affect the overall social services budget and welfare population. For example, in Cortland County, the welfare caseload is about 600 families, requiring about $5.4 million in assistance per year. The Flight Simulator indicated that a strategic emphasis on helping people keep the jobs they have or move to better ones might save the county $500,000 a year. [Bold added].

While the modeling team attempted to bring about the necessary focus to this GMB effort and provide clarity regarding the usefulness and limitations of this modeling work, the somewhat ambiguous purpose of the project limited confidence in the findings derived from this work.

What is the "best" strategy? Numerous simulations of alternative management strategies involving investments in different parts of the system highlighted that successful implementation of welfare reform depended upon a coordinated community-wide effort. Welfare reform goes beyond the reach and responsibilities of the departments of social services. The leverage appears to be at the "edges" of this system, because only at these boundaries between community organizations can the welfare system be deflated effectively. However, it is unlikely that any one strategy will produce the best results with respect to all criteria of interest, both in the short and long term. Ultimately, a concerted county-wide response involving multiple community organizations depends on effective leadership and building a shared vision of what success might mean, but full community involvement in such an effort would be no easy task.

There is much that could be done to help community leaders successfully engage in the important questions that emerged from GMB. Was the emerging list of system performance criteria complete? Are some indicators more important than others in the short or long term? Which indicators must be examined only in the context of others (e.g., rates as a function of stocks, as in ratios)? Could more robust guidelines for using the indicators be derived? Do the existing performance measures and incentive mechanisms create an adequate condition for achieving the most successful outcomes possible for the communities, and, if not, what might the "right" incentives be?

No matter how extensive the analytical effort, it is still best when clients and audiences understand clearly and thoroughly the causal links in the system and how behavior can be traced to causal structure. Even if considerable progress is made toward the development of indicators–and guidelines to use such indicators, as well as the "correct" incentive system, it is still necessary to be able to explain convincingly why a particular strategy works or not. No amount of directing clients to the right answers will ever replace the primary value of their understanding what happens to system behavior if a change in system structure or parameters is made, tracing back that solid explanation to the system structure itself.

Acknowledgements

The authors acknowledge David Andersen, Naiyi Hsiao, Robert Johnson, Tsuey-Ping Lee, Irene Lurie, and George Richardson for their many contributions to this research effort. This work was successful in large part due to the level of interest presented by the Commissioners of Social Services in the participating counties: Jane Rogers, Robert Allers and Irene Lapidez, and the genuine engagement of their management teams.

References

Allers, R., Johnson, R., Andersen, D.F, Lee, T.P., Richardson, G.P., Rohrbaugh, J., & Zagonel, A.A., 1998, Group Model Building to Support Welfare Reform, Part II: Dutchess County, *Proceedings of the 16th International Conference of the System Dynamics Society* (Québec, Canada).

Andersen, D.F., & Richardson, G.P., 1997, Scripts for Group Model Building, in *System Dynamics Review*, **13**, 2.

Bryson, J.M., 1995, *Strategic Planning for Public and Nonprofit Organizations: A Guide to Strengthening and Sustaining Organizational Achievement*, Jossey-Bass (San Francisco).

Checkland, P., 1981, *Systems Thinking, Systems Practice*, John Wiley & Sons (Chichester, England).

Edelman, P., 1997, The Worst Thing Bill Clinton Has Done, in *The Atlantic Monthly*, March.

Eden, C., 1989, Strategic Options Development and Analysis, in *Rational Analysis in a Problematic World*, edited by J. Rosenhead, John Wiley & Sons (London).

Eden, C., 1990, The Unfolding Nature of Group Decision Support: Two Dimensions of Skill, in *Tackling Strategic Problems: The Role of Group Decision Support*, edited by C. Eden and J. Radford, SAGE Publications (London).

Eden, C. & Ackermann, F., 1998, *Making Strategy: The Journey of Strategic Management*, SAGE Publications (London).

ESR, Understanding Welfare, 1998, *Empire State Report*, January.

Hodgson, A.M., 1994, Hexagons for Systems Thinking, in *Modeling for Learning Organizations*, edited by J.D.W. Morecroft & J.D. Sterman, Productivity Press (Portland, OR).

Huz, S., Andersen, D.F., Richardson, G.P., & Boothroyd, R., 1997, A Framework For Evaluating Systems Thinking Interventions: An Experimental Approach to Mental Health System Change, in *System Dynamics Review*, **13**, 2.

Lee, T.P., Zagonel, A.A., Andersen, D.F., Rohrbaugh, J., & Richardson, G.P., 1998, A Judgment Approach to Estimating Parameters in Group Model Building: A Case Study of Social Welfare Reform at Dutchess County, *Proceedings of the 16th International Conference of the System Dynamics Society* (Québec, Canada).

Lurie, I., 1997, Temporary Assistance for Needy Families: A Green Light for the States, in *The Journal of Federalism*, **27**, 2.

Reagan-Cirincione, P., Schuman, S., Richardson, G.P., & Dorf, S.A., 1991, Decision Modeling: Tools for Strategic Thinking, in *Interfaces*, **21**, 6.

Richardson, G.P., 2006, Concept Models, *Proceedings of the 24th International Conference of the System Dynamics Society* (Nijmegen, The Netherlands).

Richardson, G.P., & Andersen, D.F., 1995, Teamwork in Group Model Building, in *System Dynamics Review*, **11**, 2.

Richardson, G.P., & Pugh III, A.L., 1981, *Introduction to System Dynamics Modeling with DYNAMO*, Productivity Press (Cambridge, MA).

Rogers, J., Johnson, R., Andersen, D.F., Rohrbaugh, J., Richardson, G.P., Lee, T.P, & Zagonel, A.A., 1997, Group Model Building to Support Welfare Reform in Cortland County, *Proceedings of the 15th International Conference of the System Dynamics Society* (Istanbul, Turkey).

Rohrbaugh, J., 2000, The Use of System Dynamics in Decision Conferencing: Implementing Welfare Reform in New York State, in *Handbook of Public Information Systems*, edited by G.D. Garson, Marcel Dekker (New York).

Rohrbaugh, J., & Johnson, R., 1998, Welfare Reform Flies in New York, in *Government Technology*, June.

Sterman, J.D., 1994, Learning In and About Complex Systems, in *System Dynamics Review*, **10**, 2–3.

Thompson, F.J., 1996, Devolution and the States: Challenges for Public Management, paper presented at the Annual Meeting of the Association for Public Policy Analysis and Management (Pittsburgh, Pennsylvania).

Vennix, J.A.M., 1996, *Group Model Building: Facilitating Team Learning Using System Dynamics*, John Wiley & Sons (London).

Vennix, J.M.A., 1999, Group Model-Building: Tackling Messy Problems, in *System Dynamics Review*, **15**, 4.

Wolstenholme, E.F., 1994, A Systematic Approach to Model Creation, in Modeling for Learning Organizations, edited by J.D.W. Morecroft & J.D. Sterman, Productivity Press (Portland, OR).

Zagonel, A.A., 2002, Model Conceptualization in Group Model Building: A Review of the Literature Exploring the Tension between Representing Reality and Negotiating a Social Order, *Proceedings of the 20th International Conference of the System Dynamics Society* (Palermo, Italy).

Zagonel, A.A., 2004, Reflecting on Group Model Building Used to Support Welfare Reform in New York State, Doctoral Dissertation (Nelson A. Rockefeller College of Public Affairs and Policy, University at Albany, State University of New York), UMI Microform 3148859.

Zagonel, A.A., Rohrbaugh, J., Richardson, G.P., & Andersen, D.F., 2004, Using Simulation Models to Address "What If" Questions About Welfare Reform, in *Journal of Policy Analysis and Management*, **23**, 4.

Chapter 8

Agents First! Using Agent-based Simulation to Identify and Quantify Macro Structures

Nadine Schieritz and Peter M. Milling
Industrieseminar, Schloss
Mannheim University
nschieritz@hotmail.com

1. The Potentials and Limitations of System Dynamics Simulations in Dynamic Decision Making

The system dynamics paradigm was developed in the 1950s at the MIT, primarily by Jay W. Forrester who applies concepts from engineering servomechanism theory to social sciences [Forrester 1961, Richardson 1991]. In system dynamics real-world processes are represented in terms of stocks (for example stocks of material or knowledge), flows between these stocks, and information that determines the value of the flows [Forrester 1968]. The primary assumption is that the internal causal structure of a system determines its dynamic tendencies [Meadows and Robinson 1985]; it is not single decisions or external disturbances that are responsible for a system's behavior, but the structure within which decisions are made – the policies [Richardson 1991]. Abstracting from single events and concentrating on policies instead, system dynamics takes an aggregate view [Forrester 1961]. The reasons for using continuous simulation can be found in this aggregate viewpoint as well as in the focus on structure in general [Forrester 1961, Richardson 1991]. The mathematical description of a continuous simulation model is a system of integral equations [Forrester 1968].

The system dynamics approach focuses on the mental models of decision makers as the key information source for model building [Forrester 1994]. It therefore operates close at the stakeholders of a problem, what is according to Churchill a fundamental challenge a method has to face if it is to be used in an industrial setting [Churchill 1990]. In such a case the problem situation can not be addressed detached from the characteristics of the audience [Cropper 1990; Lane 2000]. Or, as de Geus puts it: "I have not met a decision maker who is prepared to accept anybody else's model of his/her reality, if he knows that the purpose of the exercise is to make him, the decision maker, make decisions and engage in action for which he/she will ultimately be responsible. People… trust only their own understanding of their world as the basis for their action." [de Geus 1994, p. xiv] The system dynamics modeling process provides a number of qualitative tools and techniques to ease the understanding of the model even for laypersons [Kim 1992]. Especially the integration of stakeholders in the modeling process, the so-called "group model building" accounts for the requirements an industrial application involves: "… the system dynamics community has made considerable progress in developing tools and techniques to support a group model building process. Graphical facilitation techniques, such as causal loop diagrams, stocks-and-flows diagrams and graphical functions are used in combination with guidelines for structuring and facilitating group sessions, group knowledge-elicitation techniques and appropriate consulting roles." [Akkermans and Vennix 1997, p. 3].

The group model building process brings about the advantage that insights generated during the process of model building are not only shared by all participants, but at the same time the mental models of relevant stakeholders are externalized and with that they become communicable. Misunderstandings within a group can be smoothed out; the mental models of the participants can be aligned. This can increase the commitment to a group decision [Vennix 1996]. Vennix et al. consider these aspects to be more important than the results of the actual simulation runs in many industrial settings: "… the primary objective in strategic decision making is frequently not to find a robust policy, but rather to encourage team learning, to foster consensus and to create commitment to the resulting decisions, in particular when divergent opinions are involved." [Vennix et al. 1996, p. 39] In group situations the system dynamics approach can be used as a method for structuring a discussion as well as for the development of a learning environment where assumptions and strategies can be externalized and tested [Vennix et al. 1997]. Lane evaluates the value of such an application as follows: "It is clear from experience that system dynamics offers the most benefit when it generates support and commitment amongst the group whose views are modelled." [Lane 2000, p. 16]

A consequence of the aggregate view system dynamics takes is that it often requires a higher level of abstraction what can make system dynamics models difficult to quantify. The approach deals with this problem by acting on the assumption that the structure of a system is known to the decision makers who operate within this system (however they are not able to evaluate its dynamic implications) [Milling 1991; Forrester 1991]. Moreover it is supposed that this knowledge can be extracted from the mental models of these people. If one of those assumptions is not fulfilled it will be difficult to quantify a system dynamics model or to even identify the causal problem structure. The main objective of this contribution is to analyze the potential of the agent-based approach to help out in such a situation. Beforehand, the next paragraph gives a short introduction to the agent-based modeling approach.

Contrary to the system dynamics approach agent-based studies operate on an disaggregate level. It is a key assumption of the approach that social phenomena result from local interactions of agents one level below the phenomenon; global system control does not exist [Jennings et al. 1998]. From a modeling perspective, the macro system behavior is generated by the behavior of individual entities, called agents, on a micro level [Schillo et al. 2000]; thereby the agent becomes the basic building block of a model [Jennings and Wooldridge 1998]. The concept of agency, however, is not at all well-defined. The agent-based literature identifies several different catalogues of requirements for an entity to possess agency; two prominent examples are autonomy and goal-directed behavior [Schieritz and Milling 2003]. Applications of the agent-based simulation approach are widespread. A very famous example and also one of the first models is Schellings checkerboard simulation of racial segregation [Schelling 1971]. Other applications can be found in the analysis of traffic flow [Resnick 1994, Nagel and Schreckenberg 1992], of the behavior of social insects like bees [Dornhaus 1998] or ants [Bonabeau et al. 1999, Klügl 2001], of supply chain management problems [Parunak 1998] and many more.

Due to the different level of aggregation the system dynamics and the agent-based approach have the potential to complement each other: High-level system-dynamics structures can be identified and quantified using agent-based models. The next section deals with the quantification of such macro structures using a population model as an example.

2. Quantifying System Dynamics Models Using Agents

Let the dynamics of a population be determined by the interaction of three feedback loops as depicted in Figure 1, a positive and two negative ones, which are influenced by the population characteristics *Fertility* and *Mortality*; depending on loop dominance the population grows or decays. Everything else being equal the dominance of the negative death loop (as more members die, the population is reduced resulting in less deaths) results in a decline of the population, whereas it grows as soon as the positive reproduction loop (a higher population leads to more births what increases the population) takes over. However the availability of resources slows down reproduction: As population grows, less resources are available per member what reduces the overall fertility.

For the problem to be modeled the structure is known and results in the causal relationships depicted in Figure1; moreover most of the equations can be transferred from the individual behavior of the population members in the original system to the

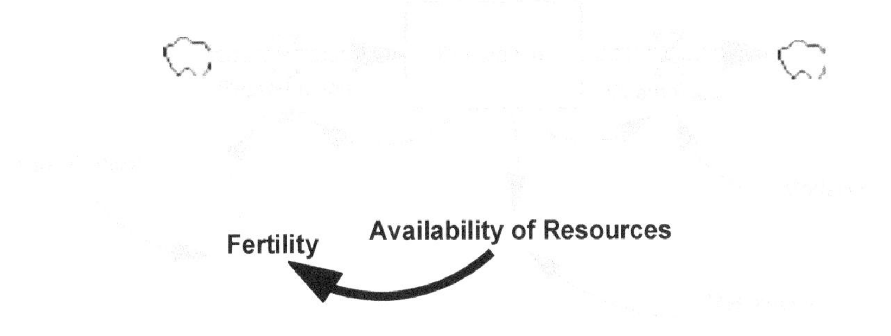

Figure 1. System dynamics model of a population with limited resources

aggregate level of the system dynamics model. Only the highlighted relationship between the available resources and the fertility needs close attention. Even though the individual policies characterizing this relationship are known (see Table 1), it can be observed that they depend on local conditions – the situation within the neighborhood of an agent. The aggregated system dynamics model, however, only includes the global availability of resources as an information source for the fertility "decision".

Table 1. Individual calculation of fertility

Individual policy for population member A
Fertility A = Max. Fertility A * (1 – Occupied Neighborhood A)

A transfer of local policies to a system level is only possible if the local situation of the major part of the agents is comparable to the global situation. In any other case the system policies have to differ from individual ones. An agent-based model of the population example is used to make this point clear. In this version of the problem it is not the population that has the characteristics fertility and mortality, but every single member of the population; the population itself is not modeled; it results from the behavior of its members. Each birth or death is then an event affecting the individual, a birth event leading to the rise of a new member (again with individual fertility and mortality) and thereby a new system element, a death leading to the exit of a member. Resources are implemented in terms of space available to the agents. Every unit of space can only be occupied by one agent and, an agent places its offspring into its neighborhood (which is defined as the Moore-Neighborhood).

Two scenarios of the policy in Table 1 are tested: The simulation *dynamic* (dark graph) allows the agents to move in their environment at random; in the *static* run this ability is disabled. The runs are depicted in Figure 2.

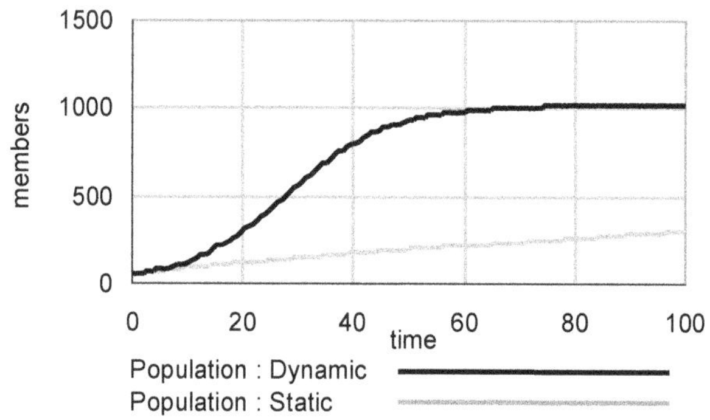

Figure 2. Dynamics of agent-based model with and without agent movement

In the beginning of the simulation many resources are available in the system compared to the overall number of agents. The simulation run *dynamic* shows exponential growth in this phase – reproduction dominates the overall behavior. By allowing the agents to move, local dependencies, which result from the fact that an agent places its offspring in its neighborhood and by that influences the future probability of reproduction, are eliminated. Without the ability to move around, however, a great portion of the local resources are used up very early in time. Therefore the grey graph shows a comparatively weak growth rate. Figure 3 gives a snapshot of the "agents' world" at time 50. Agents are more or less equally distributed in case the global situation equals the local one (dynamic run); otherwise clusters of populated areas emerge.

Given the right situation in Figure 3, the individual rules can directly be transferred to an aggregated system dynamics model; the left situation requires an adaptation as described in the following. Figure 4 shows a quantification of the functional relationship between the globally available resources of the agent model and the average fertility of all agents. It can be used as an input for the system dynamics population model.[1] In the case of a *dynamic* population (dark graph in Figure 3 and Figure 4) the functional relationship on a global level equals the linear relationship in the local neighborhood. Therefore the functional relationship described above is the bisecting line; the individual agent policies can directly be implemented into the system dynamics model.

For the grey *static* run the functional relationship on a global level is due to the reasons discussed above not linear; it does not equal the local relationships. High global resource availability is accompanied by a strong reduction of locally available resources as a result of the reproduction process; clusters of agents with low fertility are the result. With the help of a table function the relationship shown by the grey graph in Figure 4 can be implemented in a system dynamics model; it allows for the representation of a non-linear effect of resource availability on fertility due to local policies of the individuals within the system.

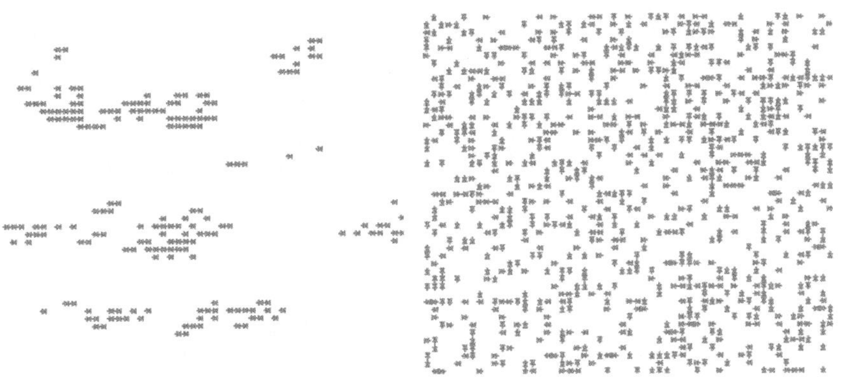

Local ≠ global situation in t = 50   Local = global situation in t = 50

Figure 3. Two agent-based scenarios with a different degree of local dependency

[1] The agent-based equation from Table 1 is formed to the following: *Fertility A / Max. Fertility* A = *1 – Occupied Neighbourhood A*.

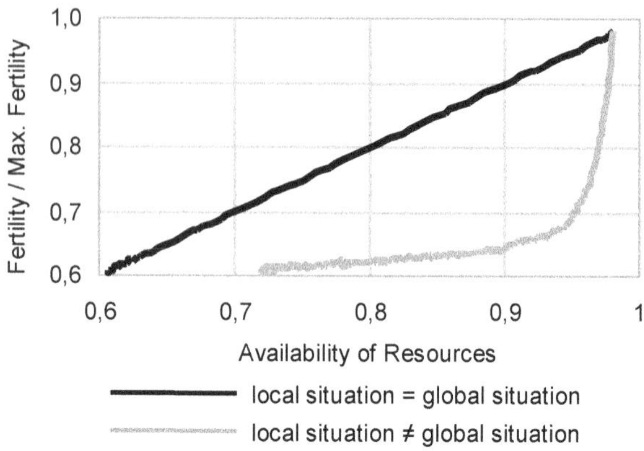

Figure 4. Relationship between resource availability and fertility

3. Agent-based Models for the Identification of Macro Structures

The use of agent-based models for the identification of "new" macro structures follows a less structured approach than the one discussed in the last section. Here again an agent-based representation of the original system takes over the role of "reality" in the system dynamics modeling process; however, now even the causal macro structure of the problem is not known. Therefore, before a quantification of functional relationships can take place, the variables characterizing the problem on a macro level have to be identified together with their causal structure. This is done by conducting simulation experiments with the agent-based model. As a prerequisite, the policies of the individual agents have to be identifiable.

The following simulation analyses aim at developing a macro theory of the evolution of a population. As the agent-based approach commonly uses variations of the genetic algorithm to model evolutionary processes, the macro structure of which is not established, an agent-based model is used for developing such a theory [Holland 1992]. The model is conceptually based on the *dynamic* scenario discussed in the previous section, however it is now two species (A and B) that compete for the available resources. The A's follow the same rules as the population described above, the B's are modified slightly as depicted in Table 2 [see also Allen 1988].

Table 2. Behavioral rules for species A and species B

	Species A	Species B
Reproduction	Maximum fertility determines the probability to reproduce.	Maximum fertility determines the probability to reproduce.
Inheritance	The characteristics of the parent generation are exactly passed on to the offspring.	The offspring is exposed to mutations concerning fertility and mortality.
Drop out of the system	Mortality determines the probability of a drop-out.	Mortality determines the probability of a drop-out.

Whenever a B agent reproduces itself, only a percentage of its offspring, represented by the parameter *PerfectRepro*, exactly inherits its characteristics, the rest is exposed to mutation that either affects its fertility or its mortality. The parameter *Deterioration* determines the percentage of the mutations that lead to worse characteristics (a lower fertility or a higher mortality), the other mutations result in an improvement. To what extent the offspring improves or deteriorates is expressed by the parameter *Mutationfactor*; the extent of change equals a percentage of the parent value (all discussed parameter values can be specified by the user). However, mutations are limited to a specific interval to avoid unrealistic situations like eternal life. The relationship between the parameters that determine the process of evolution is explained in Figure 5 using the example of a parent agent with a fertility of 0.3 and a mortality 0.2. By multiplying the probabilities along a path, the probability of a descendant with specific characteristics is obtained, e.g. the probability for agent 1 to give birth to agent 23 equals 0.3 * 0.1 * 0.8 * 0.5 at every time step. Note that the possible values of an offspring's fertility and mortality depend on the actual values of the parent and the value of the *Mutationfactor*; in the example displayed in Figure 5 the *Mutationfactor* equals 10 %.

In the simulation run displayed in Figure 6 the A species having invariable characteristics competes with population B. Initially, both populations have 50 members with exactly the same characteristics (a *Fertility* of 0.3 and a *Mortality* of 0.2).

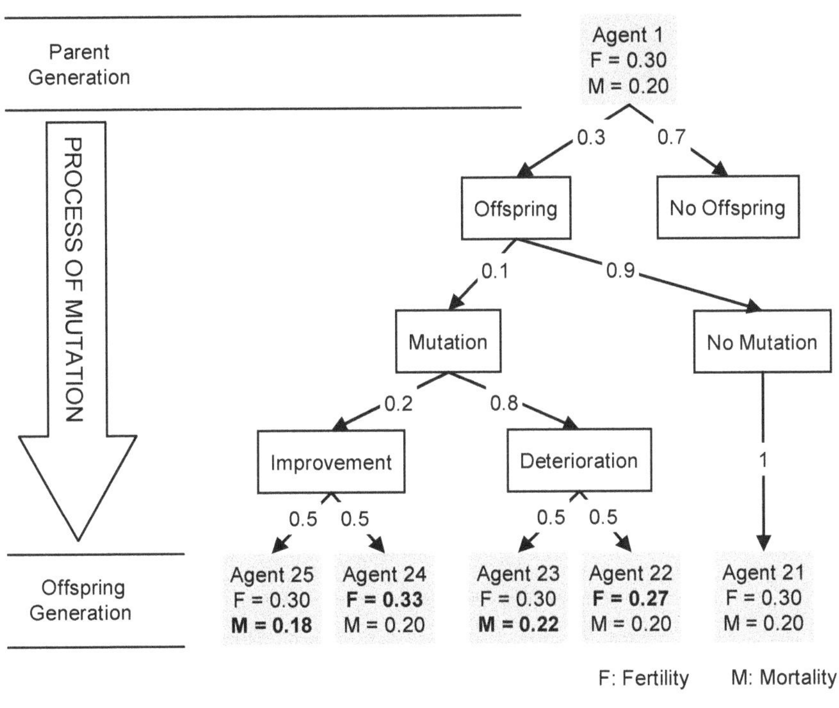

Figure 5. The process of evolution

In comparison to the simulation run above however, the offspring of B is exposed to mutation (the parameter values equal those in Figure 5). The graph shows that despite the fact that with a very high probability mutation leads to deterioration, population B becomes superior to population A after some time; in the long run, A dies out.

The behavior in Figure 6 can be explained as follows: in the beginning of the simulation deterioration dominates the dynamics of population B; a great part of the offspring is born with less advantageous characteristics than the members of population A; A leads the race (see Figure 6) However, some mutants are superior to population A. As they have a higher fertility or life expectancy or both, they manage to increase in quantity above average, at the same time the deteriorated population groups can only increase in quantity below average. Therefore the average fertility and life expectancy of the overall population B improves, resulting in an extinction of population A.

An aggregate description of the phenomenon described above takes as a starting point a model of two populations competing for resources based on the one-population-version described in the last section. Figure 7 shows a system dynamics representation of the problem.

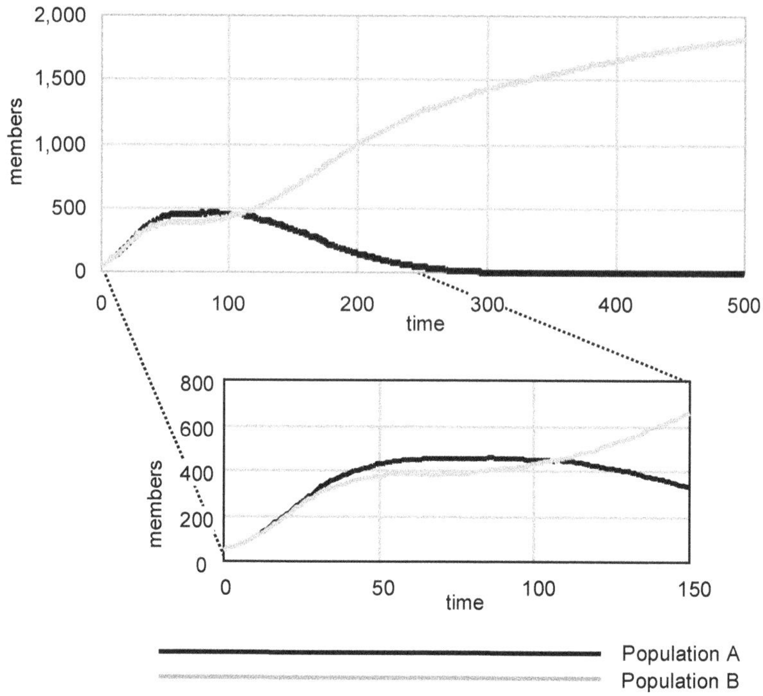

Figure 6. Dynamics of population A versus population B

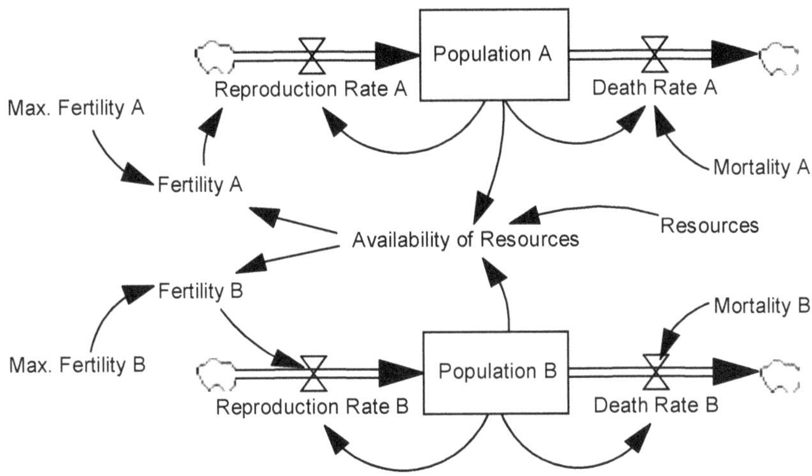

Figure 7. System dynamics model of two populations competing for resources

The aggregate representation of species B that is able to evolve has to deal with two challenges: The necessity to abstract from the events on the individual level while at the same time taking the heterogeneity necessary for the evolutionary process into account. Without this heterogeneity the mutation as well as the selection process can not be modeled. Therefore, the goal is to transfer the process of evolution from the individual to the population level and to model its causal relationships on this level. It is not the individual, that changes its behavior but the population.

The heterogeneity is achieved by disaggregating the level « Population B » and thereby dividing the overall population into subgroups: *B0* with the same characteristics as population A, *B+* with characteristics improved by the mutation factor and accordingly *B-* with deteriorated characteristics.[2] Corresponding to Figure 8 the process of mutation is modeled using flow variables between the population sub groups. A deterministic percentage of the offspring moves in an adjacent population group; the magnitude of this percentage is determined by the parameter *Percentage of Improvement* respectively *Percentage of Deterioration.*[3] It is not distinguished between the form of mutation, affecting fertility or mortality, but only between the direction of mutation – improvement or deterioration. The individual population subgroups have the same fundamental structure, only the values for the parameters *Fertility* and *Mortality* differ, depending on the mutation factor.[4]

---

[2] In contrast to the agent-based version fertility as well as mortality are equally affected by mutation. This is done for reasons of simplification.

[3] In the agent-based model the parameters *Percentage of Improvement* and *Percentage of Deterioration* correspond to the product of the percentage of the offspring that is exposed to mutation and the portion of mutations that lead to improvement respective deterioration of characteristics.

[4] Figure 8 shows a simplified representation of the population model. The entire model together with its equations can be requested by the authors.

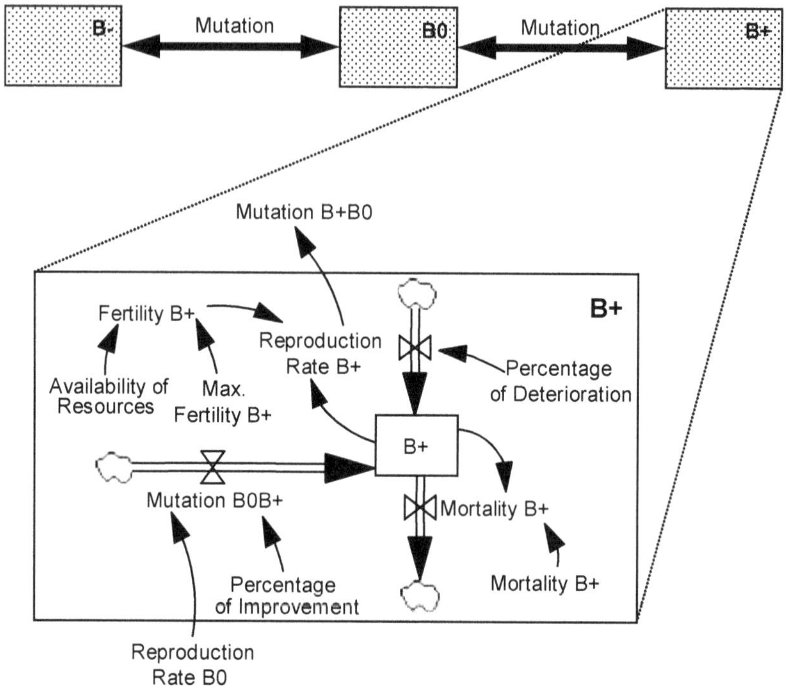

Figure 8. A system dynamics model of evolutionary population B

The model in Figure 8 corresponds to a one-to-one transference of the individual agent-based policies to an aggregate level. However, the resulting behavior of the population shows in comparison to the agent-based model qualitative differences as depicted in Figure 9.[5] Due to a reduction to three stages the process of mutation is reduced to a small part of its overall spectrum resulting in a limitation of the growth of population B in the system dynamics model, whereas B grows within the whole simulation period in the agent-based representation. A long-term improvement respective deterioration of population B, resulting from a situation where a mutated parent generation forms the starting point for new mutations is not covered by the model.

The process of mutation could be expanded by introducing two additional subgroups *B--* und *B++*, as Figure 10 shows schematically. They allow for *B-* and *B+* to act as starting points for mutation in both directions. However this only expands the process to five instead of three levels; it still has a limited spectrum, leading to the necessity to add more subgroups. A suchlike proceeding leads to an addition of more and more subgroups resulting in a situation where the system dynamics representation of the problem has the same granularity as the agent-based one; with the difference that the structural representation makes the system dynamics version more complex than its counterpart. Instead of reducing complexity, it is added by the system dynamics approach.

[5] The parameter values equal those of the agent-based model as depicted in Figure 5.

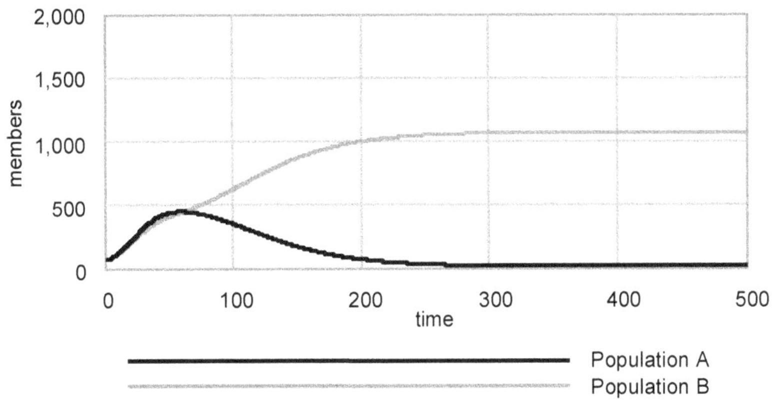

Figure 9. Dynamics of restricted process of evolution

Therefore it is indispensable that a system dynamics view of the problem abstracts from the individual policies. As already mentioned, the goal is to bring evolution to the level of the population, whose characteristics have to be changed, instead of changing the characteristics of individual population subgroups as shown in Figure 10.

On the basis of the population model composed of three subgroups in Figure 8 the essence of a structural implementation of an aggregated process of mutation lies in the introduction of an average value of the population characteristics – an average mortality and an average fertility. They are calculated as the currently existing mean of the characteristic.[6] In the following the functionality of the aggregated process of mutation is explained using the example of the variable *Mortality*.

In the beginning of the simulation the *Average Mortality* equals *Mortality B0* (e.g. 0.2), as all members of the population belong to subgroup *B0*. The *Mutation Factor* (e.g. 10 %) determines the mortality of the superior subgroup *Mortality B+* (0.18) as well as the one of the inferior subgroup *Mortality B-* (0.22). The *Average Mortality* changes caused by a movement of parts of the offspring to the superior or the inferior subgroup due to mortality. However in the long run the « normal » *Mortality B0* adjusts to the *Average Mortality* as highlighted in Figure 11 by the variable *Normality Change M.*[7]

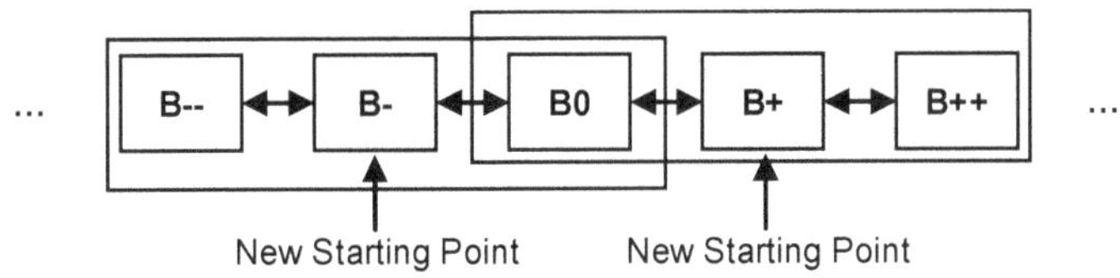

Figure 10. Adding subgroups to the population model

6 For instance the average mortality at time t is calculated as: ((*Mortality B-*(t) * *B-*(t)) + (*Mortality B+*(t) * *B+*(t)) + (*Mortality B0*(t) * *B0*(t))) / (*B-*(t) + *B+*(t) + *B0*(t)).

7 A grey box represents an information delay, meaning a change in *Average Mortality* has an exponentially smoothed effect on *Mortality B0* [Milling 1997].

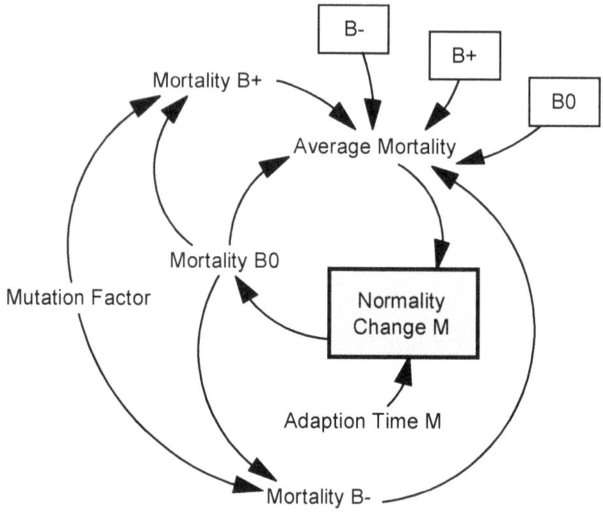

Figure 11. Causal diagram of extended process of evolution

The modified *Mortality B0* is the new starting point for the mutation process. This structure internalizes the parameter *Mortality B0* and allows a structurally created change – the parameter becomes a variable. As a result the limitation of the process of evolution to one or two stages is abolished without increasing model complexity. The point of criticism concerning the first version of the model is removed. A comparison of the behavior of the system dynamics model as depicted in Figure 12 with the agent-based model in Figure 6 shows that the fundamental mode of the agent-based behavior can be replicated.

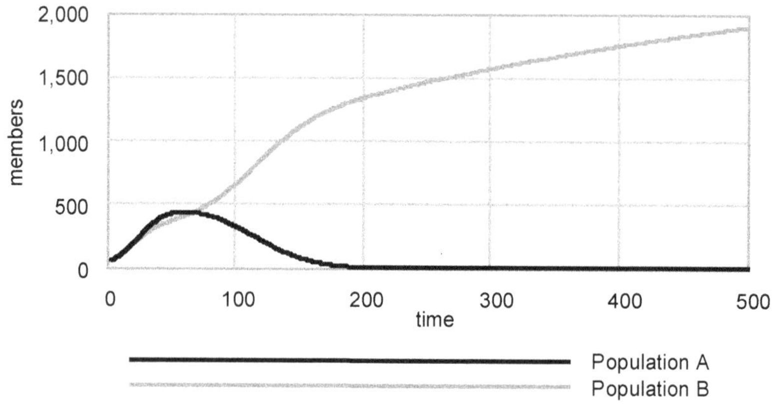

Figure 12. Dynamics of extended process of evolution

4. Conclusion: When Agents Should be First

The last two sections have used the example of population dynamics to show how a agent-based model can be used to quantify aggregate structures or to identify the causal relationship of an aggregate representation of a problem. An additional benefit of a parallel or sequential use of different methodologies for the analysis of a problem is the ability to cross-check results. However, such an approach has one major disadvantage: It requires a lot of effort. It should therefore be used only in the case of absolute necessity, namely when two requirements are fulfilled [Homer 1999]:

1. None of the available data sources, neither the mental models of the decision makers nor the available literature and numeric data is sufficient to identify/quantify the relationship under consideration.

2. The system dynamics model sensitively reacts to the relationship.

References

Akkermans, H. A. und J. A. M. Vennix, 1997, *Clients' opinions on group model-building: an exploratory study*, in System Dynamics Review, **13**, 3.

Bonabeau, E., M. Dorigo and G. Theraulaz, 1999, *Swarm Intelligence – From Natural to Artificial Systems*, Oxford University Press (Oxford).

Cropper, Steve, 1990, The Complexity of Decision Support Practice, in *Tackling Strategic Problems: The Role of Group Decision Support*, edited by Eden, C. und J. Radford, Sage (London), **29**.

De Geus, A. P., 1994, Foreword: Modeling to Predict or to Learn?, in Modeling for Learning Organizations, edited by Morecroft, J. D.W. und J. D. Sterman, Productivity Press (Portland), **xiii**.

Dornhaus, A. F. Klügl, F. Puppe and J. Tautz, 1998, Task selection in honey bees – experiments using multi-agent simulation, in Proceedings *of the Third German Workshop on Artificial Life*, edited by Wilke, C S. Altmeyer and T. Martinetz, **171**.

Forrester, J. W., 1961, *Industrial Dynamics*, Productivity Press (Cambridge).

Forrester, J. W., 1968, *Principles of Systems*, Productivity Press (Cambridge).

Forrester, J. W., 1994, Policies, Decisions, and Information Sources for Modeling, in *Modeling for Learning Organizations*, edited by Morecroft, J. D. W. and J. D. Sterman, Productivity Press (Portland), **51**.

Holland, J. H., 1992, *Adaptation in natural and artificial systems*, MIT Press (Ann Arbor).

Homer, J. B., 1999, *Macro- and micro-modelling of field service dynamics*, in System Dynamics Review, **15**, 139.

Jennings, N. R. und M. Wooldridge, 1998, Application of Intelligent Agents, in *Agent Technology – Foundations, Applications, and Markets*, edited by Jennings, Nicholas R. und Michael Wooldridge, Springer (Berlin), **1**.

Kim, D. H., 1992, *Systems Archetypes: Diagnosing Systemic Issues and Designing High-Leverage Interventions*, in Toolbox Reprint Series: Systems Archetypes, **3**, 3.

Klügl, F., 2001, *Multiagentensimulation – Konzepte, Werkzeuge, Anwendungen*, Addison-Wesley (München).

Lane, D. C., 2000, *Should System Dynamics be Described as a 'Hard' or 'Deterministic' Systems Approach?*, Systems Research and Behavioral Science, 17, **3**.

Milling, P. M., 1997, *Exponentielle Verzögerungsglieder in der Simulationssoftware Vensim*, Forschungsberichte der Fakultät für Betriebswirtschaftslehre, Universität Mannheim, Nr. 9701, Mannheim.

Richardson, G., P., 1991, *Feedback Thought in Social Science and Systems Theory*, University of Pennsylvania Press (Philadelphia).

Parunak, H. V. D., R. Savit and R. L. Riolo, 1998, Agent-Based Modeling vs. Equation-Based Modeling: A Case Study and User's Guide, in *Multi-Agent Systems and Agent-Based Simulation*, edited by Sichman, J. S., R. Conte and N. Gilbert, Paris.

Resnick, M., 2001, *Turtles, Termites, and Traffic Jams – Explorations in Massively Parallel Microworlds*, MIT Press (Cambridge 2001).

Schieritz, N. and P. M. Milling, 2003, *Modeling the Forest or Modeling the Trees: A Comparison of System Dynamics and Agent-Based Simulation*, in Proceedings of the 21st International Conference of the System Dynamics Society, edited by Eberlein, Robert L., Vedat G. Diker, Robin S. Langer und Jennifer I. Rowe, New York.

Schillo M., K. Fischer und C. T. Klein: The Micro-Macro Link in DAI and Sociology, in *Multi-Agent-Based-Simulation* edited by Scott Moss und Paul Davidsson, Springer (Berlin), **133**.

Vennix, J. A. M., 1996, *Group Model Building: Facilitating Team Learning Using System Dynamics*, Wiley (Chicester).

Vennix, J. A. M., H. A. Akkermans und E. A. J. A. Rouwette, 1996, *Group model-building to facilitate organizational change: an exploratory study*, in System Dynamics Review, **12**, 39.

PART III

System Dynamics and Agent-Based Modeling in Action

Chapter 9

Influencing and Interpreting Health and Social Care Policy in the UK

Eric Wolstenholme, David Monk, Douglas McKelvie, and Gill Smith[1]

1. Introduction - Health and Social Care: the Structures, Drivers and Performance Dilemmas

Health and social care in England are at the centre of a modernisation agenda whereby the government sets out a programme of change and targets against which the public may judge improved services. The government has made reform of public services a key plank in its legislative programme and pressure to achieve these targets is therefore immense and can be "career-defining" for the heads of the various agencies concerned.

The modernisation agenda is rooted in the NHS Plan, a ten-year milestone plan for health and social care which was initially published in July 2000 and is revised and extended as each new planning period is entered. Figure 1 describes the scope of this ten-year plan, with additional details of government investments (see Figure 2) and targets (see Figure 3). For instance, a key initiative of the NHS Plan is to take the pressure off acute hospitals by providing new services, such as diagnostic and treatment centres and intermediate care. The latter issue is one of the key subjects of this chapter affecting both national and local thinking in the UK.

[1] Eric Wolstenholme and David Monk are Directors of Symmetric SD Ltd, (eric.wolstenholme@symmetricsd.co.uk). Douglas McKelvie and Gill Smith are independent consultants.

- 10 year plan; Health Authority (HA) funding over 3 year cycle (not annually)
- New resources for intermediate care and Mental Health
- Implementation by Modernisation Agency - to be set up by autumn 2000
- A new relationship with the Department of Health (DH) based on subsidiarity (earned autonomy)
- New "concordat" with private sector (includes buying capacity to ensure wait times targets are met)
- Rationalisation of health regulation
- New processes for health complaints/suspensions
- More patient representation
- Appointments to boards no longer in hands of government
- NHS to buy capacity from private sector
- NHS Plus to sell occupational health services to employers
- NHS Lift to upgrade primary care buildings (cleaning, catering)
- National Treatment Agency (NTA) to pool resources for drug misuse
- Rationalisation of health training

Increased funding tied to targets. No change to status of Social Services - unless fail to achieve joint working (and can then be forced into a Care Trust with Health)

Figure 1. NHS Plan 2000

- £30M for clean-up of patient areas/Accident and Emergency (A&E)
- £5M development funds for trusts and HAs that are in top quartile ("earned autonomy") – total £500M. Under-performers will have to show improvement plans to earn their share
- £300M for equipment cancer, kidney and heart disease, including 50 Magnetic Resonance Imaging (MRI) scanners and 200 Computed Tomography (CT) scanners)
- 20 diagnostic and treatment centres by 2004 (day or short-stay surgery)
- 500 one-stop primary care centres
- £900M for intermediate care (hospitals to be kept for acute services) – includes 5000 rehab beds and 1700 non-residential rehab places
- 100 on-site nurseries in hospitals
- £300M by 2003/4 to introduce Mental Health service framework (in addition to £700M already announced)
- 50 early intervention teams to support young people and families
- 335 crisis teams
- 50 more assertive outreach teams over next 3 years (in addition to 170 planned for April 2001)
- 1000 graduate-level primary care workers to support General Practitioners and Primary Care Trusts (PCTs) and provide brief therapy (e.g. for depression)
- 300 staff to improve prison health services

Figure 2. Investment

"A key message.....in formulating this Plan was that it needs a small focused set of targets to drive change. Too many targets simply overwhelm the service"

- By 2002, anyone with operation cancelled on the day for non-clinical reason will get new appointment in the same month or payment to go private
- By 2004 max wait times in Accident and Emergency 4 hours; to see GP 48 hours (or primary care professional within 24 hours)
- By 2004, 7000 extra beds (2100 in general/acute wards, 4900 in intermediate care)
- By 2005 – wait times for operations down to 6 months, outpatient appt to 3 months
- Reduced mortality rates for major diseases by 2010
- Reduced health inequalities (targets TBA 2001)
- Benchmarking cost of quality care (milestones 2003/4)
- By 2010, 100 new hospitals under Private Finance Initiatives (PFIs)
- Year on year improvement in patient satisfaction (including cleanliness and food)

Figure 3. NHS Targets

Insert 1 provides updated guidance for the current planning period (2003–6). The emphasis is on growth (increased levels of facilities and staff) and transforming the patient experience (flexibility, choice, standards). It implies that there are known answers to current issues and all that is needed is focus, perseverance and attention to value for money.

In order to implement the modernisation programme, the government has set up various task forces and in order to ensure the new policies are providing the required results, the government has also strengthened the inspection and regulation framework and introduced national programmes.

The extra money coming into health and social services gives us the opportunity to make real improvements. We can expand through recruiting new staff, developing new services and creating new facilities. Even more importantly we can transform the quality of services by raising standards, tackling inequality, becoming more accessible and flexible and designing our services around the needs and choices of the people we serve.

This is about both quality and growth. The real test for success will be whether people can feel the difference and believe the services they receive are truly designed around them.

These are hugely ambitious goals. They will take time to deliver. Making progress over the next three years will be demanding and difficult and require real determination and discipline. It will need us to:

- Focus on priorities, we cannot make progress at the same pace in every area
- Extract the maximum value from every pound
- Be prepared to change old practices, be creative and take uncomfortable and difficult decisions in the drive to improve quality and respond to people using services

Insert 1. Priorities and Planning Framework 2003–6. Foreword by Nigel Crisp

2. The Potential for System Dynamics to Equip Ministers and Managers in Improving Health and Social Care

SD work in the UK health and social care field has been gaining momentum since the mid 1990s. Early work (Wolstenholme, 1993) laid the foundations for the creation of the concept of "whole systems thinking", a term which now has widespread use throughout the National Health Service (NHS). Although the early manifestation of whole systems thinking was somewhat qualitative, it subsequently paved the way for more rigorous SD modeling (Wolstenholme, 1996 and 1999; Roysten, 1999, van Ackere, 1999; Lane, 2000; Dangerfield et al, 1999 and 2000; Wolstenholme et al, 2005; Lacey, 2005). To date the method is being extensively used by the economics and operational research section of the Department of Health as well as by a number of private health and social care consultants and academics. This UK work has also been complemented by other health related system dynamics work in other countries (Hirsch et al, 2005).

There is considerable evidence in this body of work and the presentation here, that health and social care managers are actively recognising the role of system dynamics for addressing the high degree of dynamic complexity and flux present in their organizations. In particular, the work is highlighting the need to improve in line with national guidelines and the interpretation of these at a local level.

There is evidence of a willingness to engage in the (often demanding) process of externalizing mental models and also a strong recognition that performance improvement will not come through collecting ever-increasing amounts of data. Indeed there is recognition that the modeling process of itself assists the definition of appropriate data.

However, in many important places there still persists a culture of change through redefining boundaries in isolation from processes and a focus on target lead managerialism, which contrast sharply with the aims of systems concepts (Figure 4).

Issue	***Managerialism***	***System Dynamics***
Defining the problem	May be taken as self-evident (as shown by "symptoms")	Approached with caution: source of problem not seen as self-evident in complex systems
Defining the solution	Chosen as the course of action which appears the most cost-effective response to the problem	Chosen after experimenting with alternatives (checking likely responses in a complex system)
Type of solution	Framed in terms of management by objectives and targets	Focused on management of real operational processes
Reviewing outcomes	Tendency to simplistic cause/effect reasoning: e.g. X did not happen due to insufficient resources	Rigorous analysis of the behaviour of the whole system: X began to happen but caused a response in another part of the system….
Underlying principles	***Linear*** thinking, managers define actions and ensure ***compliance***, re-plan for next initiative. Tends to focus on organisation ***structure and boundaries***	***Systems thinking***, ***quantified*** view of dynamics and ***interdependency*** leading to (***innovative***) shared solutions devised by participants and ***iterative*** development of plans. Focus is on ***process and flows***

Figure 4. Managerialism versus System Dynamics

3. Influencing National Policy: Delayed Hospital Discharges

3.1. Issue

Delayed hospital discharge is a key issue which first came onto the legislative agenda in late 2001. The 'reference mode' of behaviour over time for this situation was that of increasing numbers of patients occupying hospital beds, although they had been declared "medically fit". In March 2002, 4,258 people were "stuck" in hospital and some were staying a long time, pushing up the number of bed days and constituting significant lost capacity.

The government's approach to this issue was to find out who was supposed to "get the patients out" of acute hospitals and threaten them with 'fines' if they did not improve performance. This organisation proved to be social services who are located within the local government sector and who are responsible for a small but significant number of older people needing ex-hospital ('post-acute') care packages. Such patients are assessed and packages organised by hospital social workers.

This approach was challenged by the Local Government Association (LGA-which represents the interests of all local government agencies at the national level) who suggested that a 'system' approach should be undertaken to look at the complex interaction of factors affecting delayed hospital discharges. This organisation, together with the NHS Confederation (the partner organisation representing the interests of the National Health Service organisations at a national level) then commissioned a system dynamics study to support their stance.

The remit was for consultants working with the representatives of the two organisations to create a system dynamics model of the 'whole patient pathway' extending upstream and downstream from the stock of people delayed in hospital, to identify and test other interventions affecting the issue.

3.2. Description of the Delayed Hospital Discharge Model

Figure 5 shows a highly simplified overview of the model created in the *ithink* system dynamics software. Complex behaviour over time arises in patient pathways by the interaction of the proportions of patients flowing (converging and diverging), their lengths of stays in different parts of the pathway and the capacities of each pathway, together with the actions taken by the agencies owning each piece of the pathway as patient movements conspire in and against their favour. The resultant behaviour of the whole system over time can be very counter intuitive, which is a major justification for the use of computer analysis to simplify the problem and improve understanding and communication.

Each sector of the model and its associated patient pathways was deemed necessary to fully understand the phenomena of delayed hospital discharges and is described in turn.

1.3.2.1 Primary Care

Primary Care

This sector represents only patients who are becoming ill enough to be referred for hospital treatment. Patients here are classified by GPs into three main types: medical, emergency surgical and elective surgical. It is also possible for some admissions to be avoided or diverted. If space allows, a proportion of those needing admission can instead receive certain forms of intermediate care, or a domiciliary care package. This is normally at the expense of similar services available to a patient awaiting hospital discharge, but may be provided through "new money".

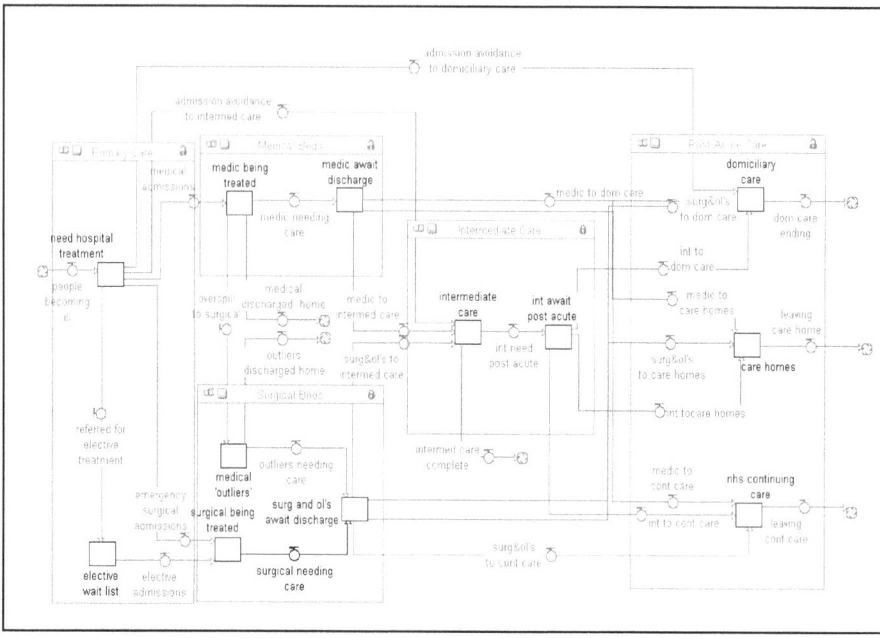

Figure 5. Simplified Version of the National Delayed Discharge Model

Medical Beds

The model uses the shorthand term "beds" to mean essentially the total number of hospital places of a particular kind. In reality, hospital capacity is constrained by a number of factors, including availability of staff and operating theatre time. Similarly, the concept of hospital "capacity" in the model is simply a number, expressed in terms of so many beds. This is set at the acute hospital's intended level of occupancy (somewhere between 85% and 95% of the number of beds available).

In the model (but not shown in this diagram), there are various pathways through medical beds representing patients with differing characteristics (those with relatively straightforward conditions and few onward care needs, those with more serious conditions and who are more likely to have onward care needs, and elderly mentally ill patients, a significant sub-group of patients within medical beds, who have very specific onward care needs). These pathways cluster the medical admissions into groups, each having different characteristics with respect to length of stay, treatment time and assessment time. This clustering is important as a relatively small number of patients with complex conditions often require complicated care packages in post acute care.

There is a fixed capacity of medical beds. The model will only allow patients to be admitted if space is available. If medical beds are under pressure numerous coping strategies are brought into play to avoid over capacity problems. First, some patients are discharged home earlier than they ideally should be. In the short term, this creates some space for new admissions. However, the unintended consequence of this action is that some patients will need to be readmitted. Second, if there is spare surgical capacity, a number of medical admissions will be admitted to and treated in surgical beds; these are referred to as "medical outliers". Again, this has the effect of dealing with the immediate problem of patients needing to be admitted, but another unintended consequence is that elective operations may have to be cancelled resulting in increases of elective wait times. Mapping is a good way of surfacing these types of coping strategies and their unintended consequences.

Once patients have been treated, if they have no onward care needs, they go home. Those with onward care needs, and who therefore cannot be discharged unless a further service is available, go into the "await discharge" stock. They will remain in that stock, occupying a hospital bed, until the correct (intermediate or post-acute) service becomes available for them.

Surgical Beds

There are three kinds of surgical admission represented in the model.

Emergency surgical patients are always admitted, regardless of how much spare capacity exists. This means that at times the hospital will exceed its target occupancy rate, which resembles reality for the current users of the model.

Those needing elective surgery go on the waiting list. They are admitted as surgical beds become available. One of the buffers in the whole system, therefore, is the elective waiting list, which increases when the hospital is full, and reduces when the hospital has spare capacity.

Medical outliers are medical patients occupying surgical beds. These admissions take place only when there are no more medical beds available but there are surgical beds.

Patients are admitted to surgical beds in the model according to this order of priority: emergency patients (always admitted), medical outliers (if medical beds are full and there are spare surgical beds), and then elective patients (if there are spare surgical beds). As with medical beds, once patients have completed their treatment, they either go home, or, if in need of onward care, wait in an acute bed until a suitable resource is available.

Intermediate Care

In this simplified version of the model shown in Figure 1, there is one entity called "intermediate care". The actual model represents a number of services, some of which are intended as alternatives to admission, and others that assist timely discharge. Some intermediate care services do both. Intermediate care is time-limited. On completing a period of intermediate care, patients / service users either go home with no further service, or are discharged with a post-acute package. The latter might involve waiting in intermediate care until the post-acute care is available.

Post-Acute Care

There are three main types of post-acute care in the model, each with a separate capacity: domiciliary care, care homes (residential or nursing), and NHS continuing care. Patients are discharged to these services either directly from acute hospital or from intermediate care. Some are referred directly to the domiciliary care service as an alternative to acute admission.

Each post-acute service has a fixed capacity. Vacancies become available according to an average length of stay (care home or continuing care) or duration of package (domiciliary care). This figure includes those whose care package ends only on their death.

3.3. The Model and Data

As is often the case in developing system dynamics models the (extensive) categories of data that are currently collected by the organisations being modelled do not neatly match the data categories required by the model. Indeed the process of matching data and model structure can create significant insights and be a revelation to managers about the data they **really** need to manage their internal processes.

Modelling facilitates an in-depth focus on data. There are some data items that are simple "inputs" to the whole system, and which can be entered into the model as such. These would include:-

- The capacity of each service
- Average treatment or service lengths of stay
- Demand data at the entry point to the whole system (i.e. numbers of people requiring acute care)
- The percentage of people using a particular service who will go on to have a need for a further type of service

However, most data items are not in themselves "inputs" to the whole system. They are indicators of how the whole system is operating but they do not drive it. A useful learning point is that, even where agencies have accurate data for a particular part of the process, it is not always possible simply to put these numbers into the model. For example, the daily admission rate to intermediate care may be known, but it is not an input. It should be clear from the diagram that once the model is running, the rate of admissions to intermediate care is a product of the rate at which people are being referred to intermediate care (mainly) from the acute hospital, and the rate at which people are leaving intermediate care. And that rate is itself dependent on the availability of some post-acute services. The admission rate is therefore used, not as an input, but as a check that the model is correctly deriving this figure.

As the model is running, variables are checked to ensure their behaviour within the model corresponds to how they "really" behave.

In general, agencies hold more data about stocks (such as how many people there are in most categories at various points in time), when it would be more useful to know about flows (that is, the rates at which people move between different stages in a process).

The model reveals an absence of data that is absolutely critical for joint planning to improve the hospital discharge process. For example, there is very detailed information about "lengths of stay" in hospital for all categories of patient but the model shows that it is equally important to know:

- length of stay up to the point where a patient is deemed ready for discharge (which would be a model input called "treatment time"), and
- The length of time spent awaiting discharge (which would be a model output that would vary according to whether there is space available in a given post-acute service)

Other useful data (currently unavailable) are the proportions of patients being discharged to the main post-acute options of domiciliary care, care homes and NHS continuing care. These categories do not neatly map against the standard discharge data available.

In the case of the delayed discharge model the best available national data was assembled.

3.4. Configuration of the Model

The model was set up to simulate a typical sample health economy over a 3 year period when driven by a variable demand (including three winter "peaks"). The capacity constrained sectors of the model were given barely sufficient capacity to cope. This situation was designed to created shocks against which to test alternative policies for performance improvement. Major performance measures in use in the various agencies were incorporated. These included:

1. cumulative episodes of elective surgery
2. elective wait list size and wait time
3. numbers of patients in hospital having completed treatment and assessment, but not yet discharged (delayed discharges)
4. number of 'outliers'

The model was initially set up with a number of fixed runs, to introduce people to the range of experiments that yielded useful insights into the behaviour of the whole system. From there, they were encouraged to devise their own runs and develop their own theories of useful interventions and commissioning strategies.

The three main polices tested in the fixed runs were:

1. Adding additional acute hospital bed capacity. This is the classic response used over many years by governments throughout the world to solve any patient pathways problem and was a favourite 'solution' here.
2. Adding additional post acute capacity, both nursing and residential home beds but also more domiciliary capacity.
3. Diverting more people away from hospital admission by use of pre-hospital intermediate capacity and also expansion of treatment in primary care GP surgeries.

3.5. Example Results from the Delayed Hospital Discharge Model

Figures 6, 7 and 8 show some typical outputs for the delayed hospital discharge model. Figure 6 captures the way capacity utilisation was displayed (actual beds occupied v total available for both medical and surgical sectors of the hospital) and shows the occurrence of 'outliers' (transfers of patients from medical to surgical beds) whenever medical capacity was reached.

Figures 7 and 8 show comparative graphs of 3 policy runs for 2 major performance measures for 2 sectors of the patient pathway - delayed discharges for post acute social services and cumulative elective procedures for acute hospitals. In each case the base run is line 1. Line 2 shows the effect of increasing hospital beds by 10% and line 3 shows the effect of increasing post acute capacity by 10%.

The interesting feature of this example output is that the cheaper option of increasing post acute capacity gives lower delayed discharges and higher elective operations whereas the more expensive option of increasing acute hospital beds benefits the hospital but makes delayed discharges worse. The key to this counter intuitive effect is that increasing post acute capacity results in higher hospital discharges which in turn reduces the need for the 'outlier' coping policy in the hospital, hence freeing up surgical capacity for elective operations.

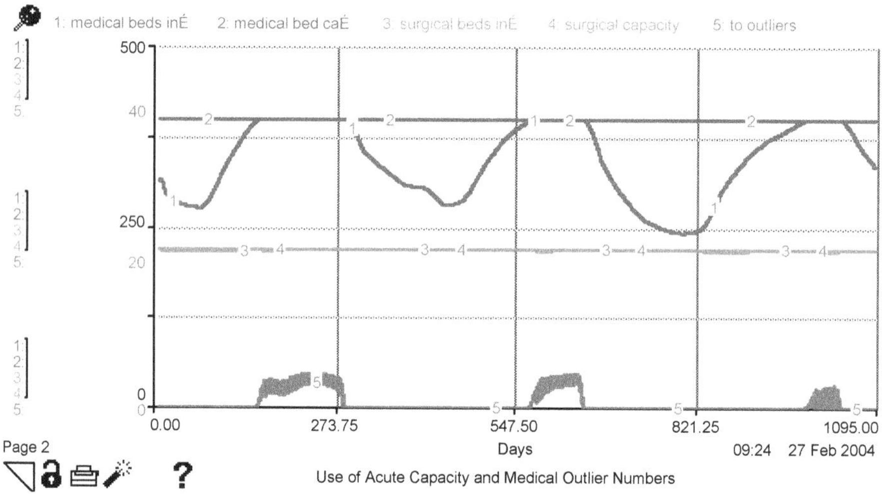

Figure 6. Medical and surgical bed utilisations in hospital and 'outliers'

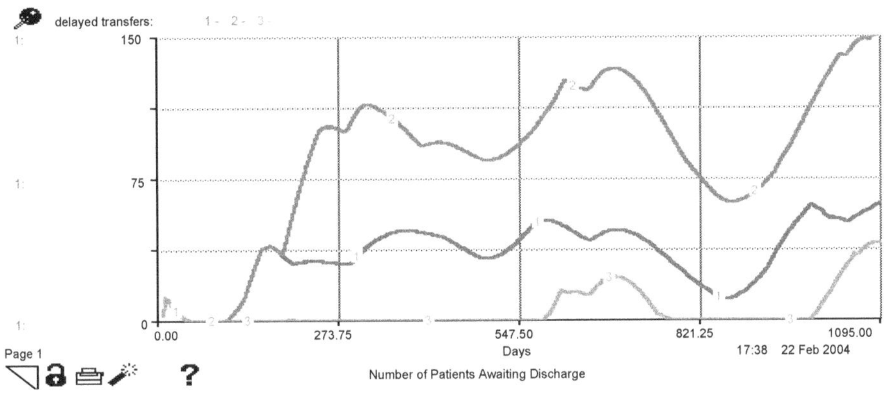

Figure 7. Delayed hospital discharges for 3 policy runs of the model

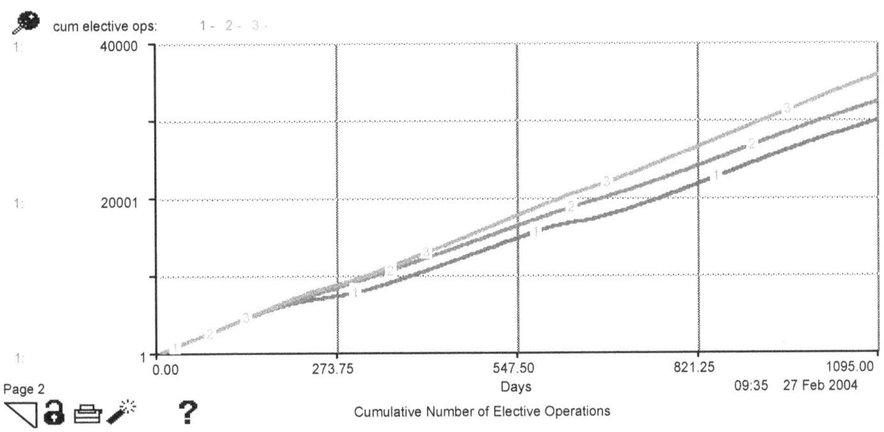

Figure 8. Cumulative elective operations for 3 policy runs of the model

Figures 7 and 8. Delayed hospital discharges and cumulative elective operations for 3 policy runs of the model.

3.6. Conclusions from the Delayed Hospital Discharge Model

Common sense solutions can be misleading

The obvious unilateral solution of adding more acute capacity can exacerbate the delayed discharge situation. More hospital admission will be made in the short term, but no additional discharges will be possible. Hence, the new capacity will simply fill up and then more early discharges and outliers will be needed.

Fines may have unintended consequences

This solution will also be worse if the money levied from social services is given to the acute sector to finance additional capacity. It will be worse still if it causes the post-acute sector to cut services. The effects of service cuts may also spill over into other areas of local government including housing and education.

There are some interventions that can help

1. Increasing post acute capacity gives a win-win solution to both health and social care because it increases all acute and post acute sector performance measures. Further, counter intuitively, increasing medical capacity in hospital is more effective than increasing surgical capacity for reducing elective wait times. This again is because 'outliers' are avoided.
2. Increasing post acute capacity gives a win-win solution to both health and social care because it increases all acute and post acute sector performance measures. Further, counter intuitively, increasing medical capacity in hospital is more effective than increasing surgical capacity for reducing elective wait times. This again is because 'outliers' are avoided.
3. Reducing assessment times and lengths of stay in all sectors is beneficial to all performance measures, as is reducing variation in flows, particularly loops like re-admission rates.
4. Increasing diversion from hospitals into pre-admission intermediate care was almost as beneficial as increasing post acute capacity.
5. If fines are levied they need to be re-invested from a whole systems perspective. This means re-balancing resources across all the sectors (NOT just adding to hospital capacity).
6. In general the model showed that keeping people out of hospital is more effective than trying to get them out faster. This is compounded by the fact that in-patients are more prone to infections so the longer patients are in hospital, the longer they will be in hospital.
7. Improving the quality of data is paramount to realising the benefits of all policies.

An interesting generalisation of the findings was that increasing stock variables where demand is rising (such as adding capacity) is an expensive and unsustainable solution. Whereas increasing rate variables, by reducing delays and lengths of stay, is cheaper and sustainable.

3.7. Impact of the Delayed Hospital Discharge Model

This model was shown at the Labour Party Conference of 2002 and generated considerable interest. It was apparently instrumental in causing some re-thinking of the intended legislation, so that social services Departments were provided with investment funding to address capacity issues, and the "fines" (re-titled "re-imbursement") were delayed for a year. Reference to the model was made in the House of Lords (see insert 2).

"Moving the main amendment, Liberal Democrat health spokesperson Lord Clement-Jones asked the House to agree that the Bill failed to tackle the causes of delayed discharges and would create perverse incentives which would undermine joint working between local authorities and the NHS and distort priorities for care of elderly people by placing the requirement to meet discharge targets ahead of measures to avoid hospital admission….. ***He referred to "ithink", the whole systems approach being put forward by the Local Government Association, health service managers and social services directors involving joint local protocols and local action plans prepared in co-operation"***

Insert 2. Reference to Model in the House of Lords

4. Interpreting National Policy – Use of Patient Pathway Models for Local Commissioning and Capacity Planning

4.1. Issue

As a result of the success of the national work, a number of local health communities (combined primary care trusts, acute hospitals and social services departments) were interested in developing similar models on a local basis, specifically for their own delayed discharge issues but more generally for commissioning (buying of services) decisions and capacity planning.

By definition this work required a model of significantly different configuration from the national delayed discharge model, which was limited to flows in and out of hospital. Local health economies are more complicated than this, with flows into social services from both hospital and from the community. However, in each locality it was found useful to use the national model as a starting template to communicate ideas. A generalised version of the emergent local health and social care economy model is shown in Figure 9 in the *ithink* system dynamics software.

The main differences between the new model compared with the delayed discharge model were:-

- the focus on older people rather than the general population
- the need for the model to focus on the population of a health economy, rather than based around the users of one district general hospital
- the need to represent a wider range of hospital beds than just medical and surgical
- the need to take account of all older people using social care services (mainly care packages at home, and residential / nursing home care) in that locality and not just those referred from the hospital
- the complexity of intermediate care and its location in the middle of a web of services
- the need to represent delays within services as well as those resulting from people waiting for a place in the "next" service

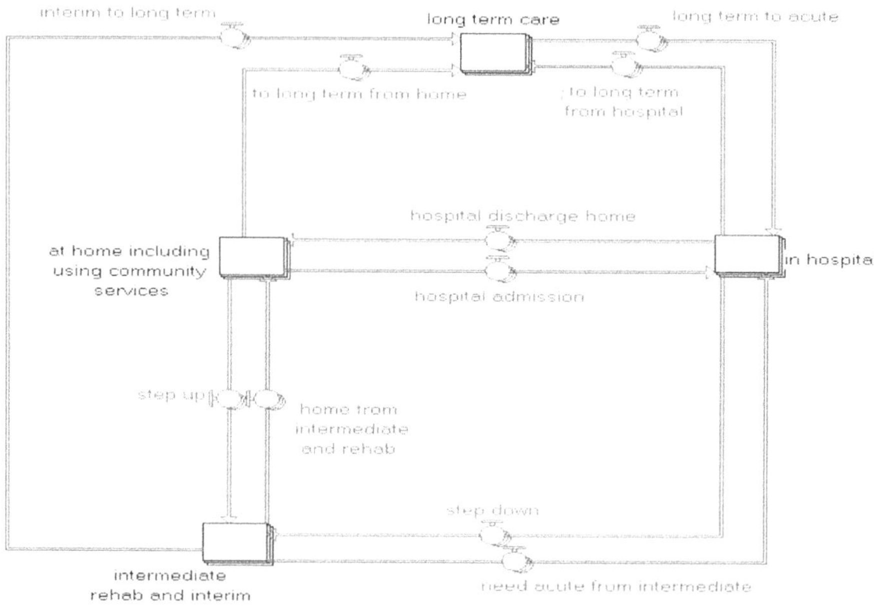

Figure 9. Overview of the local health community model

The altered shape of the diagram depicts a more circular representation of people flowing around a system, rather than the more left-to-right direction portrayed in the national work. Not all of the differences are obvious from comparing the shapes of the respective diagrams. The main difference is that the symbols in the new diagram look more three-dimensional, denoting that a number of similar flows are "arrayed" on top of each other. These arrays have two dimensions, people (divided into four age bands), and types of service (different for hospital, intermediate and Long Term parts).

The age bands adopted were:

- adults aged 18 – 64
- older people aged
  - 65 – 74
  - 75 – 84
  - 85 and over

Apart from differences in structure of the local models, there were also significant differences in the style of projects for local application. The main differences were:

- A wider range of issues to be addressed
- More detailed and dynamic complexity to be modelled
- Greater size and variation in members of the management teams

- Wider range of people to report to
- More data quality issues
- A range of possible approaches by which users access the model

4.2. Description of Model Sectors (Service Types)

The classification of services in the local modelling was agreed to be of "service type" rather than service name. These were:

At Home

Most people in the model are "at home" and a subset of these is using community services (though the main service type represented in this version is people receiving social services-funded packages of care).

Hospital

Based on an incidence rate (proportion of people becoming unwell per day), every day a number of people are referred to hospital. Although not included in the diagram; the actual model represents the stages of presentation at Accident and Emergency and further assessment (including overnight) in an acute admissions unit.
There are seven main routes through hospital.

- medical patients (mainly emergency admissions)
- patients in surgical beds, subdivided into:
    - elective patients
    - emergency patients
    - medical "outliers" occupying surgical beds
- older people's medicine
- older people's mental health
- patients in hospital beds other than the local general hospital

As an alternative to referral to hospital, the model also shows the use of intermediate care as a means of admission avoidance (labelled stepping down in the diagram).

Intermediate Care, Rehab, and Interim Care

The model represents intermediate, interim and hospital rehabilitation services as an apparently similar group of services, but each has its own particular features with different flows through (which cannot be deduced from the summary diagram). The common feature of these services is their more transitional nature.

Intermediate care is a range of services that are between home (frequently combined with a care package) and hospital. People are either admitted to intermediate care as an alternative to going into hospital, or as a stage in their discharge from hospital services. Intermediate care tends to be time-limited, for up to six weeks, but this policy can be relaxed.

Some intermediate care services provide for only one option (either admission avoidance or assisted discharge). Others can be used by people referred from home or hospital. Some intermediate care services are provided in someone's own home (including supported housing), and others are bed-based (either in care homes or in hospitals). The desired outcome from a period of intermediate care is that someone will return to full independence at home, or will manage at home with assistance from the mainstream (i.e. not intermediate care) services.

Intermediate care is closely related to hospital rehabilitation, and to older people's medicine. In the model, older people's medicine is represented in the "hospital" part, whereas hospital rehabilitation is represented in the intermediate, rehab and interim part. For very good reasons, intermediate care services often have the word "rehabilitation" in their title.

Hospital rehabilitation, in the model, is similarly located between hospital and home, but the flow is only in one direction, from hospital to home. Where it differs from intermediate care is that:-

- it is less focused on frail older people
- it is not time-limited
- it is not used for admission avoidance – it is a service for people whose recovery from an acute episode requires a therapeutic element

Long Term Care (post acute)

There are five service types in the Long Term care part of the model:-

- residential care
- nursing homes
- residential care (mental health)
- nursing homes (mental health)
- NHS continuing care

Long term care means a form of care that becomes a person's permanent home. Although people do return home from a long term placement, these numbers are small enough that they do not need to be represented in the model. However, it is possible to move between the different types of long term care in the model. It is also possible for people whose main location is in long term care to be admitted to hospital.

People flow into long term care either directly from home, from hospital, or from interim care. In some boroughs there are policy aspirations that might exclude some of these movements, e.g. people should not be admitted to a long term placement directly from hospital.

4.3. Obtaining Data for - and Learning about Data from - the Model

In a modelling project of this scale it is even more inevitable than in the national case that agencies will find that they do not have data available in a form that matches the model structure. This is because it is unlikely that any standard databases have been derived from a need to represent the stock / flow structures of a whole system of services.

Taking a whole systems perspective, the challenge is to understand how particular population groups (older people with differing degrees of dependency needs) move around between different service levels and types. This requires information about how many there are (stocks), where they are currently (stocks, but best reported over time), and the factors that govern how they move into, out of, and between service types (flows).

The "currency" of data is also important, and whether it is being used for management information or performance reporting. For example, in reporting performance, a typical measure might be the numbers that have used a particular service in the course of a year. This is a stock-like measure, but if reported in the absence of measures of turnover is much less useful than a daily (or even weekly) measure of "the number using service at this time", ideally plotted on a graph over time.

Data relating to discharge from hospital exemplify many of the differences between the kind of information that would be useful to managers (as set out in the model) and the kind of information that is actually collected. This is particularly pertinent because hospitals represent the services that have the highest throughput, and are probably most susceptible to becoming "blocked" as a consequence of the current state of other services.

There is a standard national set of data codes (although with some local variation in exactly which codes are used). Clearly, these codes represent categories of information that were once believed to be important, perhaps for accounting purposes. Sadly, they do not shed much light on the effectiveness of the whole system of services for older people. For example, the majority of people are discharged to "home / usual place of residence". Buried within that category are:

- People who returned home with no need for any further service
- People who were referred for a care assessment, but received no service (a significant number as it turns out)
- People who were discharged home with a new care package

Additional problems arise where people whose normal place of residence is a care home, from which they were admitted to hospital, are recorded in this category (because they returned to their usual place of residence). We thus "miss" a proportion being discharged to a care home. The various categories of care home in the coding framework are also frequently not up-to-date.

In essence the codes used by social service about where people come from are quite different from the codes used by hospitals as to where people go to.

In order to overcome some of these difficulties, many people worked extremely hard to extract the most meaningful data from their respective systems. Further, data research was carried out and spreadsheets were developed to transform the officially reported data into the model categories by allowing a user to vary a number of assumptions about what is going on within each coded category.

4.4. Model Configuration and Access

The model was set up to simulate time-steps of one day for five years and most model outputs were presented in the form of time-series graphs. Because the model was detailed and was designed to address a wide range of issues it had numerous policy levers and data input screens. Consequently it was relatively slow to run simulations and required considerable thought about the best medium by which to allow users access to the model.

It should be emphasised that the main purpose of this type of model is as a learning environment within which a group jointly investigates the behaviour of a complex system. The model is not a backroom tool that is fed data and gives some answers (for example, how many of these will there be in two year's time?) The power of the model is in its representation of the system's behaviour rather than in the exact figures that it produces. It is useful to know that:

- if this service gets blocked some other service will suffer
- there always seem to be a number of delays within this service but they do not impinge on other parts of the system – perhaps we can live with this
- if we divert people from here to there, this will have unintended consequences (in the form of other people who also needed to go there being blocked or having their progress slowed down)
- this kind of delay is more significant than some other kind
- if we change this capacity, we also need to change that one and monitor this other thing closely
- there is a tendency for this service to be (counter-intuitively) under-utilised when the whole system is under pressure

The optimum way to use this tool was found to be to allow users to access the model in facilitated workshops or group events, making various assumptions about how key service inputs might change over a period of five years. The model was set-up to pause every quarter, at which point it was possible to make some additional changes to the model settings (such as add or take-away capacity – open up some new services or readjust some service priorities) and then continue the run.

The most common model settings, applicable to most service sectors, were:

- The length of stay (with a wider range of settings covering the different stages in hospital) and percentage discharged to (different destinations depending on service type) settings
- Service capacity
- Allocation policies (when a service is under pressure, how admissions are prioritised based on current location of persons being referred)
- Availability and workload size of care managers
- Population (including demographic changes over the five year period) and incidence rates (e.g. rate of referral to hospital, or for community services, of people accessing from the community)

- Assumptions about the demand for elective procedures
- Extent to which intermediate care is being prioritised to assist discharge or admission avoidance

The model then displays the results of these inputs and changes, in the form of a series of graphs. Mostly these graphs show the changes in the values of stocks (or sums of stocks) over time. It is also possible to plot graphs of flows, but in a model of this complexity these tend to be used more for technical (validation or testing of the model) purposes.

4.5. Examining Some Model Outputs from Scenario Meetings

Model Outputs

Scenario planning meetings were dynamic, in that they consisted of a group of stakeholders inter-acting with the model, facilitated by the consultant. These meetings produced interruptions, "what-if?" questions, and discussions about "why did that happen?" in a relatively unstructured way.

It is difficult to capture the nuances of such an event in words. However, Figure 10 shows some examples of how the model and the workshops were used to generate insights. All graphs show changes in stock levels over a five year period measured in days.

The graphs in the top left quadrant show:

1. the numbers of people actually receiving care packages at home (dom total) compared with the capacity for this service (dom capacity),
2. the level of care management in hospital (hospsw cap) mainly dealing with patient assessment for discharge (this is a function of number of care managers multiplied by average caseload), again compared with the capacity of the service.

The graphs in the top right quadrant show:

1. the daily numbers of people in hospital accident and emergency (A&E) who are deemed to need medical or older people's medical beds
2. the total number of residents being screened in A&E per day.

The graphs in the lower left quadrant show:

The numbers of people in one particular domiciliary-based scheme for intermediate care, again compared with the capacity of that service.

The graphs in the lower right quadrant show:

The number of people being treated in hospital because they need to be there) and those delayed (for social services assessment or waiting service) in medical beds.

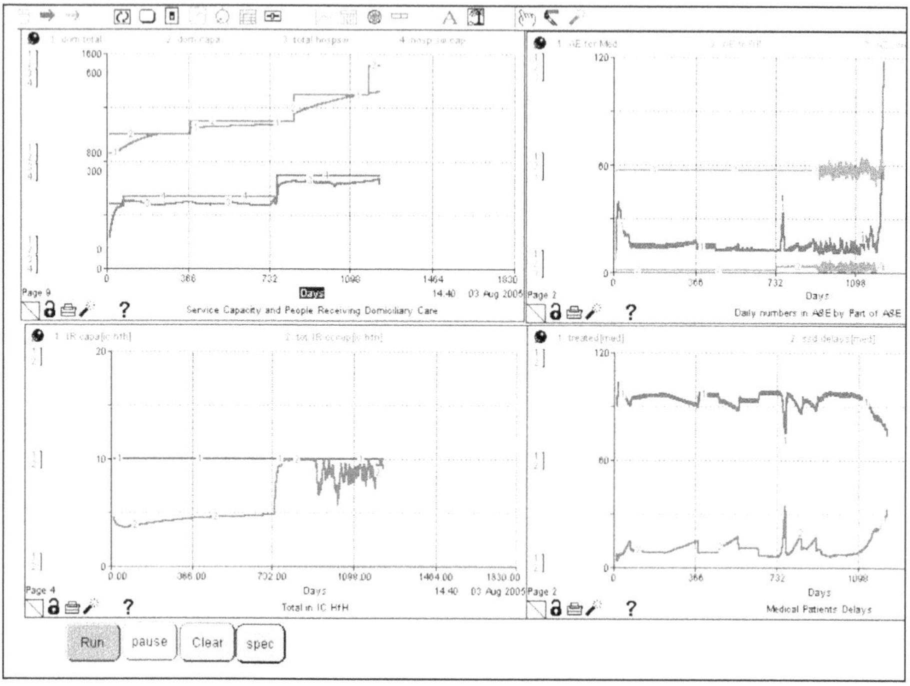

Figure 10. Example results from a Scenario Workshop using the Local Health Community Model

Experiments over the first 2 years of the run (days 0- 730)

The group first tackled the interaction between social work discharge assessment in hospital and delays in hospital accident and emergency rooms and in hospital medical treatment.

Early in the run (top left) it becomes apparent that hospital social work capacity is going to be stretched. Note the sharp rise in A and E medical delays around that time (top right) as delays occur in older people in hospital waiting for assessment to be discharged (bottom right).

As an immediate measure, an extra hospital social worker caseload is added. Immediately, the number of delays in hospital drops (but does not go away completely). Also, the potentially worrying performance in A&E stabilises. There is still some spare capacity in care packages, but not much.

Around days 60 – 200, there are still some social service delays in medical beds, even although there is spare capacity in care packages and among care managers. Note that the number of delays remains fairly constant. This is an example of what might be termed a "normal" level of delays. If referrals are typically made, say, two days before patients are deemed ready for discharge, and if the process of allocation to a care manager plus time taken to complete the assessment come to more than two days, there will always be a number of people being assessed for care beyond their statutory discharge date. This is a problem that cannot be solved by changing capacities, but only by ensuring that referrals for care assessment are made earlier in an admission, or by (somehow) speeding up assessment time. This shows the benefit of presenting data about delays as a time-series graph. Generally, delays that remain at a certain level, neither rising nor falling, are less worrying than delays that rise steadily over time.

Just after day 200, delays in hospital begin to do just that. Note that the delay line moves from being horizontal to rising just as the existing capacity of domiciliary care packages is reached. It goes on rising until the end of the year, at which point additional care packages are put into the system, and hospital delays revert back to their "normal" level. Similarly, around day 550, delays have started to rise once more. This is due to pressure on residential and nursing home places (graph not shown on this display), a problem that is resolved by increasing their respective capacities.

Experiments over the third year of the run (days 730–900)

At the end of the second year, the group tackled a different issue, which was to consider why the intermediate care service (bottom left) was relatively underused.

This was a service that mainly took referrals from people at home (and at risk of admission to hospital), people currently in A&E, and people in assessment wards. By resetting the hospital admission avoidance variables, making admission avoidance available to more people who would otherwise be admitted, this small intermediate care service quickly fills up.

But note the effect that this has on A&E (top right) and hospital treatment and delay (bottom right). With more hospital admissions being avoided, one would expect A&E throughput to increase and hospital delays to reduce. In fact, the opposite happens. Delays increase rapidly, meaning, among other things, that admissions from A&E take longer, and people begin to build up in A&E.

This turns out to have been caused by an increase in the assessment load of hospital social workers who also have to do pre hospital intermediate care assessment in preference to hospital discharge assessment. Once care managers reach full capacity, assessment gate-keeping for a range of services slows down. This causes more delays within the hospital which then feedback into A&E, the very service that this measure was intended to help.

An increase in care managers around day 730 resolves this problem. The main lesson here is that increasing service capacity without considering gate-keeping arrangements can have unintended consequences.

Experiments after the third year of the run (days 900 -)

From day 900 onwards, this group tried various other experiments with the model. The periodic rises in hospital delays were caused by further shortages in care home places (and resolved by increasing these resources). The less even graph-lines from day 930 onwards result from some experiments being done to introduce a daily variation (though the same average) to the rate of referral to A&E. Towards the end of the run, the sharp increase in hospital delays is, once again, due to a lack of assessment resources. It is in runs like this that the true usefulness of the demographic inputs is apparent. There is a gradual increase in the oldest population group over time, with a proportionate effect on demand, but no compensating increase in post-acute services.

4.6. Common Learning Points from Scenario Meetings

Types of delay:

Normal or Unavoidable Delays

Some delays just have to be expected. An example is delay in hospital discharge. The average time taken to complete a discharge assessment and provide post hospital care is simply often longer than the average time between notification of discharge and actual discharge date. Hence some delay occurs not as a result of blockages but because procedures do not quite match-up. This kind of delay will normally take the form of a straight line when plotted on a time-series graph. Such delays do not build up and, if there is sufficient capacity in the holding service, normally hospital, it is hard to consider them a problem.

Delays where Services are at Capacity

If a service is full, then the maximum number of new admissions per day is equal to the number of existing users leaving it per day. Any imbalance between arrivals and departures will result in increased numbers of people in the service and delays. The typical shape when plotted on a time-series graph is a line that is rising, with the steepness of the gradient reflecting the size of the difference between turnover and new referrals.

Obviously these delays are more serious than the "normal" type. If the number delayed within a service is rising, eventually this backlog will affect the operation of that service and contribute to backlogs further back in the whole system.

Delays where Assessment Teams are at Capacity

These delays are similar to those arising from service delays, but can be more serious with a broader and more pronounced effect. This is because assessment teams cover a range of services affecting a range of patient pathways. Hence a problem in one location can have a serious 'knock on' effect on services in many locations.

Further, the cause of a blockage in an assessment process might well not be due to change in the capacity of that particular service. The cause of the blockage might be that some other specific service has become blocked (or slowed down), thereby over-stretching the assessment team.

Even worse, this interplay between delays often leads to the perverse underutilisation of certain services when the whole system is otherwise stretched. This phenomenon is most observable in intermediate services, because they are taking service users in the middle of a patient pathway.

Intermediate Care

Most of the local groups were interested in modelling the dynamics of intermediate care services, which are seen as a major solution to reducing bottlenecks in patient flows. Modelling certainly shows why it can be hard to plan such services. They are sensitive to what is happening in just about every other part of the system. Moreover, they tend to have very small capacities relative to the size of the other services within the whole system, and are typically sub-divided into several schemes such as bed-based and community-based, hospital admission avoidance and timely hospital discharge.

The model was created with a very elaborate set of menus covering not only "who this intermediate service is for" but also how to prioritise between people coming from many different destinations, overlaid by four age bands.

A consistent learning point was that small changes in priority settings have large effects. Because the services have a low capacity, expanding the admission criteria can lead to them being immediately swamped with new referrals. The services are also intended to be time-limited, but that cannot prevent people from becoming "blocked" within intermediate care, given that a significant proportion of those in intermediate care can only leave if a domiciliary care package is available.

4.7. Conclusions from Creating Commissioning Models for Local Health Communities

Despite the challenges of having to model in more detail, with variable quality data and a wider range of issues across a management team of more variable constitution, the usefulness of describing services in basic inflow-stock-outflow terms, together with seeing the interconnectivity between such descriptions and behaviour over time, turned out to be particularly revealing. Benefits included:

1. Allowing risk-free testing of the effects of policies across patient pathways and an operational focus on ways to reduce costs.
2. Helping key individuals in different agencies developed an increasing awareness of their role in respect of the whole patient pathway.
3. Developing understanding of the way patient pathways worked by:

   - Improving the definition of service capacity and its relationship to referral rates, length of stay.
   - Explicitly defining the component states of being in hospital services as treatment, assessment and waiting for discharge.
   - Introducing the concept of equilibrium. Whilst people expected that in most services at most times the numbers being referred and accepted for service are balanced by the numbers leaving, they were often surprised at the degree of instability created by small changes, particularly to policies around intermediate care.

4. Identifying inconsistencies between maps of the way the patient pathways were claimed to work and the data collected for the pathways.

5. Understanding the most useful configuration of people and services to enhance performance. Whilst a breakdown of patients by age was used in the modelling, there was a continual discussion about the alternatives to this, such as a breakdown by 'degree of dependency'.
6. Defining the data required for patient pathway management across agencies. One major example was acute hospital data discharge codes being unmatched to the admission codes used for admissions to post acute services.
7. Facilitating discussion on defining and categorising issues and on performance measures.

5. Implications of the Work for the Process of Carrying Out System Dynamics Projects –The Challenge of Changing Thinking

System dynamics is most beneficial when, as described here, it is applied to deep-seated issues concerning long processes, which across multiple agency boundaries (Wolstenholme, 2003). Such projects are not only technically challenging but highly emotive and political. Users of the method need to recognise the context as well as content of project.

Technically, managers are unaccustomed to the fundamental nature of the questions posed by system dynamics. Questions such as "Is this care pathway complete/correct?", "Can people go anywhere else?", "What assumptions are built into routing?" and "How long do people spend in each stage?" are more fundamental than the usual ones they face every day. Further, managers do not tend to think in terms of 'flows'. They may assert that they need more resources, but often do not justify the request with an argument based on flows. Politically, multi-agency teams are also driven by professional allegiances and there can be conflicts of time and interest amongst the participants.

In order to get the best out of a system dynamics study it is important to have:

1. well-defined issues
2. consistent and representative teams of management professionals from all agencies involved, well briefed in the principles of the approach
3. well-defined expectations and well-scoped models in time and space
4. mechanisms for managing the politics of the management team and keeping senior people in touch with the project
5. ways of maintaining momentum
6. ways of reinforcing the benefits

The Symmetric approach to system dynamics used throughout the work addresses the above issues by:

- An emphasis on using modeling to create environments for learning: participative approach and involvement of the management team at every stage. No "backroom" tours de force, presenting a completed model to an astonished audience!

- A willingness to combine the use of relevant national (generic) templates with a flexible approach to the specific needs associated with local applications
- Careful discussion of model development timescales
- A commitment (however time-consuming) to help local projects with the issue of data mining. It requires significant effort to find the requisite data to animate the model. Organisations typically measure "snapshot" data (e.g. current usage of beds) rather than flows and participants may need to estimate proportions flowing through each route as well as lengths of stay

Figure 11 captures the major elements of the participative stages advocated here for successful system dynamics projects in multi-agency situations. It needs to be made clear from the outset that the work will be challenging. Successful projects require:

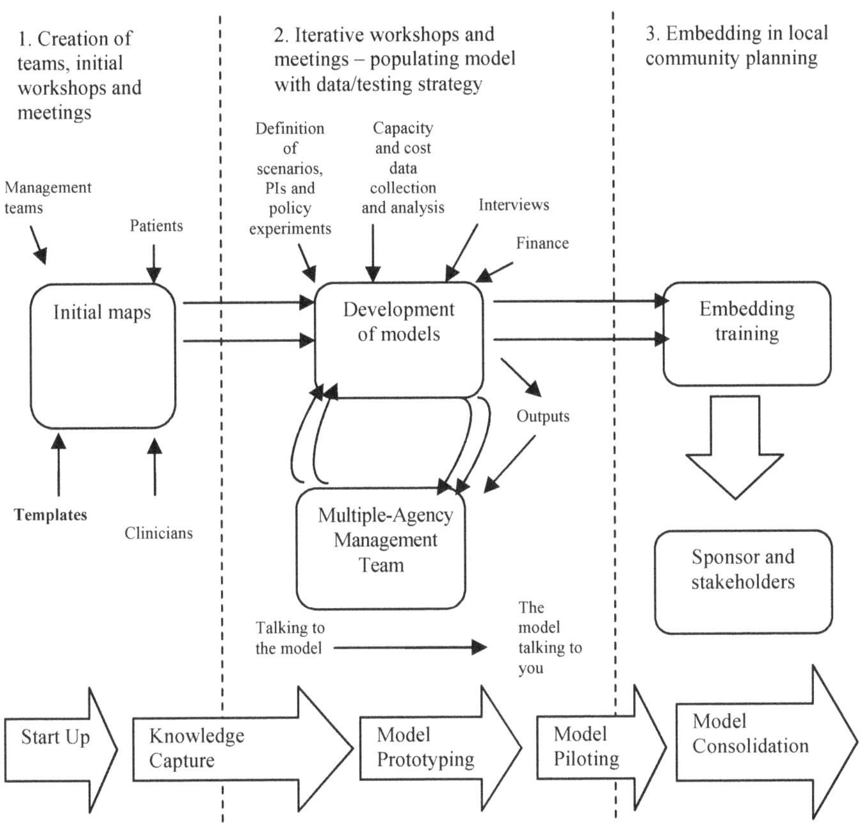

Figure 11. Stages of a System Dynamics Project – a participative approach

- partners in a multi-agency context to declare their assumptions and share their mental models of how the "system" works. There is no room for narrow views or simplistic apportionment of "blame". Participants must be able to "stand back" from single agency issues and see the broader patterns of the whole system (causal loop mapping)

- partners to agree the key issues they want to elucidate and address. These must be of significance0 to the behaviour of the whole system. It is important that they recognise the interdependence of their agencies, so that no policy or practice is considered "private" or "non-negotiable"
- painstaking work to unravel the structure (process) in a complex system and agree how to represent it in stock-flow terms, to achieve the simplest representation which reflects current realities
- recognition that models are never complete. There is always more to do.

6. Overall Conclusions

This chapter has described recent and continuing work carried out by the authors in health and social care in the UK and some guidelines for improving the success of such projects. This work has been at the forefront of both national and local policy analysis and has achieving outstanding success in influencing behaviour and decisions by its timeliness and relevance. Many managers and clinicians have been involved in sponsoring and participating in the work. These people have proved very receptive to the ideas and shown a willingness to embrace the technical and political challenges of using the ideas in a comprehensive way. They have also been very appreciative of the benefits.

Acknowledgements

The authors wish to acknowledge the help and cooperation from many health and social care clinicians and managers who contributed to the studies reported in this chapter. The results reported and views expressed for the purpose of this chapter are entirely those of the authors and are not necessarily the views of the organisations involved.

References

Ann van Ackere (1999), *Towards a macro model of Health Service Waiting Lists*, System Dynamics Review, Vol. 15, No 3

Dangerfield, B. and Roberts, C. (1999), *Optimisation as a statistical estimation tool: an example in estimating the AIDS treatment free incubation period distribution*, System Dynamics Review, Vol. 15, No. 3.

Dangerfield, B., Fang, Y and Roberts, C. (2001), *Model based scenarios for the epidemiology of HIV/AIDS: the consequences of highly active antiretroviral therapy*, System Dynamics Review, Vol. 17, No. 2.

Hirsch G., Homer J., McDonnell G. and Milstein B. (2005) *Achieving Health Care Reform in the United States: Towards a Whole System Understanding*, Paper presented at the International System Dynamics Conference, Boston, USA.

Lacey, P. (2005) *Futures through the Eyes of a Health System Simulator*, Paper presented to the System Dynamics Conference, Boston, USA.

Lane, D. C., Monefeldt, C. and Rosenhead J. V. (2000) *Looking in the Wrong Place for Healthcare Improvements: A system dynamics study of an accident and emergency department,* Journal of the Operational Research Society 51(5): 518–531).

Roysten G., Dost A., Townsend J. and Turner H. (1999) *Using System Dynamics to help develop and implement policies and programmes in Health Care in England,* System Dynamics Review, , Vol. 15, No 3.

Wolstenholme, E. (1990) System Enquiry-A System Dynamics Approach. Chichester, England: John Wiley and Sons.

Wolstenholme, E.F. (1993) *A Case Study in Community Care using Systems Thinking,* Journal of the Operational Research Society, Vol. 44, No. 9, September, pp 925–934.

Wolstenholme, E.F. (1996), *A Management Flight Simulator for Community Care,* In Enhancing Decision Making in the NHS, Ed. S. Cropper, , Open University Press, Milton Keynes.

Wolstenholme, E.F. (1999) *A Patient Flow Perspective of UK Health Service,* System Dynamics Review. Vol. 15, no. 3, 253–273.

Wolstenholme, E.F., (2002) *Patient Flow, Waiting and Managerial Learning - a systems thinking mapping approach,* Working paper, Symmetric SD Ltd.

Wolstenholme E. F., (2003), *Towards the definition and use of a Core Set of Archetypal Structures in System Dynamics,* System Dynamics review Vol. 19, No. 1, Spring: 7–26.

Wolstenholme, E.F. (2005) Monk, D., McKelvie, D. and Smith G., *Coping but not Coping in Health and Social Care – Masking the Reality of Running organizations beyond Design Capacity*, Paper presented to the System Dynamics Conference, Boston, USA.

Chapter 10

System Dynamics Advances Strategic Economic Transition Planning in a Developing Nation

Brian Dangerfield
Centre for OR & Applied Statistics University of Salford
B.C.Dangerfield@salford.ac.uk

1. Introduction

Macro-economic systems are notoriously complex. Here can be observed myriad interactions between physical production, employment, finances and government policy. In addition there are likely exogenous effects which can impinge on the system and are driven by international market movements, political upheavals or even terrorism.

The development of computers in the 1950's and 1960's, together with the systematic collection of national economic data, led to a pursuit of economic planning based upon models which attempted to make sense of this data with a view to forecasting the future of key economic variables. The emergence of econometrics as an adjunct to economic science was born. Nowadays hardly any developed nations lack an econometric model. If the central government itself have not built one, then one or more university or private research institutions will have. However, econometric models are not without their critics [Black, 1982].

The reliance on econometric models for macro-economic management is common, but it is not necessarily the best methodology for exploring, evaluating and reflecting on policy options. Meadows and Robinson [1985] review a range of methodologies and include system dynamics (SD) amongst them. They make a cogent case for the usefulness of system dynamics in national economic planning.

The application described below is one rooted in a developing economy. It derives from research conducted by the author from 2003-2005 as scientific director of a project team tasked with offering a more scientific base on which to formulate future economic and social policy in Sarawak, East Malaysia. The State of Sarawak, which lies on the northern coast of the island of Borneo, was a former British colony which joined the federated nation of Malaysia in 1963. The project is important in the sense that it is another contribution to empirical macro-economic modelling using SD.

Moreover, the project has exposed and hopefully convinced government officials as to the merits of SD as a methodology in this sphere of application. Some of the material described has been made available already whilst the research was a work-in-progress [Dangerfield, 2005].

Examples of systems-based approaches to developing economies are available in the broader operational research literature (see for instance Parikh, [1986] and Rebelo, [1986]) but it is only relatively recently that SD has begun to acquire a prominence in development planning [Barney, 2003; Chen and Jan, 2005; Morgan, 2005].

2. The Purpose of the Model

The determination of a purpose for an SD model is well grounded in the literature (see, for instance, Forrester [1961 p.137] and Sterman [2000 p.89]). The overall purpose of the Sarawak project was to provide the state with a tool to aid their future economic and social planning. But this is too broad an objective. A specific purpose needed to be defined and a period of time was spent after the commencement of the research in reviewing the various strands of thinking in the state government and, in particular, reading the key speeches of ministers to see what was preoccupying them. There would be no benefit derived from the creation of some grand planning tool if it was not consonant with the interests and ambitions of the primary stakeholders. A proposal was eventually tabled and agreement secured to develop the model with the following purpose:

> How and over what time-scale can the State of Sarawak best manage the transition from a production-based economy (p-economy) to a knowledge-based economy (k-economy) and thereby improve international competitiveness?

There had been concerns raised in ministerial speeches that the resource-based economy, which had served Sarawak well in over two decades of development, was coming under pressure from other industrialising nations, in particular China. To secure further international competitiveness, the state needed to develop more high-technology industry with higher value-added products and services. In short there was a desire to shift towards a knowledge-based economy (k-economy), [Abdulai, 2001; Neef, 1998] implying the emergence of a quaternary sector.

3. The Overall Model

The diagram shown in Figure 1 sets out, in as economical a way as possible, the overall structure of the model designed to address the issues concerning the development of a k-economy indicated above.

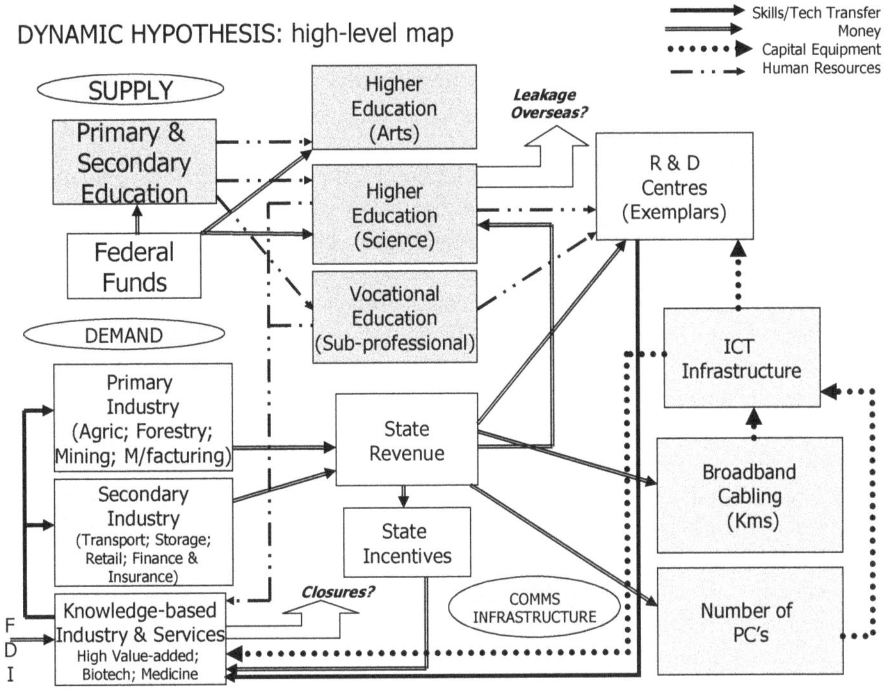

Figure 1. The high-level map used to guide the developing model (Note: FDI = Foreign Direct Investment)

There are three main aspects to be handled and an appropriate triangulation of these components is key to managing a successful transition to a k-economy:

- The supply of suitably trained human capital and entrepreneurs: the output from the education sector shown towards the top left of the diagram.

Clearly the primary and secondary education sector provides the output of students some portion of which will progress to higher education. Both arts and science specialisations are represented and there are indications that the balance here does not currently favour the enhancement of science skills which underpin the k-economy. The current graduate output ratio is weighted heavily in favour of arts courses. However, over recent years this imbalance has been tackled from the point in their school career where students choose a future pathway: Form Four (F4) level.

The vocational sector is also represented since development of a k-economy is augmented by an important group of sub-professionals (e.g. technicians) who have a crucial supporting role to play.

Funding for most education in Sarawak is provided by the federal government. However, and primarily in an effort to develop the science base, the State Government have funded certain private university developments. State funding in this manner can play an important part in expediting the flow of suitably qualified individuals who will stimulate the development of a k-economy.

- The demand side of a k-economy: those knowledge-based industries and services which are emerging (in some cases as development of primary and secondary industry) to form an ever-increasing component of the economy.

The sectors towards the bottom left of Figure 1 are split into primary (the production and resource based sector also called the p-economy), secondary (the service sectors such as tourism, finance and professional services), and the knowledge-based sector (the k-economy or quaternary sector).

The evolution of the quaternary sector is propelled by a mixture of foreign direct investment and State funding. But this alone is insufficient, for a flow of skilled human capital is essential as is the quality of the ICT infrastructure which must have attained an appropriate level of sophistication.

The emergence of a quaternary sector can appear to consume resources which might otherwise have been directed to primary and secondary industry. But it must be stressed that, contemporaneously, development of the k-economy will mean direct skill and technology transfer benefits to the existing base of primary and secondary industry. It is impossible to ignore the bedrock components of the p-economy which can, in turn, be enhanced as part of overall economic development. These sectors are currently the main providers of state revenue (via taxation and employment) and are likely to remain so.

- The state of the ICT infrastructure: in some sense this mediates the evolution of the drivers of supply and demand (bottom right of Figure 1).

The quality of the ICT infrastructure can be fairly easily measured by appropriate metrics. Two such examples are the length of the broadband data highway within Sarawak and the estimated number of PC's installed.

Again, it would be expected that the State government revenue would, in large measure, underpin the enhancement of these functions, although foreign direct investment cannot be ruled out. An ICT infrastructure of reasonable sophistication will also be necessary in order to allow the development of a number of Research and Development (R&D) Centres of Excellence, as indicated in Figure 1. The initiation of such projects is suggested in order that best practice k-economy activities can be showcased and publicised. These centres will make it clear that the State government is strongly promulgating the development of the quaternary sector through provision of funds to allow these start-up operations to proceed.

Development of a k-economy would be constrained if the supply of science graduates and suitable sub-professional k-workers are not forthcoming, which is why emphasis has been placed upon coincident (or even prior) educational changes. Initial staffing of such centres may be a problem but it is possible that, with sufficiently attractive remuneration packages, qualified Sarawak expatriates would be tempted to return.

The proposed R & D centres can be seen as crucial catalysts in the stimulation of the quaternary sector and they will offer a primary supply of people with the necessary skill sets to enthuse the creation and development of knowledge-based industry and services.

The dynamic flows to be considered are:

- Skills and technology transfer
- Money and other financial resources
- Capital equipment
- Human resources

Within the industry sectors (particularly the Primary Sector) there are also dynamic flows of goods, material and orders.

In the following section a number of policy scenario runs are described which evidence the *modus operandi* of this model: deriving qualitative prescriptions from a quantitative tool. These experiments are illustrative rather than exhaustive and centre upon policies which directly influence the basis for a transition to a knowledge based economy. A series of policy precepts are asserted. For instance, it is shown that dropouts at the level of primary and lower secondary education should be reduced. This would not only expand the general educational attainment in the economy but also yield a larger number of potential students for higher education and (particularly if they elect science and engineering) a source of recruits to high-technology firms and R&D centres. If a proportion of the lower secondary dropouts are directed into technical education at the sub-professional level, then there needs to be a corresponding increase in science and engineering graduates so as to provide the foundations for the employment of newly qualified technical staff in k-economy organisations. Technical staff alone do not establish knowledge based firms (k-firms). Finally, any policy to increase R&D spending, with a view to creating more startup schemes, needs to be phased in gradually alongside the emergence of sufficient scientifically skilled graduates to populate both the additional R&D centres as well as new k-firms. Otherwise, there is a danger of R&D centre growth crowding out the development of k-firms in the private sector.

4. Specimen Scenario Runs

4.1. Introduction

There are innumerable scenarios which can be undertaken with the model as currently configured. It consists of a total of 15 sectors (sometimes called 'sketches' or 'views') and in excess of 450 individual relationships, parameters and mappings. In keeping with the objective of the model purpose, consideration has been restricted to those changes which impact most closely on the knowledge economy.

4.2. Reduction of Dropout Rates from Various Stages of Education.

Here it has been assumed that the major crisis points for dropouts, namely after or during primary education and after lower secondary education (F3), are addressed. Over the years of the 9th Malaysia Plan (2006-2010) an assumption has been made that in the case of primary school dropouts this is reduced to just 5% by 2010 and for F3 dropouts to just 2%. Also, the improvements which have uplifted the proportion doing science and engineering at university, a near doubling from 20% to just under 40% (over 1997-2004) are incorporated.

The effects on enrolments in tertiary education are notable, allowing over 6,000 extra students per annum by as early as 2015 (see Figure 2). It must be stressed that this is without any reduction being made in those leaving after F5. Although there has been a reduction in F5 leavers in recent years, bringing the proportion down from 90% in 1995, it is still quite high as compared to normal progression rates to tertiary education in developed nations.

The beneficial effects of these extra additions to tertiary education follow the obvious links of cause and effect. Providing the ICT infrastructure is adequate then this extra stream of students will permeate through the economic system and, with many more becoming scientists and engineers, this will mean more k-firms opened and consequent improvements in GDP. The latter aspect is charted in Figure 3. It is interesting to note the delayed effects. The policy to reduce school dropouts was assumed to be initiated in 2005, but it is not possible to start to see the effects on GDP until around 2017.

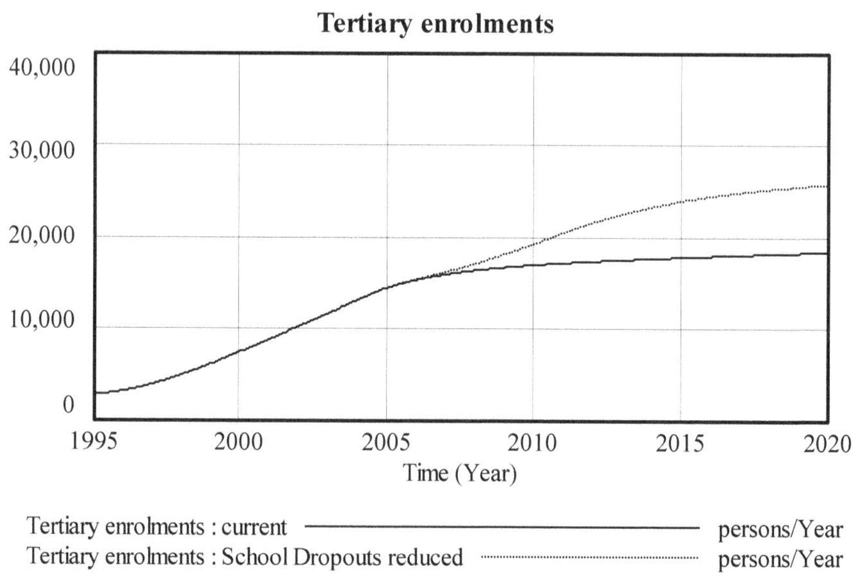

Figure 2. Increased enrolments in tertiary education after school dropouts reduced

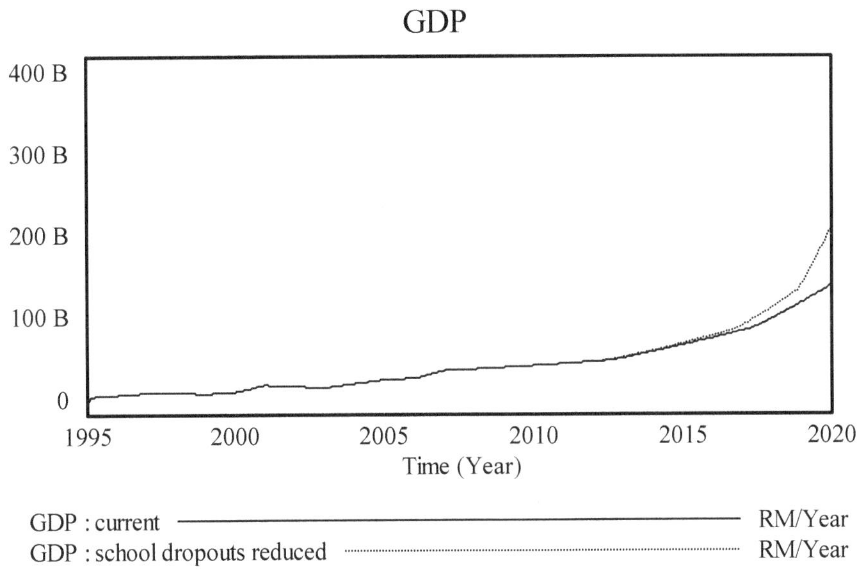

Figure 3. Effect on GDP of a reduction in school dropouts (Note: B=billions; RM=Ringgit Malaysia)

A policy precept can now be asserted: resources should be found to address dropouts at both the P1-P6 and F3 stages of education with a view to adding to the initiatives already undertaken in this regard.

4.3. Re-training of F3 Dropouts for Technical Education

Over the past, before the policies designed to tackle lower secondary dropouts was invoked, there has been a cohort of young people who have already entered the workforce with a limited secondary educational attainment. Our estimates are that as many as 150,000 persons in the Sarawak labour force would come into this category in 2005. Views were expressed that this pool represents a resource to the economy that is being wasted: they could be given further training and enter the technical labour force. This often ignored sector of the labour force is vital to the evolution of a knowledge economy. It has been assumed that each k-firm or R&D centre employing 100 engineers will also require 20 technically qualified people.

This scenario assumes that a modest increase of 500 per year of the F3 dropouts are placed into technical education. By 2010 this means 2,500 per year will be entering technical education. That has the effect of both reducing the proportion educated to only F3 (see Figure 4) and correspondingly increasing the proportion of the workforce with a technical education.

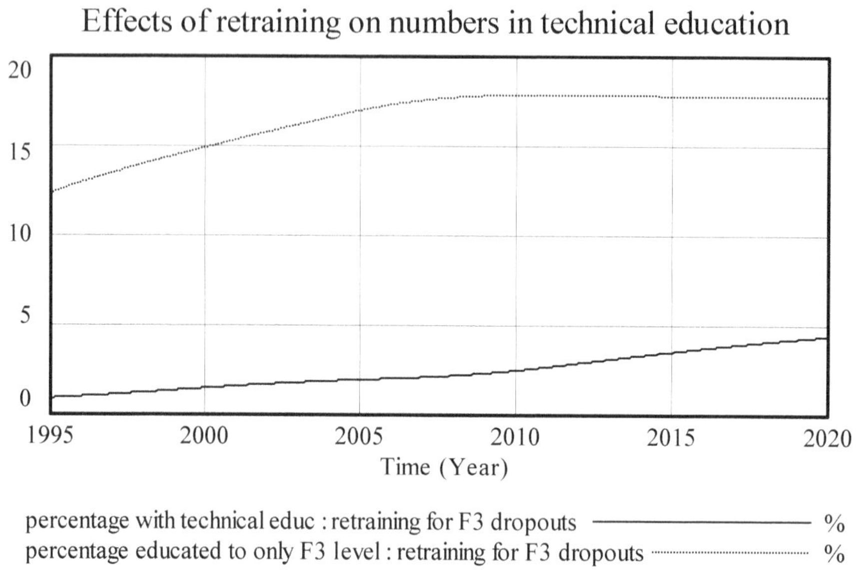

Figure 4. Percentage with technical education and with only lower secondary education

Unfortunately there is a limited impact in terms of the knowledge economy. Unless the numbers of scientists and engineers are also increased, these newly-retrained technicians enter the technical labour pool but are not employed in a high-tech environment. They can contribute to economic development in other ways (such as motor engineers) but the evolution of k-firms and R&D centres requires action across more than one front. At present the supply of labour skilled in science and engineering is not stretching the requirements for the numbers of technically qualified persons. This is not to denigrate the policy of retraining. The economic benefit to the individual is accepted (and there is an economic benefit to the State also but which is not explicitly captured in this model which addresses the k-economy). The policy precept is: retraining of F3 dropouts as technicians needs to be accompanied by increases in the numbers of science and engineering graduates if the additional supply of technicians is to gravitate to employment in the high-tech sectors of the economy.

4.4. Increasing R & D Expenditure

It is instructive to assess the effects of an increase in R & D spending. The method by which a k-economy can evolve is through increasing the number of experienced engineers by nurturing their capabilities in State-financed R & D centres. Accordingly it might be expected that to increase the proportion of development expenditure committed to R & D would be a beneficial policy.

The outcome of this policy is counter-intuitive. To see the basis for this assertion consider a policy to increase the percentage of development expenditure devoted to R & D, currently 1.5% p.a. (assumed), to 2%. This is accomplished in two ways: firstly

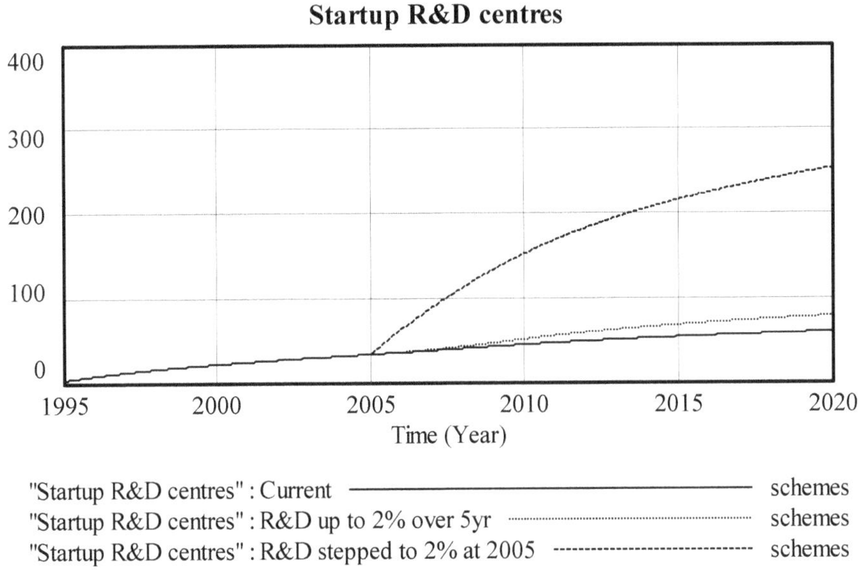

Figure 5. Effects of increased R & D spending on startup R & D centres

by gradually increasing the percentage to 2% over a five-year period from 2005-10 and, alternatively, by stepping up the percentage to 2% abruptly from 2005. The policies have a clear beneficial effect upon the number of R & D centres as can be seen from Figure 5.

But this policy attracts many of the highly skilled scientists and engineers into R & D on graduation from university. Consequently there is a surplus in that sector which crowds out the correspondingly fewer skilled persons available to become employed in k-firms. The numbers of experienced engineers are far in excess of what is required to found new k-firms as compared with the outflow of 'raw' graduates, which are then in short supply for k-firm startups.

The effect is conveyed through Figure 6. The number of potential new k-firms is depressed under the raised R & D policy because there is not the supply of new graduates to accompany the experienced ones. As with the conclusion in the previous section, action must be taken more systemically if the policy is to be effective. Besides R & D spending increases, a move to increase the supply of science and engineering graduates is also called for.

A final graph on the effects of this particular policy initiative is to portray what might be suspected from examining Figure 6. The number of firms in k-industries is indeed lower under the increased R & D spending policy (see Figure 7).

An additional insight is that the choice of the rate of change in R & D spending is also important. The policy option of phasing the change in over a period of years, rather than abruptly changing the percentage, while still being inferior to the base case with respect to the number of k-firms in the economy, at least is not as detrimental as is the abrupt change. A slower phased approach would also allow for concomitant changes to increase the supply of graduates trained in the sciences and engineering.

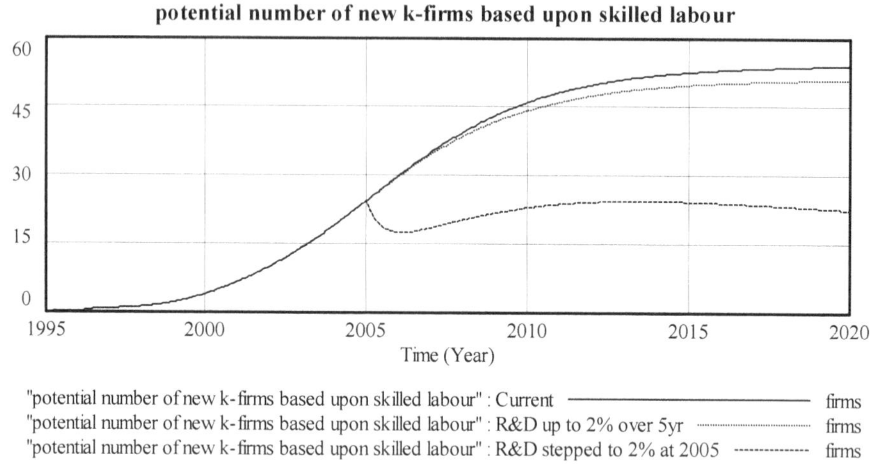

Figure 6. Effect of increased R & D spending upon the potential number of new k-firms

An additional insight is that the choice of the rate of change in R & D spending is also important. The policy option of phasing the change in over a period of years, rather than abruptly changing the percentage, while still being inferior to the base case with respect to the number of k-firms in the economy, at least is not as detrimental as is the abrupt change. A slower phased approach would also allow for concomitant changes to increase the supply of graduates trained in the sciences and engineering.

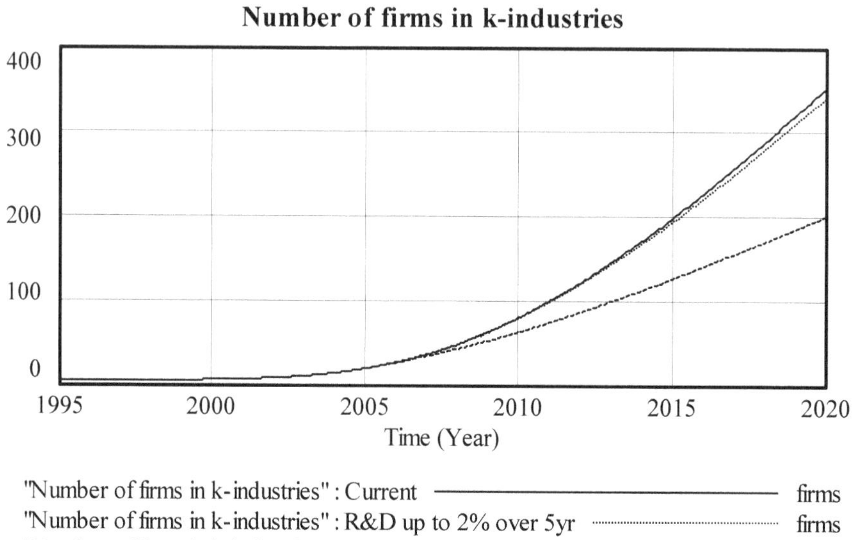

Figure 7. Effect of increased R & D spending upon the number of k-firms

The policy precept here is: increases in R&D spending, to allow more State-financed startup projects, need to be phased in over a period of years with concomitant increases in the supply of high-tech graduates to avoid a skewing of skills towards R&D centres, which thereby constrain the growth of new private sector k-firms.

5. Microworlds: Sharing and Disseminating Insights from Complex System Analysis

5.1. Introduction

System dynamics models can lever insight in complex systems such as an economy. However, it has been recognised that there is a need to promulgate those insights beyond the confines of the modelling team. To thrust a model into a policy support role means more than conducting experiments in workshop situations. The assembled audience, despite being impressed, may not be entirely convinced as the agenda may look to be well-rehearsed.

With an official as high as the State Secretary requesting the model be made available on his own computer, it was essential that a platform was delivered which would allow of easy use, but still have the potential to deliver insights. Thus it was decided to create a microworld shell around the model. This would allow anyone to run, experiment with and, crucially, reflect upon the results without interference by the modelling team. If the model was truly to leverage insight and stimulate thinking that would result in policy interventions then it was considered that the greater likelihood of this outturn would stem from personal interaction with the model by State officials rather than a guided programme offered by the modelling team. A Venapp™ environment in the Vensim™ software (at the Decision Support System version level) allows this enhancement.

The concept of the microworld in system dynamics was initially highlighted and explained by authors such as Morecroft [1988] and Kim [1992]. It has since been extended to allow an element of competition between participating teams who are allowed to intervene and make various decisions as the model progresses. In these circumstances the entire package is usually sold as a commercial item and is used in generic training exercises rather than the dedicated policy support role envisaged here.

In the figures which follow, various screen shots are depicted which reflect the way in which a State official would interact with the model and the nature of the tasks allowable in the context of the microworld. The main menu (see Figure 8) allows the user to perform four tasks: view the stock-flow diagrams, examine the past data against simulated model output, specify and perform scenario runs and analyse the causal relationships within the model. Because the model runs from 1995 to 2020, the comparison against data is restricted to, at most, some ten years (Figure 9). Scenario runs can be set up using simple slider bars (Figure 10) and the results depicted for a collection of the important variables in the model (Figure 11). Although not illustrated here, the microworld additionally makes it possible to see the formulation of the relationships expressed in the model using the 'causes' and 'uses' trees available as standard in the Vensim software tool.

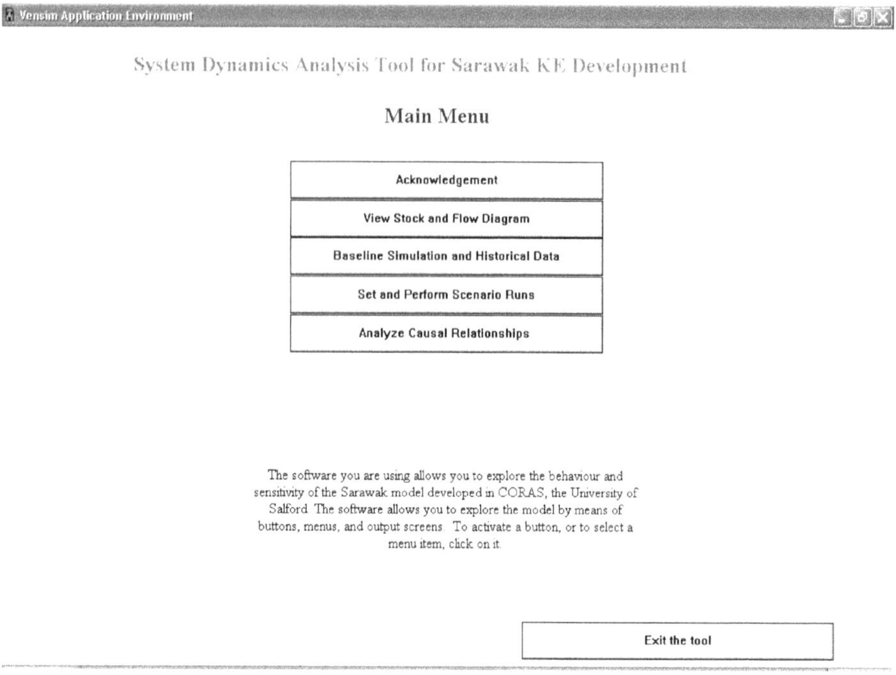

Figure 8. The opening screen depicting the Main Menu

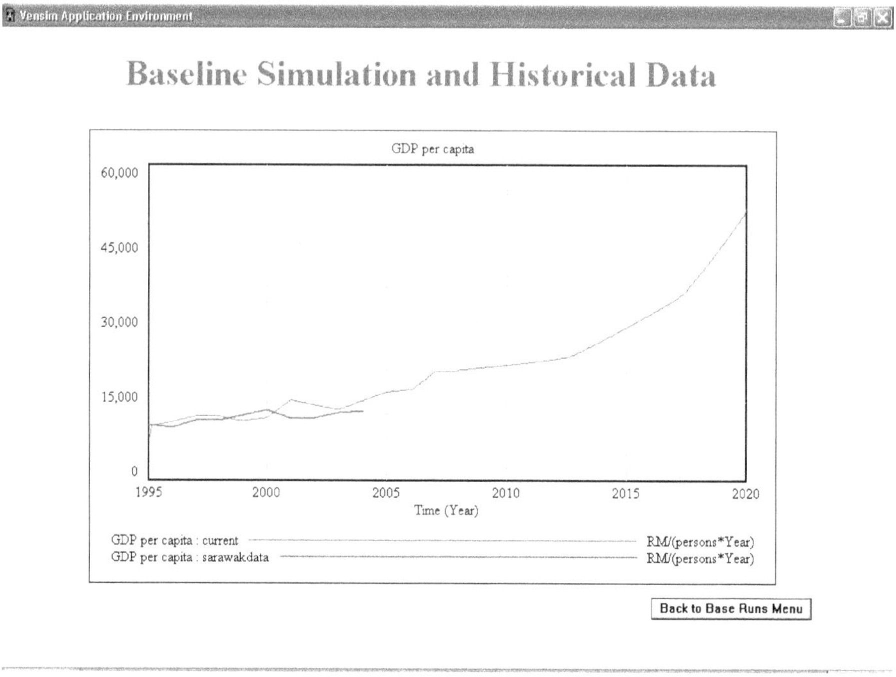

Figure 9. A screen showing the comparison with past data feature

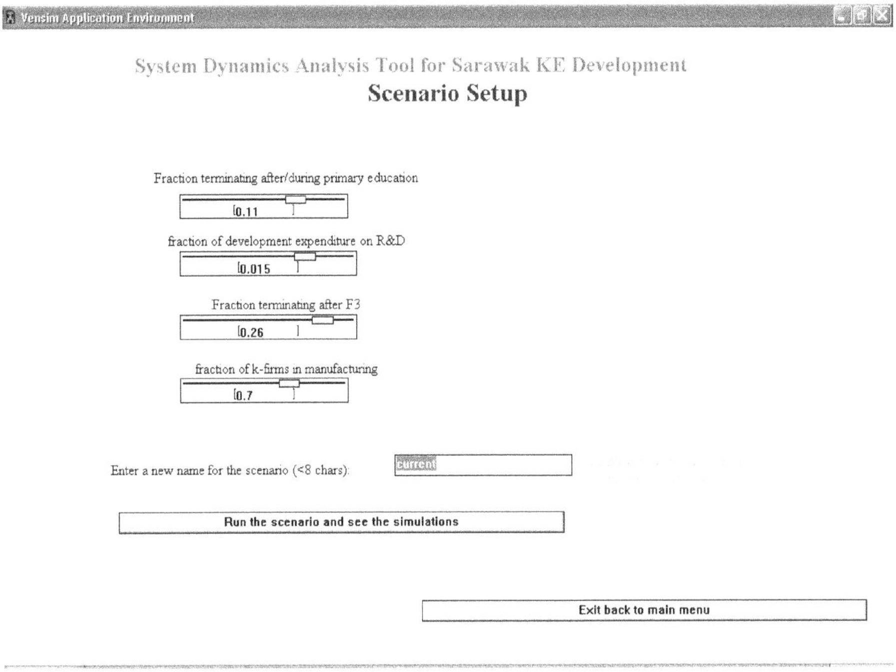

Figure 10. Slider bars allow the alteration of parameters prior to scenario runs

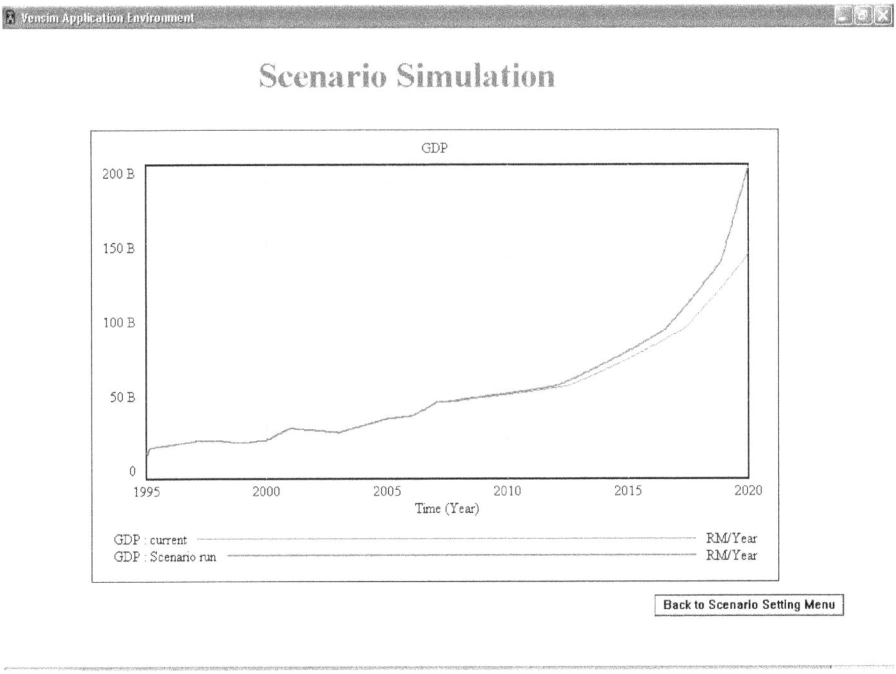

Figure 11. The results from a scenario run

6. Conclusions

The underlying objective of this modelling work was to leave the State Planning Unit in the Chief Minister's Department with a tool which would improve the quality of their thinking in respect of the new development thrust. Those who are offered the use of a model as a policy support tool are apt to say, only too easily, that 'this model has flaws'. But it is impossible in a complex system of interacting influences to even start to plan sensibly without the aid of a model specifically designed to deal with such complex networks of influences. Rarely do executives and officials apply the same forensic evaluation of their own thoughts as they are apt to do of a model; one rarely hears the comment that 'my thoughts are flawed'. Yet in the absence of a formal model, State officers would be left to plan purely on the basis of their own mental models.

The most senior State officials were strongly supportive of the work. Towards the conclusion of the project this culminated in a *Borneo Post* report of a speech by the State Secretary. It was headlined 'Government to introduce System Dynamics model'. This had an unfortunate consequence in that the officials who would likely make use of the model felt somewhat pressured. They thought the declaration would expose their inability to deliver in this more analytical environment, even with the support of those few officers who had been trained in SD and the Vensim software. The construction of thc microworld shell, together with the reproduction of reported time-series data for many of the model variables (albeit with, at most, ten data points) was designed to overcome this confidence barrier. It also meant that those without the necessary proficiencies could now interact with the model in the privacy of their own office, if desired.

At this stage it may be premature to make conclusions as to the frequency of use of the model. There may be a period of reflection before the State Planning Unit feel confident in employing the tool on a regular basis. But at the very least it would have one obvious use: as a basis for setting future objectives in relevant parts of the Sarawak component of the 5-yearly Malaysian Plans. Realistic objectives need to be set: those pitched too high result in frustration and political denigration, those set too low are portents of complacency and slower growth.

Amongst the series of tests to which SD models can be subjected, amongst the more relevant in this case is the 'system improvement test' [Forrester and Senge, 1980]. In other words, has the development and use of the model actually led to improvements in system behaviour? While this task would be easy in the case where the economic system has been manifestly under-performing, and the formulation of new policies based upon model insights has achieved a marked improvement, here we are faced with a more subtle situation. The Sarawak economy has progressed quite well over the past two decades. It has witnessed the transformation from being dependent on the agricultural sector to a production-based economy founded upon its own resource base. What has emerged out of this research, however, is a tool to aid thinking in a future situation where the forces of competition are more complex, international and arguably much stronger.

This contribution has described an empirical study conducted over 27 months and which has resulted in an SD tool for macro-economic planning being made available in a developing country. Prior to this there was little or no use of formal modelling techniques although the collection of a proliferation of statistical data was evident. As well as the availability of the tool, knowledge of the capability of system dynamics as a modelling methodology for addressing complex systems has taken root, reinforced by a technology and knowledge transfer element in the project. It is anticipated that, as new policy issues arise, the Sarawak State government will be pre-disposed to their analysis using the SD methodology.

References

Abdulai DN Ed. (2001) *Malaysia and the k-economy*, Pelanduk, Selangor.

Barney GO (2003) Models for National Planning. In *Proceedings of the 2003 International Conference of the System Dynamics Society*, New York (CD-ROM). See also www.threshold21.com (accessed May 27 2005).

Baumol WJ (1965) *Economic Theory and Operations Analysis* (2nd Ed), Prentice-Hall, NJ.

Black F (1982) The Trouble with Econometric Models. *Financial Analysts Journal*, March-April, 29–37.

Chen JH and Jan TS (2005) A System Dynamics model of the Semiconductor Industry Development in Taiwan. *Journal of the Operational Research Society* **56:** 1141–1150.

Dangerfield BC (2005) Towards a Transition to a Knowledge Economy: how system dynamics is helping Sarawak plan its economic and social evolution. In: *Proceedings of the 2005 International System Dynamics Conference*, Boston, USA. (CD-ROM). Reprinted in *ICFAI Journal of Knowledge Management*. Vol.III, No. 4, 2005. 40–56.

Forrester JW (1961) *Industrial Dynamics*, MIT Press, Cambridge MA. (Now available from Pegasus Communications, Waltham, MA.)

Forrester JW and Senge PM (1980) Tests for Building Confidence in System Dynamics Models. In: *System Dynamics: TIMS Studies in the Management Sciences*, Legasto AA, Forrester JW and Lyneis JM Eds, North-Holland, Oxford.

Kim DH Ed. (1992) Management Flight Simulators: Flight Training for Managers (Part I). *The Systems Thinker* 3(9): 5.

Meadows DH and Robinson JM (1985) *The Electronic Oracle*, Wiley, New York. (Now available from the System Dynamics Society Albany, NY.)

Morecroft JDW (1988) System Dynamics and Microworlds for Policymakers. *European Journal of Operational Research* **35:** 301–320.

Morgan P. (2005) *The Idea and Practice of Systems Thinking and their relevance for capacity development,* European Centre for Development Policy Management, The Netherlands. http://www.ecdpm.org/Web_ECDPM/Web/Content/Navigation.nsf/index2?readform&http://www.ecdpm.org/Web_ECDPM/Web/Content/Content.nsf/0/358483CE4E3A5332C1256C770036EBB9?OpenDocument (accessed February 24, 2006).

Neef D Ed. (1998) *The Knowledge Economy*, Butterworth-Heinemann, Woburn, MA.

Parikh KS (1986) Systems Analysis for Efficient Allocations, Optimal Use of Resources and Appropriate Technology: some applications. *Journal of the Operational Research Society*, **37:2** 127–136.

Rebelo I (1986) Economy-wide Models for Developing Countries, with emphasis on Food Production and Distribution. *Journal of the Operational Research Society*, **37:2** 145–165.

Sterman JD (2000) *Business Dynamics*, McGraw Hill, Boston.

Wiens EG (2005) *Production Functions* http://www.egwald.com/economics/productionfunctions.php (accessed January 28, 2005).

Appendix

Detail of Important Model Sectors

The Population Sector

Within this sector the age-based population structure of Sarawak is represented. This is an important input to modelling sustainable economic activity within the State. The flow diagram for this sector is shown as Figure A1. The main summary variables are available e.g. population age 15 and under; population 15 to 64 (taken to be the working population); population 65 and over; and the total population. The dependency ratio is also computed. This is an important demographic ratio which expresses the non-working subset of the population (the young and the old) as a percentage of the total population. The higher is this value the greater the pressure on the State's working population to support the strata that is not economically active.

Extensive census data is collected for this sector usually on a ten-year cycle and thus the formulation of the model for population flows is straightforward since the parameters are readily available. However, estimates have to be made for years between census dates and the 1995 estimates have been employed to initialise the population in each age group.

The population sub-model is presented as a flow of births which adds to population cohorts which are, in turn, depleted by deaths. A critical feature is that the population is stratified into 15 age cohorts from 0-4, 5-9, 10-14 and so on up to 65-69 and finally 70+.

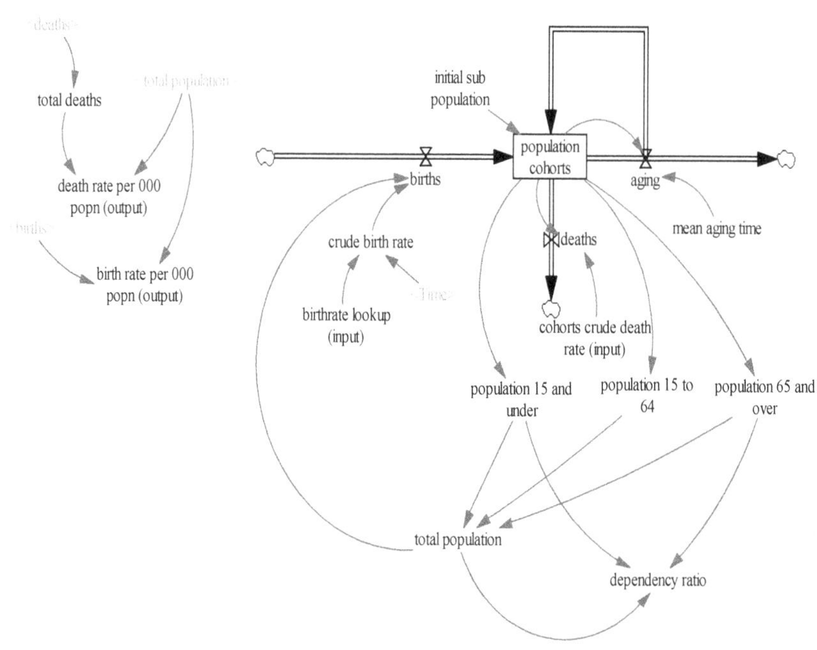

Figure A1. The Population sector sub-model

Although not shown in Figure A1, the age bands are modelled separately along an ageing chain. The Vensim software allows such a decomposition through its array facility. Although any age group can be extracted (e.g., for plotting or to use elsewhere in the model), it is usual to concentrate on the important broad age ranges and the total population, as described above.

The data which provides the parameters for this sub-model are collected in an EXCEL spreadsheet. In general, many of the model's parameters are stored in this way because Vensim allows the extraction of both parameter values and historical time series from EXCEL spreadsheets and external data files.

There are separate crude death rates for each age band (estimated from the census data) and separate initial population values for each also. By initial population is meant the initial number of persons in that age band at the commencement of the simulation in 1995.

A distinction is made between *input* and *output* for crude birth and death rates. *Input* values for crude death rates are those estimated from the data, whilst the *input* crude birth rates were provided directly as data up to the most recent year for which figures are available or are derived from a linear interpolation process out into the future to 2020. *Output* values are calculated in the model as part of the simulation and are presented as aggregates over the entire population. In part they act as a check on the data entered but also provide a useful overall metric for consideration. For instance the crude death rate in the older age bands might be expected to reduce over calendar time as a result of improved medical care. Currently, these *inputs* are held constant. This possibility, if implemented, should be reflected in the *output* value for the overall crude death rate per thousand of the population.

Education & Human Capital Sectors

Progression to a k-economy will take some time but will be propelled by the twin thrusts of investment in people and a communications infrastructure. In respect of obtaining a suitably qualified labour force it is necessary to place the centre of gravity of the model around the production of higher-educated and technically-qualified human capital. These developments underpin a suitably skilled workforce and, in view of the importance attached to this, a workforce sector has been added (see section below).

The education sector is considered here and has been split as between Primary and Secondary education (Primary 1 – Primary 6 and Form 1 – Form 5 respectively) on the one hand and Tertiary education on the other. The extent and importance of these model sectors means that they cover two separate views. Firstly the model of Primary and Secondary education is described and is represented by Figure A2.

There is a significant proportion leaving full-time education after F5. Attempts to address this are essential if the Sarawak economy is to prosper and if Malaysia as a whole is to attain developed status. An internal document suggested an estimate of 90% leaving education post-F5. However, it is believed that improvements in the proportion progressing to tertiary education have been implemented since the mid-1990's and so the figure of 0.9 is progressively reduced from the start of the model's runs. Experimentation here is obviously a highly critical aspect of our work: more students entering higher education is vital for the Sarawak's future as a potential knowledge economy.

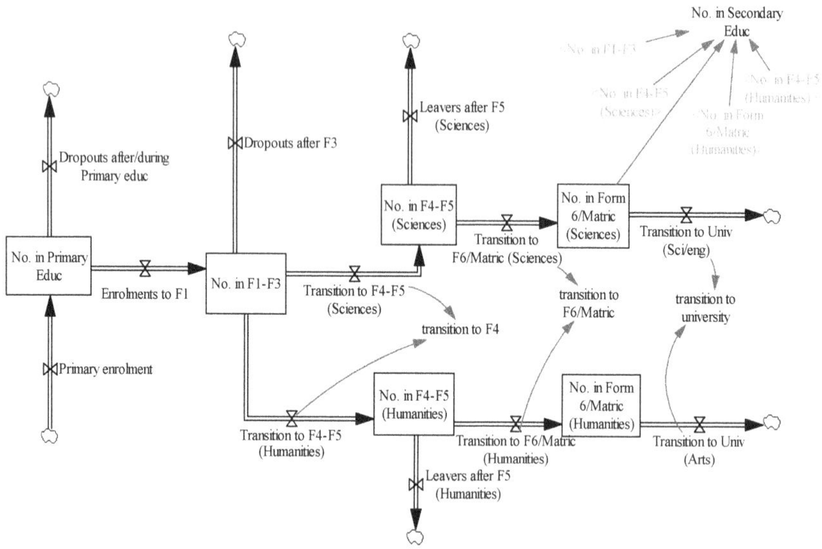

Figure A2. The Primary & Secondary Education Sector

Transition to higher education is an almost continuous progression once students have elected to carry on to Form 6 or, alternatively, elect the matriculation route. Those leaving after this point in their educational progression are negligible. No such leavers are represented in the model since the numbers are so small. Also, it has been assumed that the choice between humanities and sciences will then reflect their corresponding choices at university.

Once the student flow bifurcates before F4 there is a need to create disaggregated variables. Thus the totals entering F4, F6/Matriculation and university are separately defined. Also the total in secondary education is introduced, made up the five individual model variables which are the sub-components of "number in secondary education".

The tertiary education sector is defined as universities and also technical colleges or polytechnics. In essence it is all those institutions offering full-time education programmes to students after the level of F6 or equivalent. The model of tertiary education developed for the current purpose is presented in outline in Figure A3. The inflows to university here are the outflows from Figure A2, with the inflows to technical education arising from the leavers after F5 in the secondary education sector.

Technical education is included because, whilst knowledge-based firms require skilled engineering and scientific staff, they also need technical support staff. These people will have gained qualifications from the technical/vocational education institutions existing in Sarawak. By 1998 all secondary vocational schools were converted to State Technical Schools with around 1500–2000 students enrolled. However, the number of such schools in the whole of the State is less than ten. But it is progression on to higher technical education which is the element that underpins the knowledge economy and fulfils the necessary support role for qualified engineers.

After graduation there must exist job prospects, particularly for those qualified on the science and engineering side, for significant emigration to be avoided. If such graduates cannot obtain employment in Sarawak within one year, on average, it is assumed they will start the process of emigration. In addition to emigration a flow of repatriations is included. If the prospects and opportunities for skilled ex-patriots expand in Sarawak then it would be expected that some fraction would return to their homeland for employment and career progression. This will add to the skilled labour available to k-firms and to R & D centres.

Arts graduates can be re-trained and the existence of post-graduate conversion courses to equip them with some of the skills needed in knowledge-based employment is a possibility and is included in the sector shown in Figure A3. This initiative needs to be progressed and it assumes a capability and willingness of the universities to mount it and, furthermore, government funding may also be required. In the event that (more) such courses are launched, there is yet another source of skilled labour available to k-firms.

The promotion of Research and Development (R & D) centres is a crucial segment of the strategy to transition Sarawak towards a knowledge-based economy. The source of the labour for these centres will come from the expanding scientific and engineering graduate output, backed up by suitable technical support staff. The government have a role to play in inaugurating start-up R & D centres and must be prepared to allocate financial resources accordingly. Over time the centres will acquire collective experience: they will 'mature'. It is these mature centres which both directly and indirectly will facilitate the supply of entrepreneurs to set up k-firms. In turn these k-firms will recruit from the scientific and engineering labour pool.

The training of scientific and engineering staff will necessitate more teachers at the level of scientific secondary education and at university. This is another career

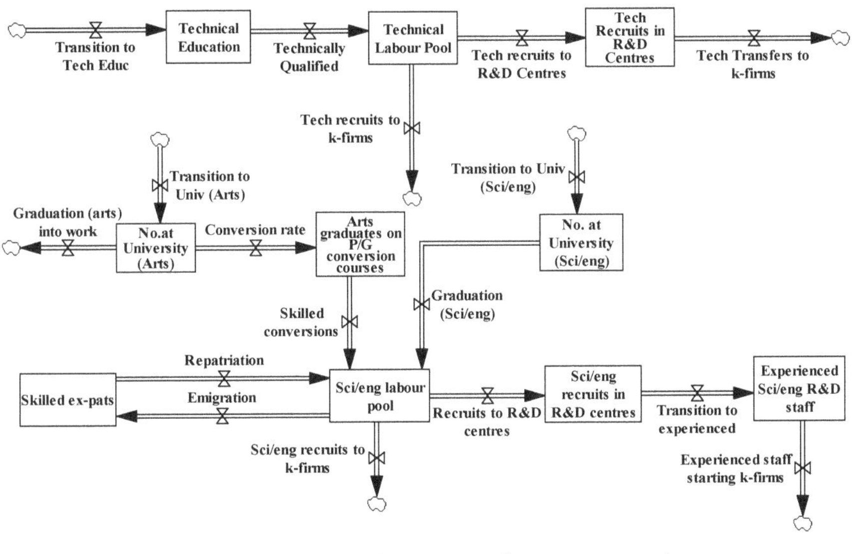

Figure A3. Tertiary education sector and consequent employment

route for appropriately qualified new graduands and one which currently is not explicitly included in the model.

Workforce Sector

The Workforce Sector presented in Figure A4 aims to portray the quality of human capital available in the Sarawak economy. It defines six categories consistent with their highest level of educational attainment. These are:

- No formal education
- Educated up to F3 (This represents the conclusion of lower secondary education and the point at which students are directed into specialist studies of either arts or science.)
- Educated to F5 (The conclusion of formal secondary education; further study is optional and dependent on ability.)
- Technically educated post F5/F6 (Certificate & Diploma students)
- Numbers with Arts degrees
- Numbers with Science degrees

The category 'No formal education' covers those who have dropped out during or just after primary education (P1–P6) or before commencing F1. Attaining just this level of education cannot be expected to help a move to a k-economy and so the description is justified in the current context. There currently appears to be a significant dropout at the F3 stage and this level of attainment is therefore specifically included. Those in the category 'Up to F5' education similarly have terminated their education after F5.

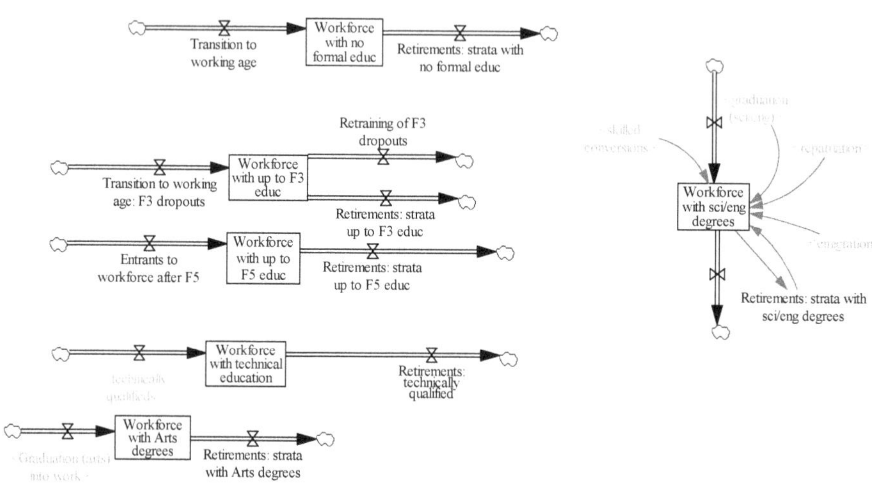

Figure A4. The Workforce Sector: main view

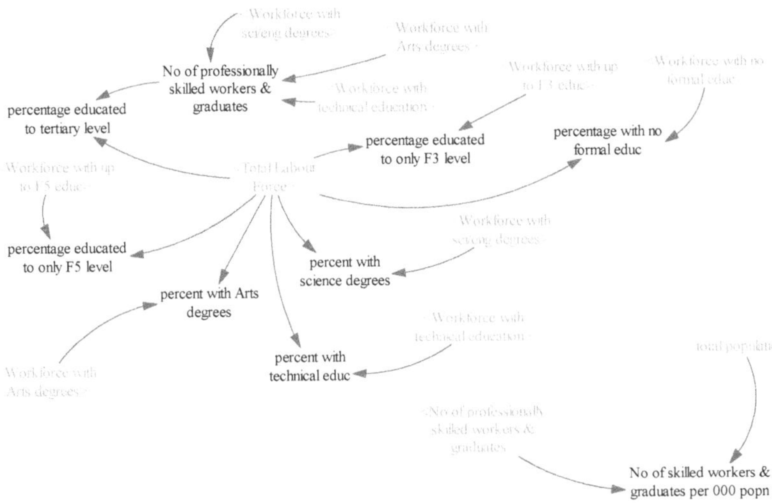

Figure A5. Various percentages and indices computed for the workforce sector

Tertiary education comprises those who have gone on to further or higher education post F5/F6. It comprises those in degree level education, either via F6 or through the matriculation route, together with those electing a technical education. The latter have an important supporting role to play in the development of a k-economy.

Accumulating all the various categories together allows a calculation of the total workforce. In addition, weighting the proportions attaining each possible level of education against the total workforce allows a computation of the 'mean years of education'. This is an important international measure of development and its improvement over the years is an indicator of Sarawak's progression to developed economy status in line with Vision 2020 for Malaysia as a whole.

In addition to the above, there are various indices computed, such as the percentage of the workforce educated to tertiary level, to F3 level, F5 level and so on (Figure A5). The proportion of professionally-skilled workers and graduates per 1000 of population is another widely-reported index and this too is easily computed. 'Professionally-skilled' is defined as those with a university degree or other post-F5/F6 qualification such as a Diploma or Certificate.

Manufacturing, Construction and Service Sectors

This section describes the representation of the manufacturing, construction and service sectors within the system dynamics model. The manufacturing sector is specified by disaggregating it into its important individual industrial components. It is not modelled globally. The separate sectors included are: 'Electrical & Electronic (E&E)'; downstream timber processing; petroleum products; palm oil and sago (starch) production. (Liquefied Natural Gas (LNG) & unrefined petroleum are categorized as part of 'mining' although these export-earning sectors are also included separately here.) The chosen groups cover 87% of the total gross value of manufacturing output in the State based on 2000 data. The construction industry is specified separately.

The sectors other than Electrical and Electronic are described below and at the conclusion of this appendix. Figure A6 shows the structure adopted for E&E, Construction and services. The diagram is a simplified version of that incorporated within the model so as to aid comprehension.

A Cobb-Douglas production function [Baumol, 1965 p.402; Wiens, 2005] is used to model E&E output and construction output. This type of function has a long history of use in economic growth models and it is flexible, being capable of exhibiting increasing, constant or diminishing returns to scale dependent upon the parameter specification. The inputs which need to be specified are (i) a constant parameter (ii) a labour index and (iii) a capital index.

The sum of the indices determines the degree of returns to scale. At present, E&E is specified with increasing returns to scale. Inputs of labour and capital are estimated based upon a simple growth function and the numbers used appear in line with existing data on labour and capital in the E&E industry. If more precise estimates are desired then it is a trivial matter to re-specify the parameters.

The influence of k-firms enhancing the output of the E&E sector is modelled by uplifting the constant value in the Cobb-Douglas function. This is in line with sound practice in economic growth theory. As the number of k-firms in the State gets bigger this will have a progressive effect on E&E output value.

The growth of services is set at 5% per annum and a compound growth function is adopted. An examination of the recent annual growth rates of the various categories of

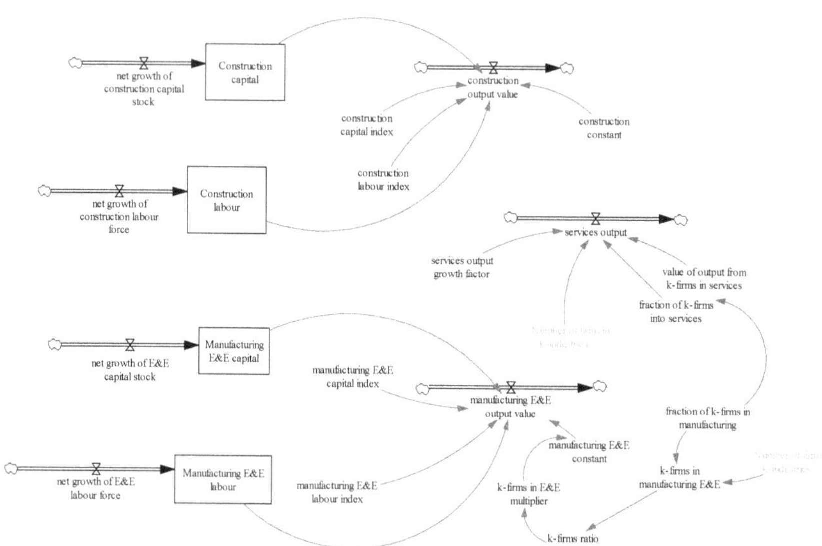

Figure A6. Simplified view of the Electrical & Electronic Manufacturing, Construction & Service Sectors

services reveals a significant variation from less than 1% to in excess of 12%. The choice of 5% per annum is indicative of the majority of sub-sectors. The emergence of a knowledge component in the service sector will enhance the output in excess of the 5% p.a. specified initially. The number of k-firms is split as between E&E manufacturing and services. Currently the ratio is 0.7 for E&E and 0.3 for services. This serves to emphasise that developments in a knowledge economy are not restricted to ICT-based manufacturing industries but can, and are expected to, permeate into a range of service sectors also.

Output from the construction industry is modelled in the same way as the E&E sector in manufacturing. That is to say a Cobb-Douglas production function is employed and the labour and capital inputs over time specified by compound growth functions. Presently this industry has indices of 0.2 for capital and 0.5 for labour, the sum equating with decreasing returns to scale.

State Finances and GDP

The formulation of the State Finances and GDP sector follows a logical path: State income allows for State expenditure. Reserves, which are built up or run down as desired, exist as a buffer between the two flows. State income is derived from the tax take on various industries such as palm oil, timber and LNG, together with an allocation from the Federal government. This is split into operating funds and development funds. For instance a large slice of operating expenditure is devoted to education, social welfare and medicine and health.

For our purposes the flows of funds from different sources can be aggregated, although expenditure is necessarily split into operating expenditure and development expenditure. This is reflected in the diagram for this sector (see Figure A7).

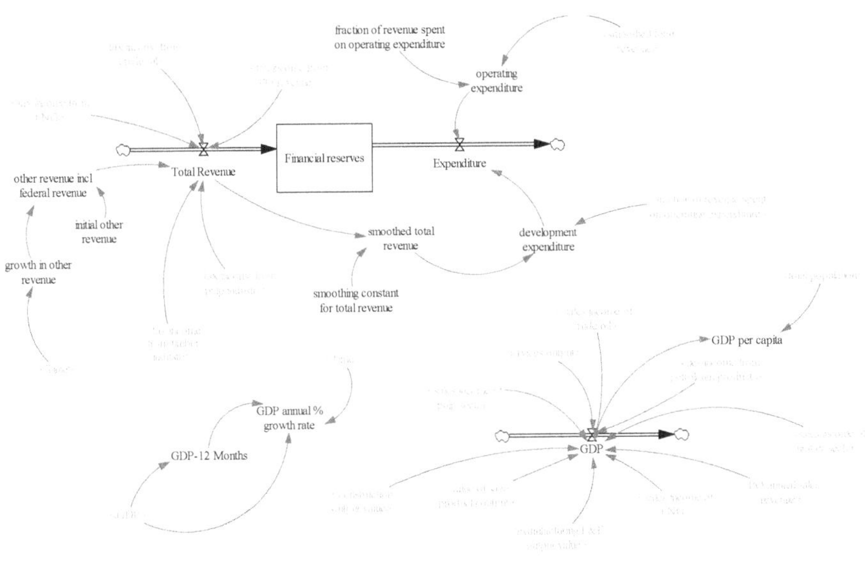

Figure A7. State Financial sector together with GDP and GDP annual growth rate

Whereas development expenditure will normally follow development revenue allocations from the Federal government, as financial reserves allow it would be prudent for the State government to take the initiative in releasing further monies for development. State R & D centres would be one such use of these funds.

The growth of 'other revenue', which covers all sources of revenue not derived from the State tax take on the industries specifically included in the model, is set to grow at 2.5% per annum from a base of 4 billion RM in 1995. It should be emphasised that the model is not aiming to faithfully reflect the full detail of financial flows into and out of the State Treasury, but rather to allow a consideration of variations in development expenditure, the primary engine of endogenous growth.

Gross Domestic Product is formulated by aggregating the output data from ten separate industries and sectors specifically included elsewhere in the model. The availability of total population data from the population sector means that GDP per capita is readily computed. It is common to find annual GDP growth rates in all countries' published data and so this measure is also included. It is worth repeating that *all* monetary data is specified in real terms based upon 1995 prices, the starting point for the all the simulation runs.

Research and Development and the K-industries Sector

The genesis of the emergence of the k-economy is to be found in research and development (R & D). Here is the instrument which any government in a developing country can deploy in order to effect an economic transformation. The sequence is normally to provide seed funding for start-up R & D centres which are staffed by qualified engineers backed up by technical support staff. These centres may not necessarily be ICT manufacturing industry based, but may embrace elements of the service economy such as health and biotechnology.

Whatever the nature of the activities ongoing in the R & D centres, they all require a supply of skilled engineers, technical backup and a high-tech ICT infrastructure. These three components are specifically included as all three together represent a necessary and sufficient condition for R & D centres to emerge (figure A8). However, the creation of k-firms requires, additionally, *experienced* high-tech engineers and scientists. Where such people are not available the k-firms need to source their expertise from outside of Sarawak. Endogenous growth of k-firms therefore is restricted in the absence of a supply of experienced engineers and scientists. Raw graduates, however appropriate their degree subject, do not equate with experienced staff.

Figure A8 shows that start-up R&D centres eventually mature over a period of years, assumed to be on average five years, and it is from the mature R&D centres that a source of experienced high-tech staff will emerge. Start-up centres are assumed to require 20 scientists and engineers, with a further 30 being employed as the centres mature. Each start-up R&D centre is estimated, on average, to cost 5 million Malaysian Ringgits (RM = $1.35m) to set up, the number of such schemes being determined by the overall annual State budget for R&D.

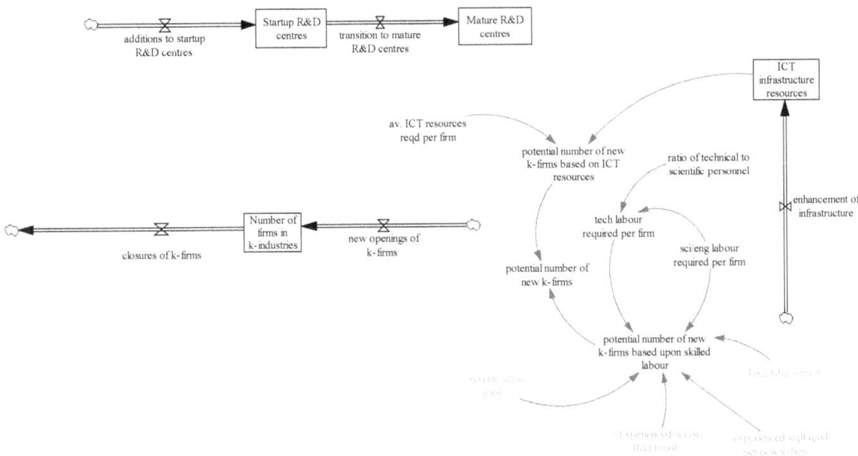

Figure A8. Research & Development Centres, knowledge-based firms and the ICT infrastructure (simplified from that in the model)

The role of the technical backup staff, both in R&D centres and, more significantly, in k-firms should not be underestimated. Within Sarawak there are indications that the supply of this element in the workforce could be constraining the State's development. It has been assumed that, for both R&D centres and k-firms, the technical labour force is 20%. In other words, there would be one technical officer for every five qualified engineers and scientists.

Remaining Sectors

For the sake of brevity the remaining model sectors are not described in detail. These sectors comprise those which make up the current resource-based economy in Sarawak. Timber is a prominent example along with the emerging pulp wood industry. Palm oil is a prominent export earner and its yield comes from oil palms planted in clearings as hardwood trees are felled. However, there is considerable attention paid to the sustainability of the hardwood species and forestry is managed closely by the government. This also applies to land made available for oil palms.

Along with the timber resources there are mining resources which contribute significantly to GDP. Oil and petroleum products is one case in point together with liquefied natural gas (LNG) which is exported to Japan. In 2003 crude petroleum and LNG together accounted for some 63% of Sarawak's exports by value (Source: Sarawak Facts & Figures, 2003/2004). In contrast timber logs and sawn timber contributed 7.5% in 2003, down from almost 28% in 1990 and reflecting the government's management of timber resources referred to above. Within the timber industry the development of plywood and veneer products (higher value-added items) has offered a counterpoint to the situation with logs and sawn timber. Plywood and veneers accounted for almost 9% of exports in 2003.

Chapter 11

Saving a Valley: Systemic Decision-making Based on Qualitative and Quantitative System Dynamics

Markus Schwaninger
University of St. Gallen, Switzerland
Markus.Schwaninger@unisg.ch

1. Introduction

The purpose of this contribution is to give substance to the notion of "systemic decision", on the basis of a pertinent case study. This case is about a weighty decision on an issue of exceedingly high complexity. The paper traces how qualitative and quantitative evaluations were combined within a systemic framework, to arrive at a decision whose superior quality could not have been attained if a non-systemic approach had been used.

According to scientific standards, decision makers are supposed to assume a rational posture when confronted with complex, dynamic situations. They should test their assumptions and corroborate their resolutions with sound arguments, basing them, if possible, on quantitative reasoning [see for example: Black 1999].

In practice, however, far-reaching decisions are often taken by relying on hunches. This approach is neglectful, however, of the nature of complex systems. These systems often exhibit behaviors which totally refute the presumption of the decision-makers. Particularly for those who rely merely on their own hunches, such results are counterintuitive [Forrester 1971]. They are usually brought about by unknown couplings of critical variables or unintended side-effects, i.e., consequences that were not anticipated.

On the side of theory, however, management science provides devices for dealing with complex issues. It has not only developed algorithms for quantitatively orientated decision-making. It has also come up with many qualitative heuristics to cope with complex issues wherever quantification is impossible or contraindicated. Essentially, those who advocate either one of these approaches have by and large remained separate, bivouacked in two "camps". Consequently, there have not been many pleas for combining quantitative and qualitative methods or methodologies. However, the growing pressure for coping with complexity appears to be leading to a better understanding of the need for integrating the two domains, and of how to bring about such integration[1]. But dealing with high complexity needs something more than a combination of methods or methodologies. It requires a different conceptual framework. I am referring to a systemic approach, or more exactly to an underlying framework based on systemic thinking.

Systemic thinking deals with systems, i.e., organized wholes made up of elements and relationships.[2] Complexity is a function of the uncertainty faced, the dynamics of the relationships, but also of the quantity and variety of the elements of the system-in-focus [cf. Rescher 1998]. The dynamic complexity of a system may give rise to new properties of that system, a phenomenon called emergence. In a nutshell, systemic thinking is holistic-integrative and multidimensional, while also being dynamic, analytic and synthetic.

In the following case it will be shown how a complex decision situation was supported by a system study which powerfully influenced the decision for the better in the social system affected. From a methodological point of view the study was remarkable in that a synthesis of qualitative and quantitative methods occurred, grounded in a systemic framework and customized to the issues at hand. It even occurred smoothly and to good effect. This was achieved on the basis of an innovative conceptual orientation, which provided to decision-makers a stronger lever than others for coping with the complexity faced when confronted with the difficult situation that will be described in the next section.

The case study will begin with a description of the context, i.e. the initial situation. Thereafter, the conceptual approach taken will be outlined. Then the steps of the decision process will follow, with decision analysis and synthesis. This will culminate in an account of the final decision. A reflective summary and outlook will conclude the paper.

2. Context

The Gastein Valley (Gasteinertal) in the county of Salzburg is one of the top tourist regions of Austria. Located at the rim of the alpine Tauern Mountains, it is connected to the north and the south via a railway coming from Salzburg in the north and continuing southward to Villach. The mountain is traversed by a tunnel leading from Bad Gastein to Mallnitz (Figure 1).

At the end of the Millennium, the Austrian state decided to build a fast railway transversal across the Alps, leading through the Gastein Valley and opening a high-capacity connection along the north-south axis Salzburg - Villach, which would also open links towards Italy (Udine) and Slovenia (Ljubljana). This was also a mandatory project, directly derived from the membership contracts of Austria with the European Union.

---

[1] Examples of approaches for integrating different methodologies are, e.g., Zeisel 1984, Mingers & Gill 1997, Rosenhead & Mingers 2001, Schwaninger 2004.

[2] The concept of *system* is a very general one, and it is subject to different interpretations.

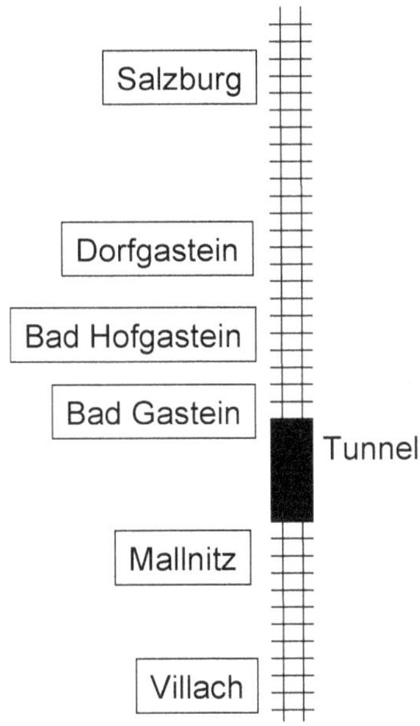

Figure 1. Scheme of the train connections to and from the Gastein Valley

The project would include a new layout of the line as a heavy duty track with two roadbeds, instead of one as in the past. It was foreseeable that such a project would have incisive consequences for the Gastein Valley. The new infrastructure would create large additional capacity which would imply the potential for a huge growth of traffic on that route. In relation to current transit frequencies, the new capacity would amount to a growth of 662 percent [Schwanhäusser 2000].

In the whole county of Salzburg and especially in the Gastein Valley, tourism is a very important economic factor [Scherrer 1998]. More than 30% of the work force of the Valley is employed by the hospitality industry. 60% of all jobs in the area depend directly or indirectly on tourism. This comes close to constituting an economic monoculture.

The tourist assets include:

- An outstanding ensemble of highly attractive natural factors (alpine landscape, flora, fauna etc.).
- Beautiful, originally preserved villages enriched by natural and cultural monuments.
- Health resorts based on a famous and powerful cure (thermal springs containing Radon and a healing gallery with the largest natural inhalatorium in the world)[3].

---

[3] The Gastein Valley is famous for the treatment of: Inflammatory rheumatic disorders, degenerative and deforming joint disorders, neuralgia, chronic pain, disorders resulting from sport and accident injuries and respiratory disorders. In addition, fertility and virility disorders, geriatric syndromes, skin diseases and disorders of the immune system and allergies are objects of the radon-based therapies.

- An advanced tourist superstructure in both qualitative and quantitative terms (9100 - mostly premium - beds in hotels and 7700 beds in other facilities).
- A comprehensive tourist infrastructure (healing gallery, thermal spa, cure and rehabilitation centre, sport medical centre, thermal mineral springs, convention centre).
- An exceptionally spacious network of promenades and hiking trails.
- 200 kilometres of ski-runs rated at all degrees of difficulty, about 60 kilometres of cross-country skiing tracks, 10 funiculars and 36 lifts.
- A set of marked socio-cultural characteristics, including a strong tradition of hospitality and ethnic customs.

Based on this combination of attributes and assets, the Gastein Valley is highly attractive, in particular for health-oriented types of tourism. Many of these market segments, such as health vacation, fitness vacation, rehabilitation and cures, are growing in the target markets, particularly in Germany.

In sum, the following critical success factors for the Gastein Valley destination were identified [Schwaninger/Lässer 2000]:

- Nature
- Health resorts (thermal springs)
- Quietness / absence of noise
- Beauty of landscape and settlements
- Tourist infra- and superstructure
- Socio-cultural factors, hospitality in particular

It is chiefly the features of peaceful quiet and health resorts which establish the basis for awarding official health-spa status to both communities, Bad Gastein and Bad Hofgastein.

3. The Need for a Solution

At the outset, the Gastein Valley was located in a virtually transit-free zone. The high volumes of transit were absorbed by the Felbertauern tunnel to the west and the Tauern highway to the east. This balance would completely change with the construction of the high capacity rail route. The levels of increase in both emissions and immissions would depend on the technical solution, i.e. the chosen construction variant.

It was clearly discernable that major changes in two dimensions would occur:

- Infrastructure (separation effects stemming from the new track and the optical and therewith esthetical changes in the landscape).
- Immissions, namely noise and - to a limited extent - vibrations.

Since the beginning of the planning phase in 1989 until the opening of a mediation process in 1998, two camps confronted each other implacably. On the one hand, a group around the Austrian Railways (ÖBB) and the Ministry of Transport, Technology and Innovation favoured an open (uncovered) and thereby less expensive tracklayout arrangement. On the other hand, a group around the local tourism industry demanded a completely closed and therewith immission-minimal variant.

There was an urgent need to resolve this conflict and eventually to move toward a consensus in order to reach a sustainable solution. This led to the initiation of a mediation process.

The mediation forum was constituted by representatives of a number of organizations, namely

- Austrian Railways (ÖBB)
- Ministry of Transportation, Innovation and Technology
- County of Salzburg
- Environmental attorneyship
- Austrian Chamber of Commerce, tourism section
- Municipality of Bad Gastein
- Municipality of Bad Hofgastein
- Tourism Association Bad Gastein
- Tourism Association Bad Hofgastein
- Civil initiative "Lebenswertes Bad Gastein"
- Civil initiative "Lebenswertes Gasteinertal - Bad Hofgastein"
- Austrian Alpine Association - Section Bad Gastein
- Civil Engineers Spirk & Partner
- Attorney Dr. H. Vana as representative of the municipalities and civil initiatives

The forum decided to schedule regular sessions and to form task forces around core issues: Tracking variants for Bad Gastein, tracking variants for Bad Hofgastein, criteria for comparing the variants, noise, etc.

The leaders of the mediation forum asked me to develop a decision base which they called "economic assessment", to serve as a foundation for the decision process. I offered to work out a system study in cooperation with Dr. Christian Laesser, the deputy director of the Institute for Public Services and Tourism, an expert on transportation and regional economics. We were put in charge and also offered all the resources available in the mediation forum. This included both the expertise of the specialists involved in the project and the many specialized studies and reports they had accumulated until then.

The decision that the construction of the High Capacity Railway (HCR) would be undertaken was definite and unquestionable. The question, therefore, was which one of the possible variants considered for that intervention was to be realized. This study would have to help the forum in taking its final decision.

4. The Conceptual Approach

The approach of the system study we envisaged had important implications:

- The study had to assess not only the economic effects of the planned intervention, but the social, technological and ecological aspects as well.
- It had to take into account the interrelationships between the variables of the multiple dimensions considered.
- It had to be dynamic, i.e. a static "photograph" of the situation after the intervention, compared to the conditions at the outset, would not suffice.
- It should incorporate not only analysis but also synthesis in order to provide a comprehensive picture.

As a consequence, this would be a study of substantial complexity. The approach promised, however, to lead to a broader view and to deeper insights as a basis for decision-making.

We started with an investigation on the spot. This included visits of Bad Gastein, Bad Hofgastein, the main infrastructural components, hotels, restaurants, natural and cultural monuments, etc. We also had meetings and led interviews with exponents of the two constituencies. Finally, we gathered written documents and set up information channels. Several local persons were designated to provide us with any further documents or data needed.

Back at our university, we conceptualized what we had learned. In the first place, we drew a conceptual and highly abstract map of the issue under study (Figure 2)

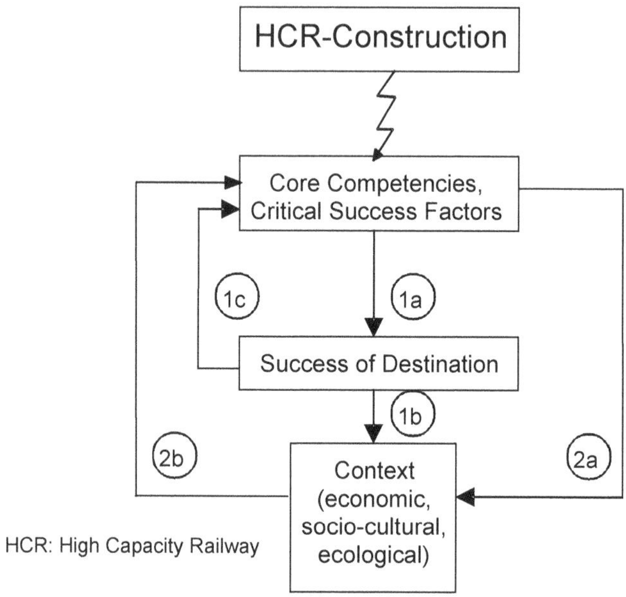

Figure 2. Object of study - conceptual map

This was also the road map for our ensuing investigation.

The construction of the High Capacity Railway (HCR) was conceived as an initial "shock" which impinged on the core competencies and the critical success factors of the Gastein Valley. From there, two causal chains were identified which were rightly to become the object of detailed studies:

1. Would the construction of the HCR have effects on the core competencies and critical success factors of the Gastein Valley destination (1a) and successively on the economic, socio-cultural and economic context (1b)? If yes, which would be the consequences and what would be their proportions? Ensuing feedbacks from core competencies and critical success factors (1c) were to be taken into account.
2. Would the consequences of the HCR construction on core competencies and critical success factors entail implications for the context (2a), and if so which ones? Also here, eventual feedbacks from the context to core competencies and critical success factors (2b) were to be analyzed.

We then proceeded to a situation analysis. This included an examination of the contextual factors of transport policy, tourist trends, Gastein Valley as a destination, tourist assets, tourist demand, core competencies and critical success factors.

This analysis allowed us to operationalize and specify the programmatic scheme (Figure 2) with a view to framing the further steps of the study. Instead of the all-inclusive concepts of core competencies and critical success factors, we were now in a position to revert to more concrete terms related to the natural ("original") and derived assets (Figure 3).

The representation used here is a common feature of qualitative System Dynamics[4]. While issues of the type studied here are usually represented by means of open causal chains, we favor the approach that relies on closed loops. These capture a characteristic feature of complex systems: Feedback, or the causal connections leading from a variable back into itself. Also delays, another characteristic feature, can be depicted in this kind of diagram.

In the present case, we can identify two self-reinforcing "Attractiveness Loops" [cf. Vester/vonHesler 1988] with further ramifications[5].

The first and primary attractiveness loop shows the influence of key variables of the original environmental package or set-up, namely landscape, nature, quietude, beauty of villages and cultural conditions. These variables, by and large, determine the

[4] System Dynamics is a widely used methodology for the modeling and simulation of complex systems, developed by Prof. Jay Forrester at MIT-Massachusetts Institute of Technology [Forrester 1961]. System Dynamics models are made up of closed feedback loops. SD is particularly apt for the discernment of a system's dynamic patterns of behavior, which may be „counterintuitive" [Forrester 1971]. It shows particular advantages for applications to social and socio-technical systems.

[5] On the methodology for elaborating these qualitative schemes, see Anderson/ Johnason 1997, Vennix 1996.

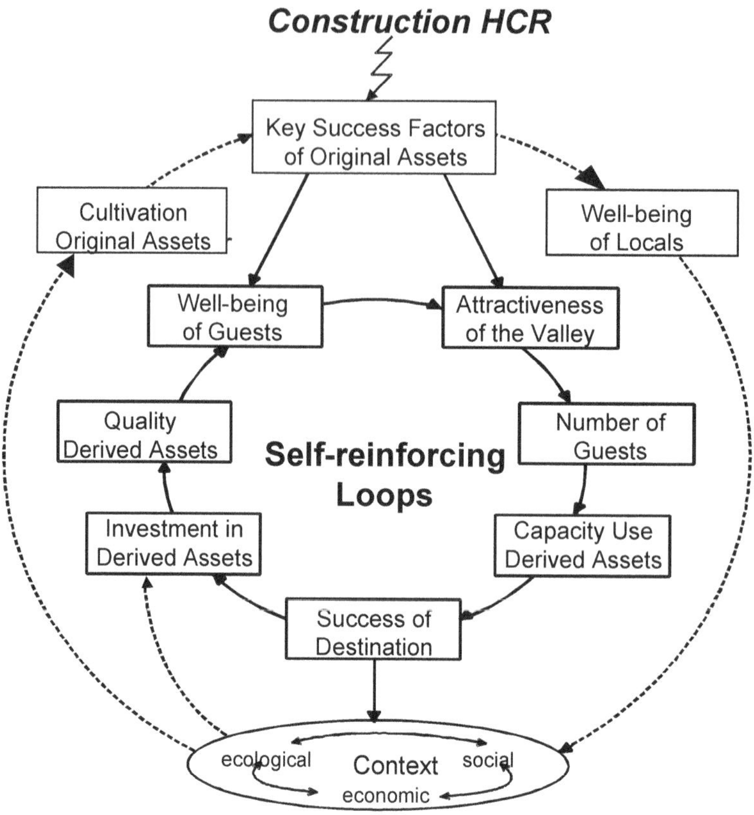

Figure 3. Causal loop diagram

well-being of guests and (in part directly, in part indirectly via the well-being of guests) the attractiveness of the valley, which is decisive for the number of changes in the number of guests. The capacity use and the economic success of the destination (both measurable in hard figures) would change as well. These changes would have consequences for the investment and thereby for the evolution of the quality of the derived set-up (infra- and superstructure). That very quality again impinges on the well-being of guests and the attractiveness of the valley.

The other, indirect and secondary loop on the outer side of Figure 3 centres on the internal context of the Gastein Valley, a context which is economic, socio-cultural and ecological. This context on the one hand is stamped by the well-being of local inhabitants. On the other hand it is influenced by the success of the destination. Via the two variables 'cultivation of the original set-up' and 'well-being of locals', the context is coupled with the key variables of the original set-up ("original assets")[6].

---

[6] Under 'original assets' we subsume mainly the natural conditions, e.g., locality, climate, landscape, flora, fauna, the aesthetics of the man-made environment, and the socio-cultural atmosphere, e.g., hospitality, tradition, customs. Under 'derived assets' we subsume the tourist infrastructure (e.g., health centers, funiculars, tourist offices, etc.) and the tourist superstructure (hotels, restaurants, etc.). For details, see Kaspar 1996.

Both loops are closed and self-reinforcing, being influenced by only one exogenous variable of interest - the construction of the HCR.

The sequence ran by and large from synthesis to analysis to synthesis. First we looked at the larger system of which the HCR is part, i.e., the Gastein Valley as a whole and its properties. Then we went about accounting for the behavior of that larger system as a function of the impacts of the HCR in different dimensions. Finally, we put these partial aspects together to explain the role of the HCR within its containing whole [cf. Ackoff 1999].

5. Decision Analysis

The study proceeded to an extensive analysis along the lines traced by the conceptual schemes (Figures 2 and 3). Accordingly, the analysis was made up of three blocks, namely:

1. Effects of the construction of the HCR on key variables of the original assets.
2. The primary causal loop.
3. The secondary causal loop.

In each one of these blocks, the pertinent variables were examined one by one. This refers to all the variables as listed in the causal loop diagram. In some instances the analysis even went down to a more detailed level; for example, in the case of the key variables (block 1) certain theoretical foundations were given, then the analysis examined quietude/noise, nature, beauty of villages and landscape, health resources/recreation factors, socio-cultural factors and hospitality in particular.

In the case of the inner, primary loop (block 2) the objects of analysis were the number of guests and capacity utilization, success of destination, investment in infra- and superstructure, quality of infra- and superstructure and well-being of guests.

The variables analyzed in relation to the secondary loop (block 3) were the well-being of locals, the economic context, the social context and the ecological context. In the case of the economic dimension, a number of partial aspects were examined, i.e., income and purchasing power, evolution of value in the hospitality industry, evolution of value in the other components of the derived tourist offer and intangible effects.

As far as these variants of the layout of the HCR are concerned, the picture was complex. For the Bad Gastein area alone, five variants of the track layout were established, and for Bad Hofgastein there were five additional ones. As the former ones could theoretically be combined with the latter ones in any way, the theoretical number of possible variants amounted to 25. Of these, the eight most plausible ones were examined in detail.

In the first place, the analysis referred to two ideal-types, namely:

Variant A: by and large open track, via a short tunnel in the region of Bad Hofgastein (Variant 1) and incomplete coverage in Bad Gastein (Variant 8). This was the track as established by the Austrian Railway Law.

Variant B: by and large closed track, via a long tunnel in Bad Hofgastein (Variant 3), a complete tunnel in the area of Bad Gastein and a cavern station (Variant 2).

These were the reference variants for the analysis[7].

For each one of the variables or partial aspects thereof, the analysis detailed first the basis for the assessment, and then proceeded to an evaluation of the consequences. In each case that evaluation was specified for the two reference variants of the HCR construction.

In the case of the variable "health resources/recreation factors" - for example, the indications treated in Bad Gastein (cf. footnote 3) - the weight of the natural factors and the impact of noise (permanent and peak noise levels) were ascertained in the section 'basis for the assessment'. In the section 'evaluation of the consequences', disturbances and effects on health and rehabilitation processes, as well as detriments to aesthetic qualities, were identified. For Variant A the transgression of noise standards as prescribed for the health-spa-status would happen; this would lead to attrition of that status. For Variant B a slight improvement of the basis for recreation and rehabilitation was foreseeable (except for the period in which the construction would be carried out).

The analysis was orientated by the conceptual schemes drafted at the outset. The logic of the analysis was essentially that of the sensitivity-analysis type. The pertinent question was: "How will the dependent variables react to different changes of the independent variables?" Much of this had to be answered in qualitative terms. These results were made readable due to the highly structured nature of the report.

On the other hand, much of the analysis was conducted along quantitative lines. The yearly costs and benefits for the reference variants were calculated in a detailed mode. The value of the investment in the railway had to be considered. Changes in tourist-generated income, regional income, tax yield etc. were calculated, changes in wages, jobs, investments and value of the tourist infra- and superstructure had to be determined as well. This made extensive data collections and also a detailed modeling necessary.

It lies beyond the scope of this paper to reproduce all the functions of that model here. However, three examples should give an idea of how qualitative constructs were translated into meaningful quantitative indices:

1. In the section about noise and its consequences, we referred to a hedonic pricing model by Pommerehne [1988], which allows one to deduce changes in the value of properties (via changes in rents) as a function of noise levels. The formula is:

$$\text{Ln } M = -0.002943 * dB^{1.3},$$

where M is the level of rents in Swiss Francs and dB the noise level in decibels.

---

[7] The variants were defined as a function of the installation of tunnels and their different lengths as well as other features of construction. E.g., Variant 8: New track parallel to the extant one (Bad Gastein; Variant 2: Continuous tunnel between Bellevue and Böckstein, and cavern station (Gastein); Variant 3a: short tunnel after Steinbach Station.

2. The construct "Success of the Destination" (primary loop) was operationalized by means of two indicators, the change in income from tourism and the value-added generated by the hospitality industry. The formulas were:

(2.1) $\Delta TGE = \Delta L * 1000$,

where TGE is the total income of the valley from tourism, L the number of overnight stays of tourists and 1000 an empirical value of the daily expenses of tourists (in Austrian Schillings).

(2.2) $\Delta WS = \Delta TGE * 0.62 * (BE - WA - ABS)$,

where WS stands for value-added, BE for the income of the local hospitality industry as 100%. WA and ABS are the respective percentages to be deduced for the cost of merchandise and depreciation. 0.62 is the hospitality industry's fraction of the income from tourism.

3. Under economic context (secondary loop), the macroeconomic income generated was calculated as follows:

$\Delta Y = \Delta TGE * \mu$,

where Y is the macroeconomic income generated by tourism and μ the multiplier [calculated with 1.37, a value ascertained empirically: Häusel 1985].

In a final section the results of the analysis were broken down to the level of the further variants, by means of a rough quantification. The comparison of the qualities of the different options came down to a set of scores on a scale between 0 (for Variant 8) and 1 (for Variant 2)[8].

6. Synthesis and Decision

The synthesis was carried out at two levels, the level of the three blocks of the analysis and the level of the total of analyses. The synthetic judgments were again underpinned by both qualitative and quantitative methods.

Firstly, at the end of each block of analysis a synthesis was undertaken which showed the larger picture, for example in case of the first block the overall change in attractiveness of the valley. Variant A would have disastrous consequences for each one of the components of the original offer as it also would for their totality. It would be a "devastating plan, ruinous for the whole valley" [König 2000], where the loss of the status as a health spa would only be one partial effect. The fateful process would start with an almost immediate loss of - modestly calculated - at least 15% (i.e. 300'000) of the guest nights, essentially due to the additional noise. And it would have strong and undesirable side-effects.

Variant B was in sum the one with the relatively least negative consequences. Some advantages, compared with the status quo, concerning noise and landscape in the area of Bad Gastein and Bad Hofgastein would still be compensated by two disadvantages: The increased traffic volume would affect Dorfgastein where no tunnel would be

[8] For example, the Variant 3a mentioned earlier got a score of 0.5.

provided. Furthermore, during the period of construction, unrest, noise and complications were anticipated.

Similar syntheses were carried out concerning the primary and the secondary loops. Both were identified as vicious loops, and in the case of Variant B the "vice" was even found to be disastrous.

The synthesis at the level of the whole had to bring the partial aspects together in clear-cut and easily comprehensible categories. The qualitative approach consisted in putting together all the partial, descriptive results of the analysis in a synoptic manner and lodging them in a special chapter. The quantitative approach resulted in a synthesis of economic results in another special chapter as well.

In addition, a small, highly aggregated, quantitative system dynamics model[9] was built on the basis of the conceptual scheme in Figure 3. This model[10] generated a result which the economic analysis had not been able to deliver, namely an overall account of the dynamic evolution of the valley's economy in the years to come. The most expressive result of the respective simulations is shown in Figure 4.

The main lesson delivered by this curve is that the results would be significantly worse than the ones calculated by the economic analysis: That scenario was calculated on the basis of a 15% reduction in guest nights (from two million to 1.7 million) within the first three years, at the outset for the case of Variant A. This was still in accord with the results of the economic calculus. The SD model, however, showed a decline of guest nights, from the initial value of 2 Million, by 34% within the first decade and a further reduction up to 52% by the end of the second decade (see Figure 4).

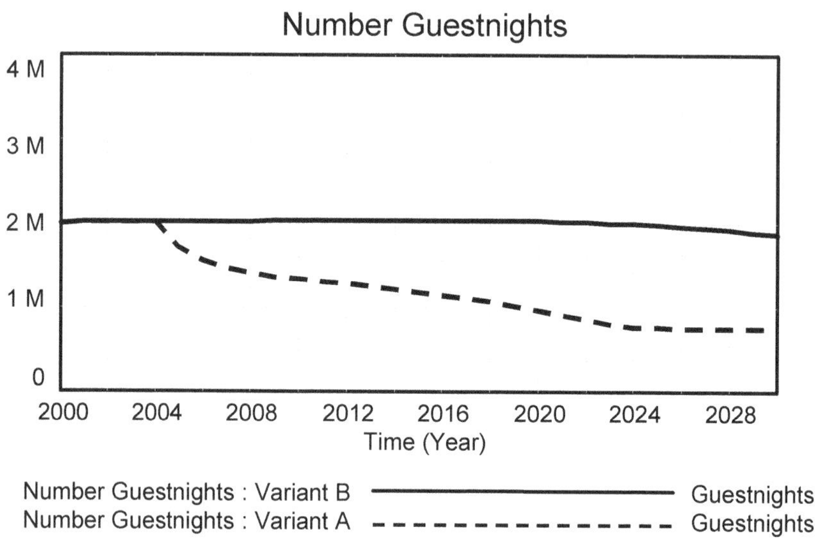

Figure 4. Evolution of the number of guestnights over 30 years (Scenario for Variants A and B)

[9] System Dynamics simulation is of the continuous type; the mathematics is based on differential equations [cf. Sterman 2000].

[10] The model is not reproduced here, but is scheduled for separate publication.

Finally, here are the results of our investigations, which were presented as a basis for the pending decision:

First and foremost, our study showed that the intervention was a counter-systemic interference, whichever variant would be chosen.

Secondly, our report culminated in the following recommendation[11]:

"The minimum of immissions achievable with the most gentle variant is equal to the maximum of what is still tolerable."

In other words, the study had led to the conclusion that out of the many options for the HCR layout studied the most environment-friendly variant (Variant B) was superior to all the others not only in ecological terms but also, and to a comparable degree, in macroeconomic and social terms.

The study had found, among others, the following concrete result: The optimal variant (Variant B) required an additional investment for its realization. According to the calculations that were carried out, the period needed for the amortization of the pertinent amount was found to be no more than 0.9 to 1.6 years[12]. It became clear that the most expensive variant was indeed a very good business proposition for the Austrian Republic.

The most intriguing observation that we made during the entire process concerns the fact that our report had an integrative impact on the mediation forum. We presented our results to the forum's plenary sessions twice. The first instance was in the middle of the study period, when the overall direction was clear but only a few of the details were apparent, and the second one, when the report had been finished. During the time we worked for the forum, we observed a successive convergence of the opinions among the participants. At the end of the process there was a strong consensus, with no significant opposition.

In 2002 the definite decision to adopt Variant B was taken by the forum. The pertinent project was then integrated into the General Traffic Plan ("Generalverkehrsplan") of the Austrian Ministry of Transportation, Innovation and Technology. The plan was to be implemented by the Austrian Railways, while the Austrian State would provide the necessary financial resources.

7. Conclusions

The case study reported on here brings several features to the fore, features which characterize a systemic approach to studying complex and dynamic issues.

The system under study exhibited exceptionally high complexity. The whole valley's socio-economic existence was challenged by the planned construction of an HCR system.

---

[11] In the original: "Das mit der schonendsten Variante erzielbare Minimum an Immissionen ist gleich dem Maximum des noch Tolerierbaren" [Schwaninger/ Lässer 2000: 95].

[12] If the additional System Dynamics-based scenario (Figure 4) would hold, then the payback time for Variant B would be 4 months only [Schwaninger/ Lässer 2000: 81].

The intervention and the ramifications of its effects as studied in this project made up a multidimensional web of relationships susceptible to analysis. That web produced a dynamic pattern which only a synthetic approach could fully elicit.

This study was a systemic one, which is expressed in several features: First of all, it was multi-dimensional, holistic and integrative. This is remarkable insofar as studies of this kind tend to be limited to one dimension, whether economic, psychological or social and so forth; and they are also often static. This study was multi-dimensional in that several dimensions - in particular the economic, social and ecological dimensions - were examined. It pursued a holistic approach as it focused on the larger whole and integrated the partial aspects of the different dimensions into a larger picture.

Furthermore, the study was based on a dynamic view. The dynamic patterns elicited, e.g. the long-term evolution as drawn by the System Dynamics model, were not only a useful but also a necessary ingredient for an understanding of the situation.

Finally, this study coupled analysis and synthesis. This combination stands in contrast to the monochromatic dominance of analysis and the lack of synthesis widely observable in studies of a comparable kind. However, this combination of analytic and synthetic reasoning is a basic feature of systemic thinking, a feature which is vital to a deeper understanding of a complex issue [Ackoff 2003].

It is cogent to argue that the results of a less systemic study, i.e. one not showing such multi-dimensionality, dynamics, and synthesis, etc., could not reach the depth and richness achieved in this case for both the investigation and its conclusions. Besides the richness of the conclusions presented in the final report, one must, e.g., take into account the valuable insights conveyed by the systems diagrams, the closed loop diagram (Figure 3) in particular. This richness offers a better basis for decision-making in a complex, dynamic environment than the one-dimensionality of more reductionist studies.

Both the analytical and the synthetic components of the study made use of qualitative as well as quantitative approaches. Neither one nor the other was sufficient in itself. The combination of both proved to be necessary. The qualitative analysis was extensive, yet the partial conclusions drawn on the basis of that analysis were highly differentiated and not easy to integrate into a conclusive picture. Contrary to many other cases, however, all of them pointed in the same direction. This fact by and large facilitated their integration. In addition, the qualitative investigation widened our horizon in that it brought detailed and sometimes hidden structures to the fore, which otherwise would not have been taken into consideration.

At any rate, the quantitative methods were an essential complement of the qualitative ones. They were not conceived as a mysterious calculus, leading to a figure about costs and benefits at the end. On the contrary, the result was a clear quantitative account which translated the qualitative deliberation into economic figures. Assumptions were specified throughout; the whole calculus was transparent and capable of being duplicated.

In addition to the relatively static calculus, a dynamic model was constructed which showed the long-term dynamics of the socio-economic evolution of the valley. That model showed that the results of conventional calculations underestimated the fatal consequences of the planned intervention. The insights triggered by the simulation results raised collective awareness within the forum.

As is usually the case with this kind of study, not all aspects can or should be quantified [Vennix 1999]. Therefore, the calculations were complemented by the additional qualitative arguments. If we had worked only qualitatively, it would have been much more difficult to take a decision, which is generally the problem with "thick descriptions" [Geertz 1993]. Had we worked only quantitatively, however, the basis for the decision would have been narrower and incomplete.

Altogether, this was a case of applying systemic thinking to a complex situation and thereby generating superior results in comparison to what might have been achieved with traditional approaches.

Most important, though, is what remains once the method has been accounted for. The study was a vital ingredient in the decision-making process, and ultimately proved to supply a crucial argument for choosing the most beneficial, although most demanding, of the variants. The implementation of any other variant would have had a negative impact on the valley in social, economic and ecological terms. Depending on the variant, that impact would have ranged from dangerous to disastrous.

In this sense it is no exaggeration to claim - *cum grano salis* - for the case in point: A valley was saved by means of a system study.

References

Ackoff, R.L., 1999, *Ackoff's Best. His Classical Writings on Management*, Wiley (New York).

Ackoff, R.L., 2003, *Redesigning Society*, Stanford University Press (Stanford).

Anderson, V., & Johnson, L., 1997, *Systems Thinking Basics: from Concepts to Causal Loops*, Pegasus Communications (Waltham, MA).

Black, T.R., 1999, *Doing Quantitative Research in the Social Sciences: An Integrated Approach to Research Design*, Measurement and Statistics, Sage (London).

Forrester, J.W., 1961, *Industrial Dynamics*, MIT Press (Cambridge, MA).

Forrester, J.W., 1971, Counterintuitive Behavior of Social Systems, in: *Technology Review*, Vol. **73**, January 1971, No. 3, pp. 52–68.

Geertz, C., 1973, *The Interpretation of Cultures: Selected Essays*, Fontana Press (London).

Häusel, U., 1985, *Die regionale Inzidenz von drei Infrastrukturobjekten: regionalökonomische Auswirkungen der N2-Osttangente Basel, des Biozentrums Basel und des Kantonsspitals Basel*, Dissertation, University of Basel (Basel, Switzerland).

Kaspar, C., 1996, *Die Tourismuslehre im Grundriss*, 5th edition, Haupt (Bern, Switzerland).

König, Ch., 2000, *Hochleistungsstrecke Gasteinertal der ÖBB; Mediationsverfahren, Kurortestatus, Erholungswert der Landschaft; Stellungnahme, 8.2.2000*, Land Salzburg - Landessanitätsdirektion/Umweltmedizin (Salzburg, Austria).

Mingers, J., & Gill, A. (eds.), 1997, *Multimethodology. The Theory and Practice of Combining Management Science Methodologies*, Wiley (Chichester).

Pommerehne, W.W., 1988, Measuring Environmental Benefits: A Comparison of Hedonic Technique and Contingent Valuation, in *Welfare and Efficiency in Public Economics*, edited by D. Bös, M. Rose, & C. Seidl: Springer (Berlin/ Heidelberg/ New York).

Rescher, N., 1998, *Complexity. A Philosophical Overview*,: Transaction Publishers (New Brunswick, U.S.S./London, U.K.).

Rosenhead, J., & Mingers, J. (eds.), 2001, *Rational Analysis for a Problematic World Revisited: Problem Structuring Methods for Complexity, Uncertainty and Conflict*, Wiley (Chichester).

Scherrer, W., 1998, *Wirtschaftsfaktor Tourismus, Antworten auf fünf zentrale Fragen zum Tourismus im Bundesland Salzburg*, Land Salzburg (Salzburg, Austria).

Schwanhäusser, W., 2000, *Eisenbahnbetriebswissenschaftliches Gutachten über das Leistungsverhalten der Tauernbahn,* unpublished report (Salzburg, Austria).

Schwaninger, M., 2004, Methodologies in Conflict: Achieving Synergies between System Dynamics and Organizational Cybernetics, *Systems Research and Behavioral Science*, Vol. **21**, pp. 1–21.

Schwaninger, M., & Lässer, Ch., 2000, *Volkswirtschaftliche Auswirkungen des Neubaus einer Hochleistungs-Bahnverbindung der ÖBB auf das Gasteinertal*, Gutachten, Institut für Öffentliche Dienstleistungen und Tourismus, Universität St. Gallen (St. Gallen, Switzerland).

Sterman, J.D., 2000, Business Dynamics. Systems Thinking and Modeling for a Complex World, Irwin/McGraw-Hill (Boston, Mass).

Vennix, J.A.M.., 1996, *Group Model Building: Facilitating Team Learning Using System Dynamics*, Wiley (Chichester).

Vennix, J.A.M., 1999, Group Model-building: Tackling Messy Problems, *System Dynamics Review*, Vol. **15**, No. 4, pp. 379–401.

Vester, F., & von Hesler, A., 1988, *Sensitivitätsmodell*, 2. Auflage, Regionale Planungsgemeinschaft Untermain (Frankfurt am Main, Germany).

Zeisel, J., 1984, *Inquiry by Design. Tools for Environment-Behavior Research*, Cambridge University Press (Cambridge. U.K.).

Chapter 12

Building Long-Term Manufacturer-Retailer Relationships through Strategic Human Resource Management Policies: A System Dynamics Approach

Enzo Bivona
Faculty of Political Sciences
University of Palermo
enzobivona@sciepol.unipa.it

Francesco Ceresia
Faculty of Educational Sciences
University of Palermo

1. Introduction

The present chapter is the result of a research project conducted by the authors together with a manufacturer operating in the high-tech industry. It is based on the hypotheses that in order to support retail stores successfully, a manufacturer has to design (a) internal policies based on Human Resource Management (HRM) practices aimed at increasing retailers' employees sales effectiveness, and (b) external-oriented policies capable of fostering potential customers' acceptance of company product benefits. The present study demonstrates the effectiveness of the System Dynamics methodology at supporting decision makers to explore alternative scenarios in a strategic planning setting [Morecroft 1984] and at fostering managerial learning processes [Sterman 1994; 2000] about how to build successfully strong, long-term manufacturer-retailer relationships. Results from strategic scenario sessions have been also used to design an empirical test of the identified successful policies on a significant sample of company's retail stores.

In today's economy all manufacturers need to pay attention to how to build strong and long-term relationships with their own distribution chain. In fact, it has been demonstrated that short term policies aimed at providing retailers with immediate benefits (e.g., product price discounts) may inhibit the development of fruitful long term relationships [Liker and Choi 2004].

These issues have been debated in the field of Distribution Channel Management [Dwyer et al. 1987; Anderson and Narus 1990; Ganesan 1994; Yilmaz et al. 2004]. In particular, it has been emphasized that manufacturers cannot ignore designing long-term growth-oriented policies, strategies aimed at improving the quality of their relationship with retailers and increasing their mutual satisfaction [Brown et al. 2000; Geyskens et al. 1999].

Analysing successful relationships between Japanese and North American companies, Liker and Choi [2004] remarked "that immediate benefits of low wage costs [or aggressive pricing policy] outweighed the long-term benefits of investing in relationships". In particular, Liker and Choi underlined that Toyota and Honda built great supplier relationships by implementing a set of different, but coherent, policies aimed at:

- investigating how their suppliers work;
- supervising their vendors;
- developing their suppliers' technical capabilities;
- sharing selected information intensively;
- conducting joint improvement activities.

Such practices highlight the central role of the human resource sub-system in building long term manufacturer-retailer relationships. This requires the manufacturer to identify a set of coherent practices that will enable its retailers' human resource management to increase their personnel effectiveness in promoting the manufacturer's mission. This has been identified as the domain of the Strategic Human Resource Management (SHRM) [Wright and McMahan, 1992].

This chapter presents an analysis based on a real case aimed at helping *Jeppy Ltd* [1] management to build successful policies to enhance the relationship with its retail chain. Through several meetings, which included the management, a System Dynamics model was built to explore alternative scenarios and related strategies. In particular, the use of a System Dynamics approach allowed us to identify an effective set of internal policies based on HRM practices (recruitment, training and goal setting policies), as well as external efforts aimed at increasing the awareness of potential customers about the company's product benefits.

In section II of this chapter, an analysis of the main contributions of Distribution Channel Management is outlined. In section III, the role of Strategic Human Resource Management is also considered.

Section IV of the chapter is the case-study. The evolution of the business, problem issues, feedback analysis of managerial growth policies, unperceived potential limits to

[1] The name of the firm has been disguised.

growth and policy design to remove business limits to growth are discussed here. It is worth remarking that problem issues and feedback analysis of adopted company growth policies are the results of several meetings with the involvement of the management.

Finally, an analysis of the main stock and flow structure of the System Dynamics model used to explore alternative scenarios and results of two simulation runs are provided.

2. Distribution Channel Management

Research in the field of Distribution Channel Management underlined the relevance of enhancing the quality of relationships between manufacturers and resellers [Dwyer et al. 1987; Anderson and Narus 1990; Ganesan 1994]. As noted by Anderson and Narus [1984], manufacturers should focus on strategies aimed at increasing resellers' satisfaction of the adopted distribution channel system [Brown et al. 2000; Geyskens et al. 1999]. In particular, as remarked by Yilmaz et al. [2004], manufacturing companies may increase the level of resellers' satisfaction by acting on four main areas:

- delivery (e.g., how well the manufacturer fulfils [retailers] resellers' procurement requirements);
- operation (e.g., manufacturers' contribution to [retailers] resellers' inventory management, store design);
- personnel (e.g., manufacturers' support of [retailers] resellers' personnel competence, courtesy and responsiveness);
- financial and sales (e.g., manufacturers' support of [retailers] resellers' sales and profits).

By effectively managing the above sub-systems, manufacturers may strengthen retailers' commitment. As a consequence, retailers might be more prone to invest in the manufacturer's business and to share its mission, on the one hand, while manufacturers could be more inclined to invest in distribution channel management, on the other.

Such a phenomenon is likely to generate a virtuous circle, supportive of a long-term relationship between the manufacturer and its retail chain [Fein and Anderson 1997; Ross et al. 1997].

As remarked in the introduction, analyzing successful relationships between Japanese and North American companies, Liker and Choi [2004] emphasized the ineffective role of short-term benefits, such as, for instance, product price discount. As the Toyota and Honda case shown, to build great supplier relationships it is necessary to implement a set of different, but coherent, policies aimed at:

- investigating how their suppliers work;
- supervising their vendors;
- developing their suppliers' technical capabilities;
- sharing selected information intensively;
- conducting joint improvement activities.

Such a set of policies can also be applied in managing the relationships between manufacturer and retailer.

In particular, it is worth acknowledging that in the above suggested policies, a crucial role is played by suppliers' or retailers' human resources management practices. In fact, designing training programs aimed at developing retailers' technical capabilities and at improving their performance, on the one hand, and implementing feedback mechanisms to monitor retailers' activities and, eventually, to suggest new plans to be adopted on the other hand, have proved successful strategies.

Based on such statements, the next paragraph will provide the reader a brief review of the SHRM approach and recurrent HRM practices often applied by companies.

3. Strategic Human Resource Management

The SHRM approach has been defined as a planned set of practices aiming at developing human resource skills and competences which allow organizations to gain their goals [Wright and McMahan 1992].

In particular, in order to recognize how to apply effectively a SHRM approach, Wright [1998] underlines four main issues on which organizations must focus.

The first issue concerns the role of human resources, as one of the main levers on which to act in order to build a sustainable competitive advantage. The second is related to the drawing up of a coherent and articulated set of programs, actions and activities that may facilitate, on the one hand, personnel to acquire desired skills and competences, and on the other hand, organizations to create an internal environment able to fully exploit personnel abilities to gain a sustainable competitive advantage. The third refers to the necessity to systemically develop HRM practices. In fact, such practices have to be consistent with corporate strategies.

Finally, to allow the organization to reach its own objectives, the fourth focus is on the strategic role the personnel unit must play. This issue emphasizes the pragmatic dimension of the SHRM approach.

3.1. Human Resources Management Practices: which one?

The literature on SHRM underlines the relevance of HRM practices in fostering companies' competitive advantage. However, organizations also need to identify which specific HRM practices have to be activated to improve companies' competitive positions.

Many researchers have suggested that HRM practices enable organizations to improve performance and achieve a competitive advantage that is likely to be substantially more enduring and more difficult to duplicate [Becker and Gerhart 1996; Delery and Doty 1996; Ferris and al. 1999; Guest 1997; Legare 1998; Pfeffer 1994; 1998].

In particular, Pfeffer [1998] emphasized the following seven HRM practices:

1. employment security;
2. selective hiring of new personnel;
3. self-managed teams and decentralization of decision making as the basic principles of organizational design;
4. comparatively high compensation contingent on organizational performance;
5. extensive training;
6. reduced status distinctions and barriers, including dress, language, office arrangements, and wage differences across levels;
7. extensive sharing of financial and performance information throughout the organization.

Pfeffer's results are also supported by the study conducted by Hiltrop [1999] in 115 multinational and 204 domestic companies located in Western Europe. Through interviews with human resource managers and personnel officers, Hiltrop tried to explore the recurrent HRM practices adopted by firms. As a result of this study, he identified the 11 most effective HRM practices for both employees' productivity and company performance. Furthermore, he detected that such practices are likely to attract and retain talent. In particular, he identified the following HRM practices:

1. employment security;
2. opportunities for training and skill development;
3. recruitment and promotion from within;
4. career development and guidance;
5. opportunities for skill development and specialization;
6. autonomy and decentralization of decision-making;
7. opportunities for teamwork and participation;
8. equal benefits and access to perquisites for all the employees;
9. extra rewards and recognition for high performance;
10. openness of information about corporate goals, outcomes and intentions;
11. pro-active personnel planning and strategic HRM.

Although an exhaustive discussion of the HRM practices adopted by organizations to improve company performance is far from the purpose of this work, the above analysis may help the reader to better understand the reason why we identified the suggested HRM practices in the project's policy design phase.

In the literature, criticisms have been raised of the use of unselective HRM practices. In particular, Guest [1997] remarkes that few studies adopt a systemic HRM approach. In addition, empirical analyses do not often investigate the structural relationships between HRM practices and their conjunctive effect on firms' performance.

This work tries to sketch the structural relationships between HRM practices and company performance on the basis of an empirical study conducted by the authors. The next section of the chapter shows the case-study, the main feedback structure of the investigated issues, the main stock and flow structure of the system dynamics model and a scenario analysis.

4. The Case-study Analysis

4.1. Introduction to the Case-study: Jeppy Ltd

At the end of the 1980's, a few months after their marriage, Samantha and Marc decided to launch a new venture. The business idea of Jeppy Ltd revolved around a simple concept: to create a technology able to increase human capabilities and to improve people's quality of life. Such a vision started from an observation of the diffusion of innumerable electronic products used for various purposes in houses, offices and other settings. Even though most of these products embody a common electronic part, they are used for distinct applications at a time. This is because each producer designs a product as a closed system. For instance, when you buy a washing machine, an air conditioner and a computer you are often unaware that you are buying some common electronic components three times. As a consequence, you are forced to pay three times for the same component. In addition, these three products are very often not able to interchange information.

In order to support people in managing simultaneously various applications keeping costs low, in both home and working activities, Jeppy Ltd proposed a new *Unitary Technology* (UT) concept, based on an unlimited set of components (so called bricks), that can be linked to electronic products and each other. It's the same philosophy of the LEGO bricks game where, by combining various components, it is possible to build endless products.

4.2. The Evolution of the Business

At the beginning of the 1990's, in order to create high quality products Samantha and Marc signed a partnership with major high-tech producers located in Asia. Such an agreement also allowed them to sustain low production costs.

In order to keep company product prices low and to reach potential customers interested in high-tech products, Samantha and Marc decided to sell Jeppy's product through high-tech product retailers. As Samantha and Marc were conscious that the market was not ready to understand and, hence, to buy "UT" products (in spite of the high product quality/price ratio set), they also decided to launch traditional products (so called "Trojan horse" products) onto the market.

Even though the latter appeared to end consumers as traditional ones (i.e., computers, videorecorder, DVD, etc.), they also embodied Jeppy's UT components. This was the initial commercial strategy pursued by Jeppy Ltd.

Jeppy Ltd products were sold by high-tech retail stores all over the domestic market. As the products had a very high quality/price ratio, due to the use of the new technology, in a few years, Jeppy's sales revenues reached a satisfactory level.

At the end of the 1990's, during a management meeting, commercial managers were very happy with company results as they had been able to reach the desired level of sales revenues. However, the company founders, Samantha and Marc, had some concerns relating to the company mission. Analyzing business figures, they detected that the concept of UT, offered by Jeppy Ltd, was not appreciated by end consumers. In other words, it seemed that customers bought Jeppy's products without any awareness of the bricks logic behind them.

In particular, Samantha and Marc pointed to two phenomena:

- end users do not tend to connect Jeppy products with others;
- Jeppy's UT products show a dramatically decreasing sales pattern over time.

As they recognized that their initial strategy, based on the "Trojan horse" products concept, had failed to diffuse among customers the concepts and benefits of unitary technology products, they decided to reshape their strategy.

At the beginning of the third millennium, the management launched their own retail chain, to explicitly promote the concept of unitary technology products. In order to effectively introduce to potential customers such a revolutionary product concept, the company decided to provide retailers' employees with an initial training course.

4.3. Problem Issues

At the beginning of 2001, Jeppy Ltd management decided to open a growing number of retail stores in the domestic market. By the end of 2004, Jeppy Ltd had 166 stores (see figure 1).

In figure 1, in the observed period, company sales revenues reached about Euro 30 Million. However, by analyzing the quantity of products sold over the selected four years (see figure 2), it is worth remarking that the so called "Trojan horse" products showed a continuous growth trend. On the contrary, the company products that explicitly embodied the unitary technology concept showed a weak growth in the first two years, and then a dramatic decline during the last period (see figures 2 and 3).

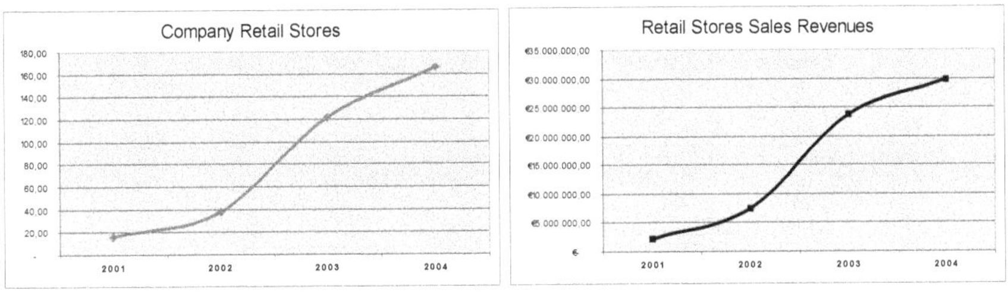

Figure 1. Retail stores and related sales revenues dynamics (2001-2004)

Figures 2 and 3 show the negative trend of the company's UT products sales. In fact, at the end of 2004, the latter made up less than 1% of the total company revenues. Such figures stimulated the following key questions:

- Why do sales quantities of "Unitary Technology" products show a weak growth and a decreasing pattern?
- What are the main causes of such a drastic reduction?
- Why do sales quantities of "Trojan horse" products show a continuous growth trend?
- What is the real contribution of the company retail chain to the achievement of the company mission?

Based on such key issues, the management decided to launch a project to detect the causes underlying the company's strategy failures in achieving its mission and to design policies to overcome business difficulties. After being contacted by the management of the company and exploring the problem issues, the authors suggested the management to build a System Dynamics simulation model. The simulation model would be used to investigate and better understand the business system behavior and to evaluate new strategies emerging from managerial debates and discussions [Morecroft 1984].

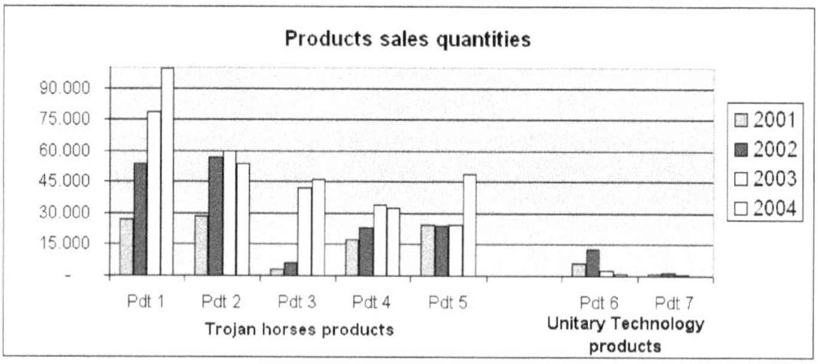

Figure 2. Company products sales quantities (2001–2004)

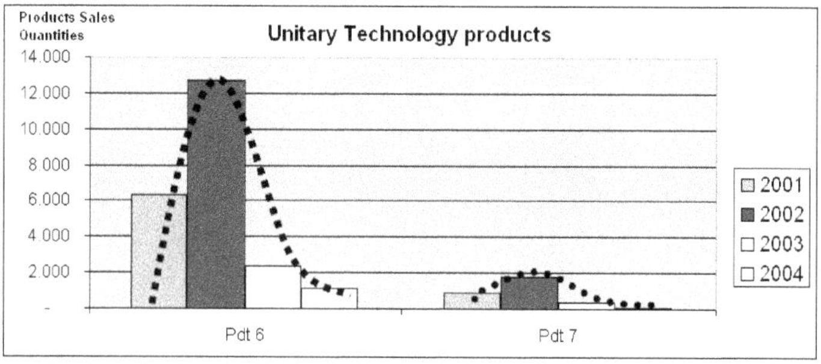

Figure 3. Unitary technology products sales quantities (2001–2004)

4.4. A Feedback Analysis of Main Management Policies to Overcome Limits to Growth and to Foster Business Development

The project actively involved the management and, in particular, the company's Organizing manager, who played a critical role in the success of the study. He strongly supported us in creating the "managerial" environment to develop scenarios based on the strategies emerging from the use of the system dynamics model. He also arranged meetings with key executives and provided data and information for the simulation model.

Once key variables and related behaviors were defined over time (figures 1, 2 and 3), the research focused on detecting cause-and-effect relationships underlying company growth policies and disclosing unperceived managerial potential limits to growth. Such an approach facilitates the construction of a system dynamics model capable of supporting strategic planning and decision making processes [Senge 1990; Kim 1989; Isaac and Senge 1994].

In particular, the above key issues have been explored by using qualitative diagrams. This was done through five meetings with the company management. With the use of such diagrams, company key performance indicators, policy levers and main cause-and-effect relationships among business variables have been made explicit [2].

In order to fill the gap between the desired and the actual value of company sales revenues, the management planned to invest in marketing and opening new retail stores. As a consequence, the management expected an increase in both customers and sales revenues. An increase in company revenues would have had to reduce the gap between the desired and the actual value. Once company revenues reached the desired target, marketing investments would be oriented to maintain such level of sales. This policy is represented in the balancing feedback loop in figure 4.

Figure 4. Balancing feedback loop leading to company sales revenues growth through marketing investments

[2] The methodology used to facilitate managers' mental model elicitation is based on the group model building approach [Vennix 1996].

The other policies adopted by the company to meet the desired sales revenues are based on investments aimed to open new retail stores, to provide retailers product discounts and an initial training to their employees. According to managers' vision, such investments in the retail chain would have had to allow the company to increase UT products sales. In fact, in order to communicate effectively to potential customers the concept and related benefits of UT products, the increase of retailer employee skills and motivation was perceived as very critical. Managers agreed that to increase UT products sales it is possible to adopt two different policies:

- a basic training course of retailers' employees on the UT concept and related benefits
- a high product cost discount, compared to other resellers.

In fact, managerial expectations were based on the fact that an increase in retailer employee skills and motivation would have fostered the effectiveness of communicating UT product' benefits and potential customer acceptance. Furthermore, as the UT product quality/price ratio was very high, an increase in potential customer acceptance should boost the number of potential customers that decided to buy UT products.

Such phenomenon could have increased company sales revenues to the desired target. Once such a target had been achieved, the management could invest in new resources in the development of the retail chain to keep such market position or to set higher sales targets (see balancing loop B2 reported in figure 5).

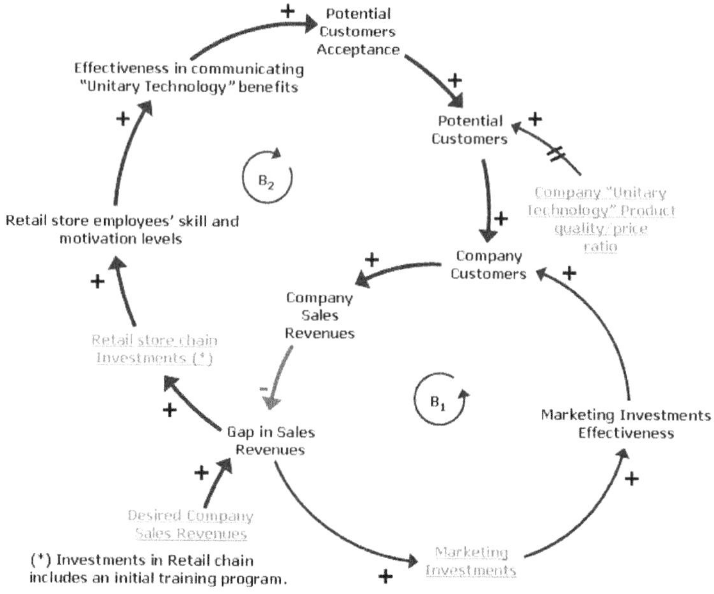

Figure 5. Company retail chain investment policy aimed to reach the desired level of sales revenues

4.5. Unperceived Potential Limits to Growth of Management Policies

The feedback loops in figures 4 and 5 show management mental models underlying main business growth processes and related policies to foster UT product sales. Such balancing feedback loops permit the company to apply controls to achieve an upward target growth objective (see figure 6).

It is worth remarking that the management also recognized two other positive phenomena that may contribute to foster company growth over time. Firstly, the new resources derived from higher revenues could be devoted to increase both marketing and retail chain investments. Secondly, as a consequence, sales quantities would increase and new resources could be spent to fuel business development.

Comparing expected and actual UT products sales quantities leads to the identification of a gap that underlines the limits of management mental models in explaining the outcomes of business growth policies (see figure 6).

What were the main causes underlying such a gap? What were the main limits to growth unperceived by the management?

In order to the investigate main causes of discrepancies between desired and actual company product sales quantities, the authors also decided to interview some key actors of the retail chain stores.

By interviewing personnel working in a sample of Jeppy's Retail stores, we detected three main key aspects that act on retailers investments effectiveness (see figure 7).

According to retails' employees, customers' cultural attitude constituted the strongest limit to growth of UT product sales. The common habit of perceiving technological products as "closed worlds" was the main obstacle to accepting the innovative idea that is possible to create endless applications able to work through a common technological device. Such a phenomenon very often pushed retail stores' employees to promote the so called "Trojan horse" products since they are easy to sell.

Figure 6. Comparison between expected and actual behaviors of "Unitary Technology" product sales quantities

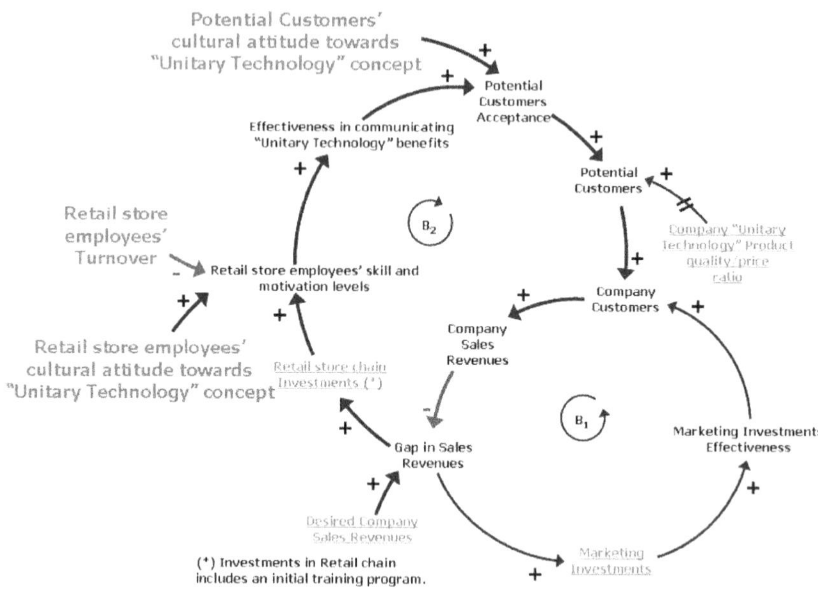

Figure 7. Main unperceived variables preventing company investments effects

On the contrary, efforts spent in promoting UT products very often do not generate any sales and, hence, de-motivate personnel who adopt a reactive approach to reach their personal sales target [3].

The second limit was related to a low retail store employee cultural attitude toward UT concept. In fact, through discussions with front-office employees, we could discern their difficulty in effectively communicating to potential customers the benefits of UT products. This seemed to us a consequence of employees' low motivation, due to lack of effective and continuous training activities. Although, there was not a full managerial consensus, most of the people involved in our meetings with the company were not surprised at our findings.

Furthermore, difficulties in promoting UT products were also amplified by the high turnover recorded in the stores' personnel. This can be seen as another limit preventing retail chain investment effectiveness through training programs.

4.6. Designing Policies to Cope with Limits to Growth

Based on the causal loop analysis shown in the previous pages, we suggested the management to act on the above commented limits to growth (retailers employees' cultural attitude towards UT concept; retailers employees' turnover; potential

[3] A similar phenomenon was detected by Morecroft [1984], who identified a natural hierarchy in the allocation of time of salespeople. "the highest priority goes to reactive sales efforts, ... The rest of the time is split between proactive sales effort, ... and general market maintenance. Second highest priority goes to proactive effort, unless the product sales objective has been completely satisfied."

customers' cultural attitude towards UT concept) by introducing on the one hand, customized retailers personnel recruitment, advanced training and goal setting policies, and on the other hand new investments in developing potential customers' acceptance of the UT concept.

As shown in the previous section (the theoretical framework of the SHRM proposed in literature) personnel recruitment, advanced training and goal setting policies are among the main practices in this field [Pfeffer 1998; Hiltrop 1999].

In particular, in the Jeppy case, an appropriate recruitment policy would have allowed the company to identify and hire the best people for the job. In fact, as stressed in literature [Guion 1997], a properly developed assessment tool may provide organizations a way to select successfully sales people and concerned customer service representatives. We suggested the management to introduce a selective retailer' personnel recruitment policy able to verify in advance human resource potential and their initial level of skills and competences in promoting UT products.

An advanced training activity was also suggested, as it would have allowed the company to communicate effectively to retailers' employees not only the organization's mission, strategies, values and so on, but also practical ways to better transfer to potential customers UT product benefits. This practice is also likely to stimulate and motivate employees, who often recognize training as a real appreciated form of organizational benefits [Hiltrop 1999]. Furthermore, we suggested the management to periodically run workshops in the retail stores, to demonstrate the use of UT products.

A goal setting policy was also recommended to the company management. Such practice is conceived in the literature as one of the most powerful motivational approaches [Locke and Latham 2002]. It assumes that when difficult, specific and clear goals are assigned to workers, the latter increase personal effort to improve task performance [Locke and Latham 2002]. To be effective, a goal setting policy has to embody a feedback control mechanism, aimed at verifying and supporting employees in achieving goals. In fact, employees need feedback about their progress in relation to the achievement of their own goals. If they do not know how they are doing, it is very difficult for them to adjust the level or direction of their efforts to match goal requirement [Bandura and Cervone 1983; Matsui et al. 1983]. In particular, in the analyzed case-study, a goal setting policy was necessary to increase retailers' personnel motivation and to provide employees well defined and stimulating targets. As a consequence, personnel could be able to make some clear priorities and effectively manage trade-offs (for instance, employees' efforts spent to promote UT products may absorb higher time associated with lower results if compared to employees' efforts spent to promote traditional products). Additionally, goal setting is also likely to reduce retailers' personnel turnover; as a consequence it could allow retailers – all conditions being equal – to maintain personnel with a higher level of competences for a longer period of time. Furthermore, goal setting policies can positively influence commitment, which in turn is likely to foster employees' learning processes [Seijts and Latham 2001].

Finally, in order to diffuse the concept and related benefits of UT products among potential customers and to increase their acceptance, such investments must be periodically supported by public events, conferences and targeted news in monthly reviews.

To pursue the company mission and the desired level of sales of UT products, we suggested the company management to explode retail chain's investments in a coherent set of policies that do not need to be conceived as alternatives, but complementary. In fact, during meetings with the management aimed to explore and validate alternative strategies, we demonstrated, through the use of an insight system dynamics model, that decisions aimed at implementing, for instance, a goal setting policy is likely to fail without proper personnel recruitment and advanced training activities.

Before building the quantitative simulation model, we shared with the management the main feedback structure reported in figure 8. It embodies the cause-and-effect relationships generated by the suggested policies used to overcome business limits to growth.

All the suggested policies are able to foster both the "Effectiveness in communicating Unitary Technology benefits" and "Potential Customers' cultural attitude towards "Unitary Technology" concept". As a consequence, on the one hand, retail stores' personnel was able to communicate UT products advantages effectively, and on the other hand an increase in potential customers acceptance could be likely to give up the number of potential customers. This could contribute to a significant increase in both the number of customers that will buy UT products and company sales revenues. As a result of this phenomenon, the gap between desired and actual company sales revenues could decrease.

Figure 8 shows the feedback loops B3, B4, B5 (human resources management practices) and B6 (external investments) described above. According to figure 8 all suggested policies give rise to balancing loops which help to boost the actual value of

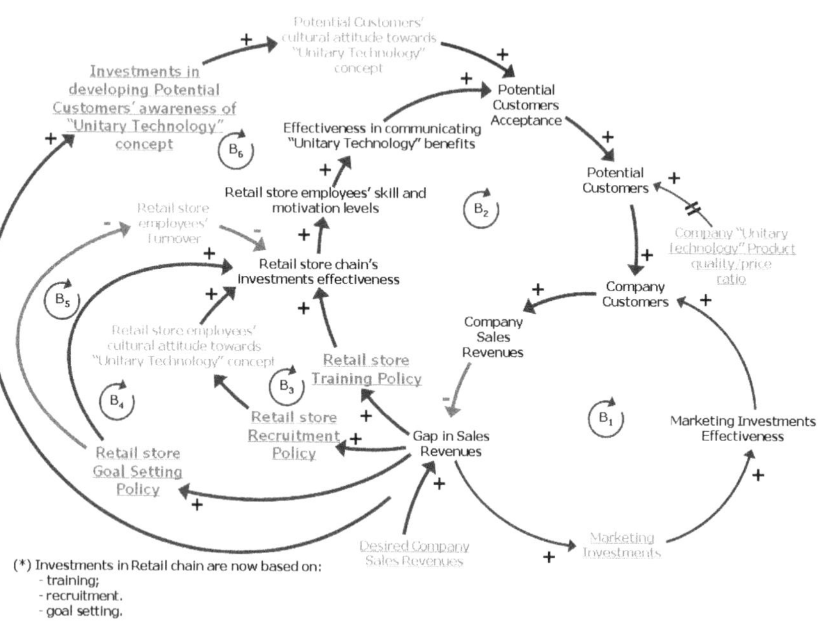

Figure 8. Suggested policies to overcome company limits to growth

company sales revenues toward the desired one. Even though the analysis also disclosed positive feedback loops (for instance, the word-of-mouth phenomenon that could take place when the number of customers who adopted UT products becomes relevant and could contribute to increase the number of UT potential customers), such relationships have not been included in figure 8, nor discussed due to their low relevance to the investigated phenomena.

4.7. An Analysis of the Main Stock and Flow Structure of the System Dynamics Model

Based on the above commented feedback structure, the authors built an insight system dynamics model. Such a model helped to capture company key-variables past behaviors (in particular, company products sales quantities) and to explore the effectiveness over time of the suggested strategies (retail stores' employees' recruitment, advanced training and goal setting policies, and investments in developing potential customers' awareness of UT concept). The system dynamics model covers three main sub-systems:

1. Company Customers related to "Trojan horse" and "UT" products sales;
2. Company retail store;
3. Human Resources Management.

The peculiarity of the third sub-system stock and flow structure will be discussed in the following pages [4].

The stock and flow structure shown in figure 9 has been adapted on the basis of the Skill Inventory Model proposed by Winch [2001]. In particular, it is possible to observe a "physical" structure related to "retail stores employees" and two main co-flow structures associated with retail stores employees' skill level and retailers employees' motivation level.

In particular, the company aims to maintain a number of retailers' employees proportionally to the number of stores. Employees' turnover is likely to affect the retailers recruitment policy, as it drains the number of employees.

Retailers employees' skill level may decrease due to the employees' skill obsolescence rate and leaving rate. Such a stock may increase not only through new recruits associated with a higher skill level (due to a selective recruitment policy), but also by continuous training programs provided to employees.

Figure 9 also delineates the main relationships affecting retailers employees' motivation level. In particular, such stock variable may decrease due to both a normal employees' motivation outflow and a leaving rate. Retailers employees' motivation level may grow through customized recruitment and goal setting policies. It is worth mentioning that a goal setting policy is likely to indirectly affect retailers employees' skill and motivation levels as it reduces employees' turnover.

---

[4] Model equations are available from authors.

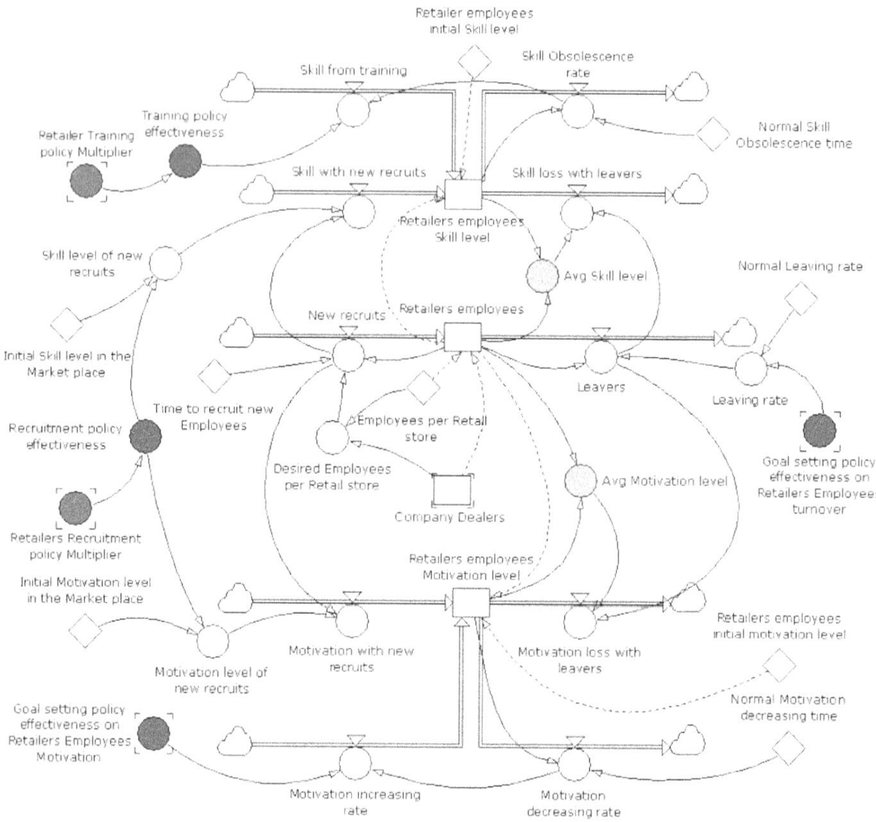

Figure 9. Stock and flow structure related to Human Resources Management sub-system

4.8. Scenario Analysis

In order to support the management in exploring and testing the effectiveness of our suggested policies to overcome business limits to growth, a system dynamics simulation model was used to analyze different scenarios.

In particular, retail stores' employees recruitment, advanced training and goal setting polices and investments in developing potential customers' awareness of UT concept have been hypothesized according to three possible options: low, medium and high.

Figure 10 shows key company variables related to the actual policies undertaken by the management. The simulation covers 10 years, from 2001 to 2010. In order to replicate the behavior of past business variables, the model has been initialized with company data provided by the management [5].

---

[5] In particular, the number of retail stores in the first 4 years of the simulation run has been generated by an external input.

As it is possible to observe in figure 10, such decisions may lead to a decreasing pattern in retailers employees' skill and motivation levels [6], which affect the effectiveness in communicating UT benefits to potential retailers' visitors. As a consequence, UT customers and related company sales dramatically collapse. Such phenomenon is also a consequence of a lack of investments fostering potential customers' culture attitude towards UT products that causes a fall down in potential customers' acceptance.

It is worth noting that management policies implemented in the initial four years (from 2001 to 2004) focused mainly on increasing the number of retailers. As figure 10 shows, at the end of 2004, the company had 166 retail stores. Such a growth is the main driver of the initial company's UT sales. In fact, as the number of stores remains unchanged, from 2005 it is possible to record an acceleration in the decreasing rate of company UT products sales. Such a phenomenon can also be observed in the dynamics of UT products' customers variable.

The analysis of company key-variables dynamics generated by the SD model provided a satisfactory fit with company past results. Furthermore, as the management was also confident on the key-variables behavior generated by the SD model, it has been used to assess the effectiveness of alternative scenarios; two of which will be commented here. In these two scenarios, the suggested policies have been implemented from 2005 to 2010.

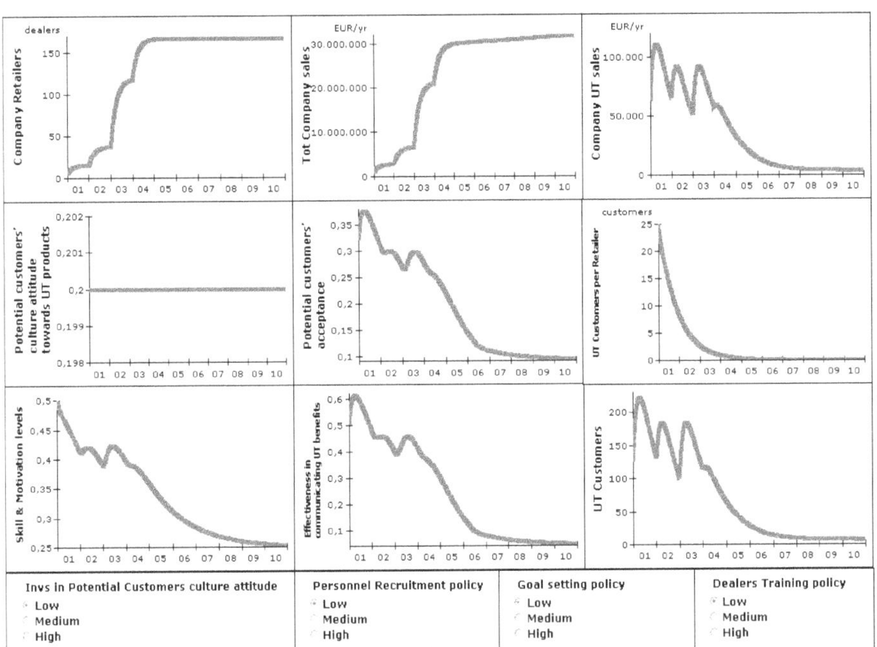

Figure 10. Main business variables dynamics (base run)

[6] Such variables are modeled as indexes (min. 0; max 1).

The first scenario (reference) is based on high investments aimed at fostering potential customers' culture attitude towards the UT products and unchanged policies on retailers' human resource management practices.

The second scenario (current) is based not only on decisions made to increase potential customers' culture attitude towards the UT concept, but also on high investments in customized retailers' employees recruitment, training and goal setting policies.

Figure 11 portrays both reference and current runs. In particular, it is worth remarking that company decisions aiming at only fostering potential customers' culture attitude towards UT products are not sufficient to boost the number of UT customers and sales revenues. This phenomenon is strongly related to the low level of retailers employees' effectiveness in communicating UT benefits. In fact, such key-variables are affected by retailers employees' skill and motivation levels, which in this first scenario do not vary.

On the contrary, the current run is likely to generate the desired effects on both external and internal company systems. In fact, such decisions are responsible for the increase: on one hand affecting potential customers' culture attitude towards UT benefits, and on the other hand affecting retailers employees' skill and motivation levels. The higher the employees' skill and motivation levels are, the greater employees' effectiveness in communicating UT benefits will be. This fosters a further increase in potential customers' acceptance, which in turn affects the number of potential customers that are more prone to buy UT products leading to an increase in Company UT customers and sales revenues. Finally, an increase in UT sales revenues allows the company to meet its desired level of revenues (see figure 6).

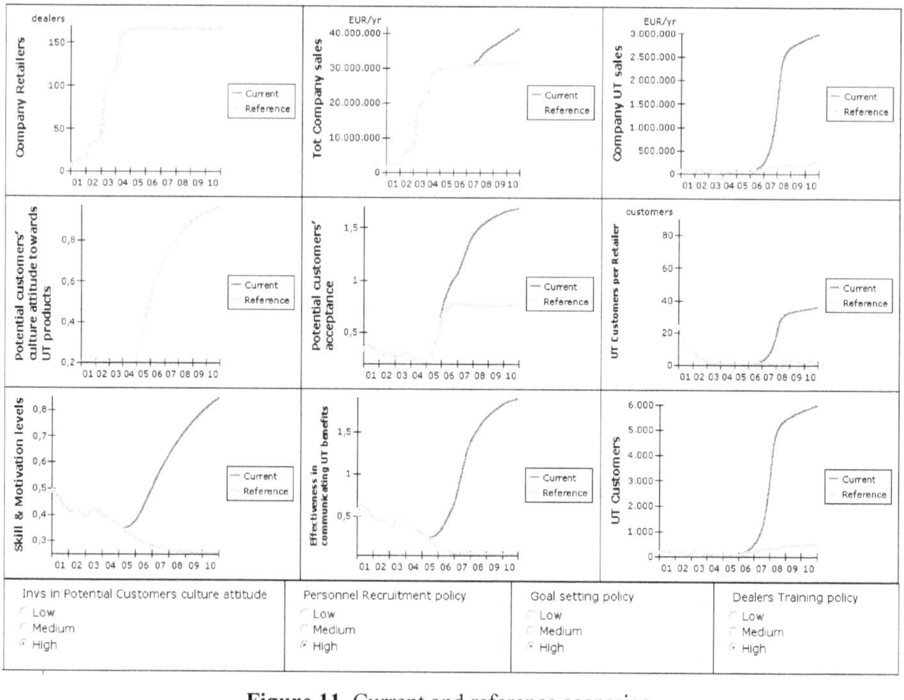

Figure 11. Current and reference scenarios

5. Conclusions and Further Research

This chapter has demonstrated that in order to successfully design long-term policies to foster manufacturer-retailer relationships, manufacturers must not make decisions exclusively oriented to generate immediate benefits [Liker and Choi 2004]. In fact, such policies may disclose future company failure.

The System Dynamics methodology displayed effectiveness in exploring and understanding the complex and dynamic phenomenon related to the manufacturer and retail chain relationship. In particular, such an approach allowed the management of the company to systemically design and assess policies based on human resource management practices (recruitment, training and goal setting policies) and external efforts to increase potential customers awareness of company product benefits. The above suggested human resources practices may represent an innovative vehicle for the development of studies to adopt a systemic HRM approach, as suggested by Guest [1997].

References

Anderson, J.C., & Narus J.A., 1984, A model of distributor's perspective of distributor–manufacturer working relationships, *Journal of Marketing, (25), pp. 23– 45.*

Anderson, J.C., & Narus, J.A., 1990, A model of distributor firm and manufacturer firm working partnerships, *Journal of Marketing, (48), pp. 62–74.*

Bandura, A., & Cervone, D., 1983, Self-evaluative and self-efficacy mechanisms governing the motivational effects of goal systems, *Journal of Personality and Social Psychology, (45), pp. 1017–1028.*

Becker, B., & Gerhart, B., 1996, The impact of human resource management on organizational performance: progress and prospect, *Academy of Management Journal* 39 *(4), pp. 779–801.*

Brown, J., Dev, C.S., & Lee, D., 2000, Managing marketing channel opportunism: the efficacy of alternative governance mechanisms, *Journal of Marketing, (64), pp. 51–65.*

Brown, J., 1981, A Cross-Channel Comparison of Supplier-Retailer Relations, *Journal of Retailing, (57), pp. 3–18.*

Delery, J. E., & Doty, D. H., 1996, Modes of theorizing in strategic human resource management: Tests of universalistic, contingency and configurational performance predictions. *Academy of Management Journal*, 39(4): 802–835.

Dwyer, F.R., & Oh, S., 1987, Output sector munificence effects on the internal political economy of marketing channels, *Journal of Marketing Resellers, (24), pp. 347–358.*

Fein, A.J., & Anderson, E., 1997, Patterns of credible commitments: territory and brand selectivity in industrial distribution channels, *Journal of Marketing, (61), pp. 19–34.*

Ferris, G.R., Hochwater, W.A., Buckley, M.R., Harrel-Cook, G., & Frink, D. D., 1999, Human Resource Management: Some New Directions, *Journal of Management.* (*25), pp. 385–415.*

Ganesan, S., 1994, Determinants of long-term orientation in buyer seller relationships. *Journal of Marketing, (58), pp. 1–19.*

Geyskens, I., Steenkamp, J.E.M., Kumar, N., 1999, A meta-analysis of satisfaction in marketing channel relationships", *Journal of Marketing Resellers, (36), pp. 223–38.*

Guest, D.E., 1997, Human Resource Management and Performance: A Review and Research Agenda, *International Journal of Human Resource Management (8), pp. 263–276.*

Guion, R.M., 1997, *Assessment, Measurement, and Prediction for Personnel Decisions,* Lawrence Erlbaum Associates (Mahwah, New York).

Hiltrop, J. M., 1999, The Quest for the Best: Human Resource Practices to Attract and Retain Talent, *European Management Journal*, *(17), pp. 422–430.*

Isaac, M., & Senge, P., 1994, *Overcoming limits to learning in computer-based learning environment*, in: Morecroft, J.W., & Sterman, J., (ed.), Modeling for Learning Organizations, Productivity Press (Portland), pp. 270–273.

Kim, D., 1989, *Learning Laboratories: Designing a Reflective Learning Environment*, in Milling, P., & Zahn, E., (ed), Computer-Based Management of Complex Systems, Springer-Verlag (Berlin).

Legare, T.L., 1998, The human side of mergers and acquisitions, *Human Resource Planning (21), pp. 32–41.*

Liker, J., Choi, T., 2004, Building Deep Supplier Relationships, *Harvard Business Review, December, pp. 104–113*

Locke, W. A., & Latham, G. P., 2002, Building a practically useful theory of goal setting and task motivation: A 35-year odyssey, *American Psychologist, 57, pp. 705–717.*

Matsui, T., Okada, A., & Inoshita, O., 1983, Mechanism of feedback affecting task performance, *Organizational Behavior and Human Performance, (31), pp. 114–122.*

Morecroft, J., 1984, Strategy Support Models, *Strategic Management Journal, (5), pp. 215–229.*

Pfeffer, J., 1994, *Competitive Advantage through People: Unleashing the Power of the Work Force*, Harvard Business School Press (Boston).

Pfeffer J., 1998, *The human equation*, Harvard Business School Press (Boston).

Ross, W.T. Jr, Anderson, E., & Weitz, B.A., 1997, The causes and consequences of perceived asymmetry of commitment to the relationship, *Management Science, 43, pp. 680–704.*

Seijts, G.H., & Latham, G.P., 2001, The effect of learning, outcome, and proximal goals on a moderately complex task, *Journal Organizational Behaviour, (22), pp. 291–307*

Senge, P., 1990, *The Fifth Discipline: The Art and Practice of the Learning Organizations*, Doudleday (New York).

Sterman, J., 1994, Learning in and About Complex Systems, *System Dynamics Review, (10), 2–3*

Sterman, J., 2000, *Business Dynamics. Systems Thinking and Modeling for a Complex World*, Mc Graw Hill (Boston).

Vennix, J., 1996, *Group Model Building*, Wiley (Chichester).

Winch, G., 2001, Management of the 'Skills Inventory' in times of Major Change, *System Dynamics Review, (17), pp. 151–159.*

Wright, P. M., & McMahan, G. C., 1992, Theoretical Perspectives for Strategic Human Resource Management, *Journal of Management (182), pp. 295–320.*

Wright, P., 1998, Human resources-strategy fit: Does it really matter?, *Human Resource Planning, (21), pp. 56–57.*

Yilmaz, C., Sezen, B., & Tumer Kabadayı, E., 2004, "Supplier fairness as a mediating factor in the supplier performance–reseller satisfaction relationship", *Journal of Business Research (57), pp. 854– 863.*

Chapter 13

Diffusion and Social Networks: Revisiting Medical Innovation with Agents

Nazmun N. Ratna*, Anne Dray, Pascal Perez, R. Quentin Grafton, David Newth, Tom Kompas

*Crawford School of Economics and Government
ANU College of Asia and the Pacific
The Australian National University, Canberra, Australia
nazmun.ratna@anu.edu.au

1. Introduction

In this chapter we reanalyze *Medical Innovation* by Coleman, Katz and Menzel (1966), the classic study on diffusion of Tetracycline, which at that time was a newly introduced antibiotic. Their pioneering study elaborated on how different patterns of interpersonal communications can influence the diffusion of a medical innovation in four medical communities in Illinois.

The motivation for our reanalysis is to capture the complex interactions involved in the diffusion process by combining Agent-based Modeling (ABM) and network analysis. Based on the findings in *Medical Innovation*, we develop a diffusion model called *Gammanym*. The topology of networks generated in *Gammanym*, and its evolution, are analyzed to evaluate the network structure influencing the diffusion process.

We describe the original study and the rationale for our study in the following section. Section 3 describes the modeling framework, modeling sequences and methods. Simulation results under different scenarios are analyzed in Section 4. The structure of the social networks depicted in our model, and its evolution are analyzed in Section 5. We explore the significance of types of network integration after normalizing adoption curves in terms of numbers for professionally integrated and isolated doctors in Section 6. The paper concludes with discussions and a review of the implications of the simulation results.

2. Diffusion and Networks

2.1. Diffusion of Tetracycline

Tetracycline was launched in November 1953. The success of its release was uncertain for the pharmaceutical companies because of an already established market for broad-spectrum antibiotics. Four US Midwestern cities: Peoria, Bloomington, Quincy and Galesburg, were selected by Coleman et al. who wanted to set their study in a group of three or four cities in one contiguous area that was not in the shadow of a large medical center, but where the cities differed from each other in terms of available hospital facilities: the number of teaching hospitals and general hospitals and population. The Coleman et al. sample constituted 148 general practitioners, internists, and pediatricians in active practice, of which 126 (85 percent of the sample) were interviewed. In an attempt to evaluate the importance of social networks, each of them was asked about their close associates (e.g., friends, colleagues and advisors) in the medical community. In order to measure the time of adoption, a prescription audit in the local pharmacies was carried out for 125 doctors (121 general practitioners, internists, and pediatricians, and 4 listed as surgeons or proctologists) over a 16-month period following the release of Tetracycline for general sale. The prescriptions of the doctors were audited for three successive days at approximately monthly intervals (Coleman et al. 1966: 194).

Coleman et al. (1966) study identifies two broad categories of variables influencing the diffusion process. The first category represents personal traits or individual variables, affecting individual receptivity. Individual variables are: 1. type of practice, 2. medical background, 3. contacts with out-of-town institutions, 4. media behavior, and 5. orientations and attitudes. The second category defines social variables influencing the adoption process as a result of social or professional ties to other members of the community. The influences of professional networks were evaluated on the basis of four factors that enable exchange of information in professional settings: 1. hospital affiliation, 2. shared office, 3. advice seeking, and 4. discussion. The friendship structure in the original study, emerged from the informal discussion among doctors in a social setting and doctors were asked to name three doctors whom they see most often socially.

The original study revealed strong evidence for social contagion, i.e., doctors' decisions to adopt Tetracycline were strongly influenced by the people they were connected to either socially or professionally. It appeared that the 'integrated' doctors differed very little from their isolated colleagues at the beginning of the introduction of Tetracycline, but their rate of adoption accelerated to produce an increasing gap between 'integrated' and 'isolated' doctors (Coleman et al. 1966).

2.2. Social Contagion vs. Media Exposure

Burt (1987) (:1327) used the Coleman et al. study to derive three major conclusions: "1. where contagion occurred, its effect was through structural equivalence[1] not

[1] "Structural equivalence is the degree that two individuals occupy the same position in a social system" (valente 1995:55). A structural equivalence model of diffusion postulates that individuals are influenced to adopt an innovation by imitating the behaviour of others to whom they are structurally equivalent.

cohesion; 2. regardless of contagion, adoption was strongly determined by a physician's personal preferences, but these preferences did not dampen or enhance contagion; 3. there was no evidence of a physician's network position influencing adoption when contagion was properly specified in terms of structural equivalence".

Strang and Tuma (1993) apply an event-history framework on the Coleman et al. (1966) data to analyse a class of diffusion models incorporating spatial and temporal heterogeneity. Their result contradicts Burt, as their models indicate that, "Contagion in medical innovation is not a simple product of structural equivalence. Cohesive ties based on advice giving and discussion also contribute to diffusion, as do structures of similarity in physicians' orientation towards their work" (Strang and Tuma 1993: 638).

Valente (1995) tests his threshold/critical mass (T/CM) model on *Medical Innovation* data. The T/CM model requires a threshold measure and a determination of the critical mass, "a system level measure of the minimum number of participants needed to sustain a diffusion process (Valente 1995: 79). The study indicates that the opinion leaders, who have greater exposure to external influence, play a dominant role in the diffusion process. Another study by Van Den Bulte and Lilien (2001) provides strong support for external influence in the diffusion of tetracycline by incorporating a data set on advertisement volume. As the empirical results were unable to detect statistically significant contagion effect once advertising effect is controlled for, the authors conclude that the data do not show that diffusion was driven by contagion operating over social networks, and that earlier analyses confounded the contextual effect of the aggressive marketing effort by the pharmaceuticals with social contagion (Van den Bulte and Lilien 2001).

2.3. Rationale for Gammanym

We reanalyze the *Medical Innovation* study by applying ABM for two principal reasons: the non-inclusion of dynamics of network structure and complexity in any of the previous studies, and a limited exploration of the impact of external influence or media exposure from previous reanalyses of the Coleman et al. data.

The major limitation of all the previous studies is their static exposition of network structure, which falls short of representing the evolving process. These dynamics can be described by the *dynamics of the network*, or the evolving structure of the network itself such as the making and breaking of network ties. The other set of dynamics refers to *dynamics on the network*, created due to the actions taken by individuals. The outcome of their individual actions is influenced by what their neighbors are doing and, therefore, the structure of the network. In the real world, both kinds of dynamics exist simultaneously (Watts 2003). Until now, however, neither *Medical Innovation* study nor the subsequent reanalyses of the original dataset have incorporated any of those dynamic features. Our study therefore complements the extant work.

Our work also makes a contribution in that the complexity generated in the diffusion process has not been examined by any previous studies. In Coleman et al. (1966) study, the extent of influence was evaluated for pairs of individuals. Individual network (discussion, friendship or advice) was, therefore, perceived as a set of discrete/disjointed pairs. Given the existence of overlapping networks and consequent influences on doctors' adoption decisions, the complexity of actual events can not be captured by pair analysis. Indeed, the authors themselves considered that analyzing in pairs was the major limitation of the study[2]. ABM enables us to address this limitation by considering the whole network as a unit of analysis.

Apart from the study by Van den Bulte and Lilien (2001), external influence in the diffusion process has also not been analyzed rigorously for the *Medical Innovation* data set. External influence is created from different forms of advertising and intensive coverage by representatives from the drug companies. In the *Medical Innovation* study, doctors acknowledged a 'detailman' or a pharmaceutical representative, and also direct mail from the drug companies, as the major sources of initial information about Tetracycline. The original study however did not contain any information on marketing effort by drug companies. An attempt to collect data on marketing effort was constrained by the fact that back issues of the *Journal of the American Medical Association* (JAMA) did not have any advertising supplement, as they were removed before binding for storage (Van Den Bulte and Lilien 1999). In our study, we examine the impact of external or media influence by performing sensitivity analysis. The simulation results allow us to explore the actions of the pharmaceutical companies, which has not been examined in a MAS (Multi-Agent System) setting before.

3. Modeling Framework

Using the Smalltalk programming language, we develop *Gammanym* with the CORMAS platform under VisualWorks environment. The codes and the detailed UML diagrams for *Gammanym* are available in the CORMAS website: http://cormas.cirad.fr/en/applica/gammanym.htm.

First, we discuss the spatial representation and the rationale for specified attributes in *Gammanym* for each objects/agents with reference to the original study. The decision making process by the social agents is then described and the section concludes with a discussion on modeling sequence.

3.1. Spatial Representation and Passive Objects

Based on the distribution of office partnership and number of hospitals in the original study, we represent the medical community in an 8 X 8 spatial grid. The unit cell captures three different locations for professional interactions: Hospitals, Practices and a Conference Center, which are created as passive objects. In the original study, on average 47 percent of doctors were alone in an office, 20 percent were in clinics,

---

[2] "To analyse pairs of individuals instead of single individuals may seem like only a very modest step towards the analysis of networks of social relations. It would be more satisfactory, and truer to the complexity of actual events, if it were possible to use longer chains and more ramified systems of social relations as the units of analysis. But so little developed are the methods for the analysis of social processes, that it seemed best to be content with the analysis of pair relationships", so write the authors themselves (Coleman et al. 1966:114).

17 percent were working with two colleagues and 15 percent were sharing an office with one colleague. We captured the categories of office partnership into three practice types[3]; *Private* (alone in office), *Center* (shared office with two partners) and *Clinic* (working with four colleagues). The doctors in the *Gammanym* model are thus distributed among 46 private, 11 centers and 4 clinics. *Gammanym* specifies two hospitals, as all the cities, on average, have two hospitals. Conference center, the third passive entity, provides the context in which a much larger group of professionals can interact with each other. Although the original study did not specify a conference center as such, our rationale for this specification is to signify the importance of contact with out-of-town institutions (Coleman et. al. 1966: 44).

Our model, *Gammanym,* has sixty-one practices, two hospitals and one conference center, randomly located over an 8 X 8 spatial grid. Random allocation of these practices over the grid reflects that spatial representation is not sensitive to distance. In other words, the doctors' decision to go to hospitals or the conference center does not depend on the location of their practices.

3.2. Social Agents

Gammanym depicts two kinds of social agents - *Doctor* and *Laboratory*. Initially located in their respective practices, *Gammanym* creates 99 doctors, the principal agents in the diffusion process. A laboratory, on the other hand, influences doctors' adoption decisions by sending information through multiple channels: medical representatives, journals and commercial flyers.

Located and Communicating Agents: Doctors

In *Gammanym*, doctors are specified with the attributes generating network effects only. Individual traits have impacts on the adoption decision. Nevertheless, we opt for this simplification on the basis of the correlation coefficient estimated in *Medical Innovation.* Four network variables, shared office, advice seeking, discussion and friendship, showed a stronger association to the date of first use of Tetracycline than any other individual variables, with the single exception of total volume of prescriptions for the class of drugs. Holding the volume of prescriptions constant, the association between integration and adoption increased (Coleman et al. 1966: 92). Thus, we explore if all the doctors are homogenous in terms of their individual attributes or 'degree of predisposition to adoption' (Burt 1987), to what extent does

---

[3] Elaborating on significance of office partnerships to generate network effects, the researchers in the original study specified its three broad categories (Coleman et al. 1966: 73): alone in office, shared office, clinic. Shared office was further classified on the basis of number of colleagues being one, two, and three or more. Shared office for three or more colleagues was not incorporated as only 3 doctors in City B was identified for this category. We specified 5 doctors to practise in a clinic based on the representation of office partnerships in City D (Coleman et al. 1966: Figure 15). A majority of doctors in City A were practising alone (53 percent); in contrast to which 50 percent of doctors were affiliated with clinics in City C. None of the doctors in City B was practicing in a clinic. See *Medical innovation* (P 73-74) for detailed analysis.

integration matter for adoption decision? In our model, a random friendship network, and professional networks are created through discussions with office colleagues, or through hospital visits or conference attendance, or all of the above.

Constrained by the unavailability of the original dataset, we opt for a random friendship network for *Gammanym*. The doctors are, thus, initialized with random number of friends and *counter for friends*[4]; both ranging from 0–3. In the original study, indices of similarity[5] of pairs were analyzed in order to identify the attributes the doctors share with the colleagues they choose as friends, discussion partners or advisors. We treat the indices with reservation, because of their limited statistical relevance with only 111 friendship pairs (Coleman et al. 1966: 143). The contacts with friends are not spatially defined, as *Medical Innovation* contains no information in this regard. Each doctor communicates with his/her friend(s) and evaluates mean adoption rate when the counter for friend is 4.

Professional interactions are spatially defined to signify discussion networks[6]. We do this to signify the importance of tacit knowledge or non-codified knowledge, which requires face-to-face contacts for its transmission (Appold 2004; Balconi, Breschi and Lissoni 2004; Becker and Dietz 2004; Doutriax 2003; Fritsch and Franke 2004; Grafton, Knowles and Owen 2004; Howells 2002). The doctors, therefore, consider the others as discussion partners if they are situated in the same cell. After each visit to hospitals or conference center, doctors return to their practices. Office partnership is, therefore, of central importance for networking as the doctors in *Gammanym* made most of their interactions in their practices.

Communicating Agent: Laboratory

In *Gammanym*, we incorporate one pharmaceutical company as a communicating social agent, termed as the Laboratory (LAB from here on). The LAB adopts a mixed marketing strategy with three different channels to send information about the new drug:

i. Detailman (pharmaceutical representative) visiting practices;
ii. Flyers, available at the conferences;
iii. Journals, sent to doctors' practices.

---

[4] An attribute for the doctors to specify the frequency of interactions with their friends.

[5] The index varies from a value of +1, where all doctors choose others like themselves, to -1, where doctors choose only others who are unlike themselves. Based on the indices of similarity (in parenthesis for each factor), the order of importance for choosing friends was: religion (0.403), professional age (0.195), place of growing up (0.146) and medical school (0.167).

[6] Our intension to distinguish among discussion and advice networks was constrained, primarily, by unavailability of original data. The criteria for advice networks, was further complicated by the fact that none of the common background factors seemed to work for choosing the advisor, as the indices of similarity for religion, professional age, place of growing up and medical school are respectively 0.109, 0.042, 0.077, 0.018 (Coleman 1966:143). Interestingly enough, analyses by Burt (1987) and Stang and Tuma (1993), despite using the original dataset, were not specific in distinguishing the structures of those two networks either.

57 percent of the doctors asked to reconstruct their stages of diffusion, identified detailman as the first source of information. Hence, we opt for a blanket exposure of all doctors to detailman. In *Gammanym* the detailman visits all the doctors at their practices. At each time step the LAB keeps records of the practices visited by the detailman and randomly chooses one of the remaining practices or available practices, only if at least one of the doctors is present at the practice. As the time step signifies one week, so the detailman is able to contact all the doctors during his visit.

In the original study, drug house mails or direct mail advertising were identified as the second major source of information. Assuming similar influences by direct mail advertising by the pharmaceutical companies and journal advertisements, we specify journals as the second option for the LAB. The LAB issues journals only when the number of newly adopted doctors in the previous time step, i.e., the last increment, is less than half of the average number of adopted doctors. Journals are sent to all the practices and thereby ensure a blanket exposure to all doctors.

From the perspective of the LAB, inclusion of flyers is crucial as it adds another dimension to the marketing mix by targeting a large group of doctors at the same time. To avoid the notion of blanket exposure to all doctors, we specify the criterion that the LAB sends flyers based on the number of previous conference participants. Receiving information from a flyer is therefore conditional upon the number of available flyers, and not all the doctors attending the conference receive the flyers.

3.3. Decision-making Process

Doctors' decisions to adopt a new drug involve interdependent local interactions among different entities in *Gammanym*. Diffusion scholars have long recognized that an individual's decision about adoption is a process that occurs over time, consisting of several stages (Coleman et al. 1966; Rogers 1995). We specify five stages of adoption: i. Awareness or knowledge, ii. Interest, iii. Evaluation/mental trial, iv. Trial, and v. Adoption/acceptance. In our model *readiness* is specified as the attribute signifying the above stages of adoption. All doctors are initialized with *readiness* number of 4 that is decremented when they receive an alert from different sources.

In *Gammanym*, the doctors are homogenous in terms of individual predisposition to knowledge. Thus, the criterion for alert generation is same for all doctors irrespective of their adoption stages or sources. For doctors working in centers and clinics, they evaluate the mean adoption rate of their office colleagues while situating at their practices. Otherwise (during visits to the hospitals or conference center), each doctor randomly chooses five of their acquaintances and evaluates their mean adoption rate. Thus, discussions with other doctors, either friends or colleagues at practices, conferences, or hospitals generate an alert when the mean adoption rate is 0.50 or above. In the case of the LAB, on the other hand, an alert is created each time a doctor receives information from the detailman, flyers or journals.

Doctors' *readiness* is gradually reduced with alerts from all the aforementioned sources i.e. doctors move to the next stage of adoption when alert is greater than or equal to one. When the readiness reaches zero (Adoption/acceptance stage), doctors adopt the new drug.

3.4. Modeling Sequence

Gammanym is divided into four phases: i) managing professional interactions; ii) external influence; iii) decision making process; and iv) networks formation. At each time step, *Gammanym* resets the attributes of the practices. Thus, the doctors are at their respective practices at the beginning of each simulation.

Phase I entails the methods for doctors' professional interactions. Primarily, the doctors interact with the office colleagues at their practices. Hospitals are another location for professional interactions, where they have their monthly visits. The doctors are initiated with an attribute, counter for hospital, ranging from 0 to 3. The doctors randomly choose one of the hospitals and move to the designated cell when the counter for hospital is 4. The third location for information exchange is the conference centre. The invitations to conferences are sent randomly to 30 doctors. Doctors receive the invitation and move to the conference centre if they are available at their practices at the time step when the invitations are sent. Phase II and Phase III are discussed earlier in Section 3.2 and Section 3.3 respectively. Phase II depicts the mixed marketing strategy adopted by the LAB. Phase III is the decision-making process based on readiness. Phase IV constitutes the methods for network formation.

Gammanym generates 137 matrices, the network matrices for professional and friendship ties at each time step, and the adoption matrix at the end of simulation. Network matrices contain the number of interpersonal interactions for each doctor at each time step. The adoption matrix, on the other hand, specifies the adoption status (adoption vs. non-adoption) at each time step for all the doctors. These matrices are the links for our novel approach of integrated modeling as the analysis on network topology and evolution of networks in Section 5 is based on these matrices.

4. Simulation Results

In this section our discussion traces the shape of the cumulative diffusion curves under three scenarios. The three scenarios are specified to evaluate the degree of influence by different factors in the diffusion process:

1. Baseline Scenario; with one 'seed' or initial adopter, one detailman and one journal;
2. Heavy Media Scenario; with one initial adopter but different degrees of external influence, by increasing the number of detailman to five and the number of journals to four;
3. Integration Scenario; one initial adopter, without any external influence from the LAB.

All three scenarios have been run over 68 weeks or 17 months, which was the time length for the original study. As several random functions are included in the algorithm, each scenario is repeated 100 times in order to estimate the output's variability. For each of the cases, the seed or the innovator is chosen among the doctors who are practicing at centers, i.e., doctors who have two colleagues[7]. The cumulative diffusion curves (CDC), representing the percentage of adopting doctors at each time step are shown in Figure 1. All three curves are derived after averaging over 100 simulations.

Baseline scenario (Figure 1) with one innovator and one detailman generates a logistic or S-shaped curve, similar to those found in cases of mixed influence diffusion models (Ryan and Gross 1943; Mahajan and Peterson 1985; Rogers 1995; Valente 1995). In this scenario, an initial phase of slow diffusion exists until the first inflection point at the 24-time step where 23 percent of doctors have adopted the new drug. Thereafter, the rate of adoption speeds up as more doctors are exposed to someone who has already adopted. The rate of adoption then gradually begins to level off as fewer doctors remain in the population who are yet to adopt. Overall, 92 percent of doctors adopted the new drug by the end of 48 weeks, or within 12 months. With one innovator in the system, diffusion seeds through the system as the doctors' adoption decisions are simultaneously influenced by their exposure to the new drug through actions by the LAB, as well as interpersonal communication.

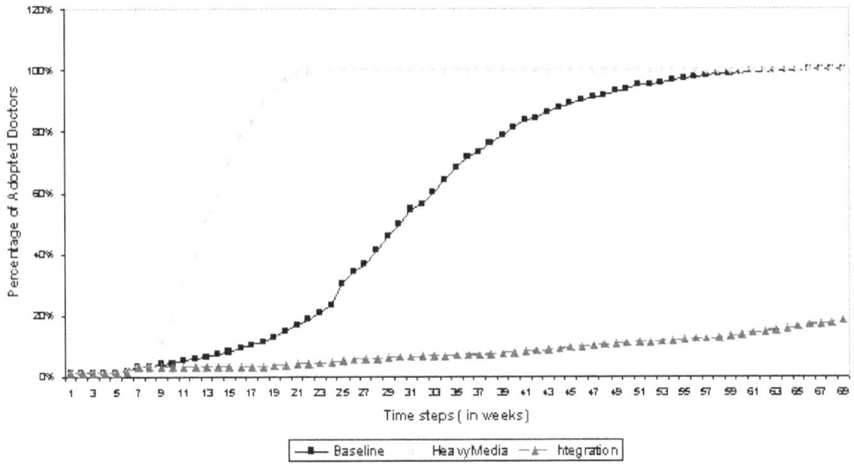

Figure 1. Cumulative diffusion curves for three scenarios: Baseline, Heavy Media and Integration

[7] We explored with initial adopters with different levels of social and professional integration. The cumulative diffusion curves for each of the cases represent the same process with point of inflection at 36th time step and levelling off at 50th time step irrespective of the integration status of the initial adopter.

The steepest diffusion curve (Figure 1) represents a heavy media scenario, where 50 percent of the population adopts the new drug at the end of 12 weeks. The rate of diffusion increases up to 16 time steps and decreases afterwards as only 18 percent of doctors at that time have failed to adopt and remain unaffected. At the 25 time step, the CDC levels off as all the doctors have adopted the new drug.

The integration scenario represents an extremely slow diffusion process (Figure 1). As the only means to have an alert is to be in contact with the initial adopter, only 18 percent of the population adopts the new innovation at the end of 68 time steps.

5. Network Analysis

In this section we will examine: 1. the topology of the interaction networks that are generated by the agent based simulations; 2. how the networks evolve over time; and 3. the way the uptake evolves across these networks.

5.1. Network Topology

We first calculate the degree distribution from the interaction matrices generated at each time step from four possible alternatives: i. regular lattices; ii. random graphs (Erdös and Rényi 1959); iii. small world networks (Watts and Strogatz 1998); and iv. scale-free networks (Barabási 2002).Degree is defined as the number of edges connected to a given node. The results were averaged over 100 replications of the simulation. Figure 2 compares the degree distribution resulting from *Gammanym* and an Erdös-Rényi (1959) random graph. From this figure we can see the networks created from the simulation are indistinguishable from a random graph. We also perform a χ^2 test against an Erdös-Rényi random graph with the same number of nodes and edges. There was no statistically significant difference between the degree distributions of *Gammanym* and the random graph when α=0.01. We do not claim that the structures of the network in the four medical communities in 1950's Illinois were random, but that the structure of the network developed in *Gammanym* appears to be random.

Many social systems display two statistical properties; first they display a high degree of clustering when compared to random graphs; second they have an average shortest path-length similar to that found in random graphs. These two properties are known as the small-world properties. To determine if the interaction networks generated by *Gammanym* have small world properties, we calculate the clustering coefficient and path-length statistics.

The clustering coefficient (CC) is a measure of the number of friends of friends type relationships found within the network. For a given node N_i with k_i neighbors, E_i is defined to be number of links between the k_i neighbors. The clustering coefficient is the ratio between the number of links that exist between neighbors (E_i) and the potential number of links $k_i(k_i - 1)/2$ between the neighbors.

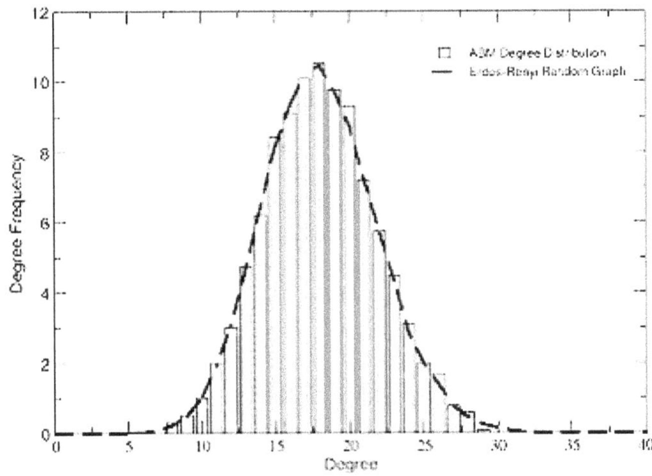

Figure 2. Degree distribution of the interaction networks in *Gammanym* and Erdös-Rényi random graph

The average clustering coefficient is:

$$CC = \frac{1}{N}\sum_{i=1}^{N}\frac{2E_i}{k_i(k_i-1)}$$

The average shortest path length (PL) is defined as:

$$PL = \frac{1}{N(N-1)}\sum_{i=1}^{N}\sum_{j=i+1}^{N}PL_{min}(i,j)$$

PL_{min} is the minimum distance between nodes i and j.

A network is said to have small world properties if, compared to an Erdös-Rényi random graph, the following conditions hold: $PL \approx PL_{rand}$ and $CC >> CC_{rand}$. The comparison between the interaction networks generated by the simulation model, and an ensemble of random graphs reveals that $PL_{model} = 1.98$ and $PL_{rand} = 1.99$ also $CC_{model} = 0.172$ and $CC_{rand} = 1.68$. This shows that the networks depicted in *Gammanym* simulation do not have the small-world properties.

5.2. Evolution of Networks

In this section we analyze the evolution of networks in *Gammanym*. As agents interact with each other, new connections form between them, while others are reinforced. Figure 3 shows the evolution of connectivity between agents within the simulation. The connectivity is defined as $L/N(N-1)$, where L is the total number of interactions within the system and N is the total number of agents within the simulation. Connectivity values are averaged over an ensemble of 100 runs. From Figure 3, it can be seen that the connectivity within the system grows as a function of $n \log n$ over time. An interesting property in Figure 3 is the sudden jumps in connectivity associated with conference events. The most notable of these occurs at time step 55 and appears to capture an element of many real world 'networking' events, where people come together to exchange ideas and experiences.

Initially the agents are only connected to doctors that they directly interact with, within their practices, defined as *contact groups*. Hence, the initial state of the system consists of a set of disconnected contact groups. This means that the initial flow of information within the system is limited to only small isolated pockets of interactions. However, as connectivity increases, the size and the nature of these contact groups changes.

In order to study the nature and structure of the contact groups, we define a *cluster* as a set of doctors belonging to the same contact group. We explore the structure and evolution of the professional network and then the combination of the professional and friendship networks[8] jointly, over the 68 weeks. At each time step within the simulation

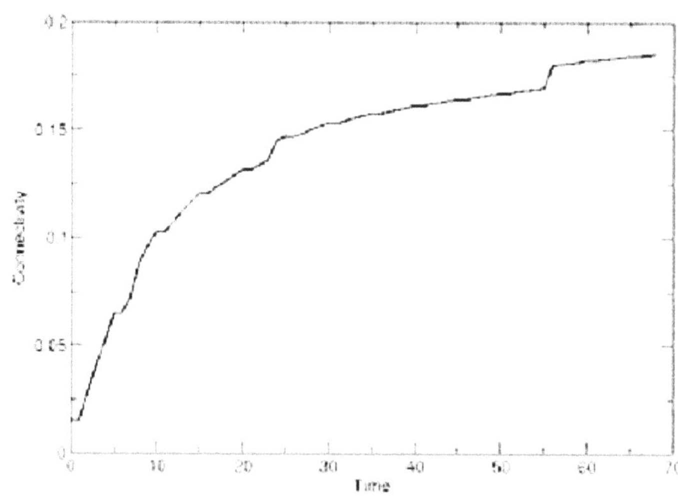

Figure 3. Evolution of connectivity within professional and friendship networks

[8] In *Gammanym* the specification for interactions with friends (see Section 1.3.2) implies that the friendship networks are static in nature.

we count the number of clusters of agents and calculate the average cluster size and the average shortest path length between agents within the system (Figure 4). All statistics are averaged over an ensemble of 100 runs. Figure 5 is a graphical illustration of network at the beginning i.e. time step =0 (Figure 5 (A, C)) and after the formation of the giant cluster at time step 10 (Figure 5 (B, D)).

In case of evolution of professional networks, we see that initially the simulation contains approximately 61 clusters (Figure 4 A), as defined by 61 practices over the spatial grid (Section 3.1). As new interactions occur, the number of clusters quickly decays. The average cluster sizes quickly explode as the agents become consumed by the giant cluster (Figure 4 B). The system begins to behave as one giant cluster after about 7–10 time steps (Figure 5 B). . We note that the average shortest path length between nodes initially increases rapidly (Figure 4 C), as more and more nodes become connected to the giant cluster. When the system becomes connected, such that there exists a path between all agents, the shortest path-length between any two nodes is, on average, relatively long. This is because of the sparsity of interaction matrices and the network containing many long paths. However, as the system increases in connectivity the path-lengths become smaller until there is approximately two degrees of separation between any two agents within the system (Figure 4 C). In summary, the system initially consists of a number of disconnected contact groups or clusters, but quickly evolves to form a single connected component or cluster while the size and structure of these clusters define how far and how quickly Tetracycline is adopted.

Figure 4 (D-F) represents evolution of clusters over time, when both professional and friendship networks are considered. Given the static nature of the friendship networks, the evolution for both the networks essentially signifies a similar process

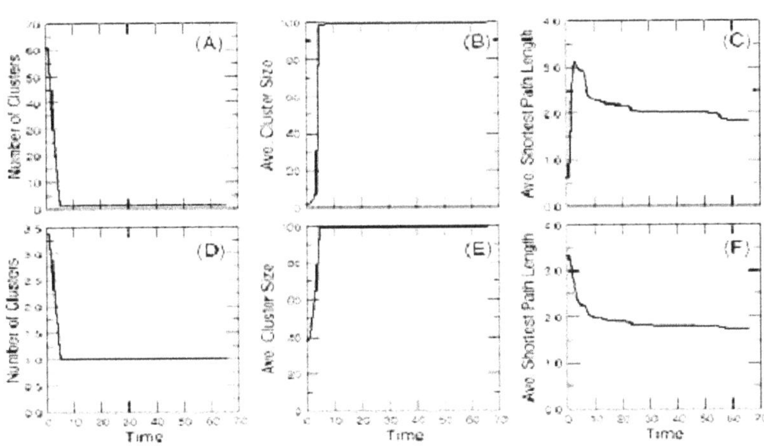

Figure 4. Network Statistics for professional networks- (A) Number of clusters; (B) Average cluster size; (C) Average shortest path length; Network statistics for professional and friendship networks- (D) Number of clusters; (E) Average cluster size; (F) Average shortest path length

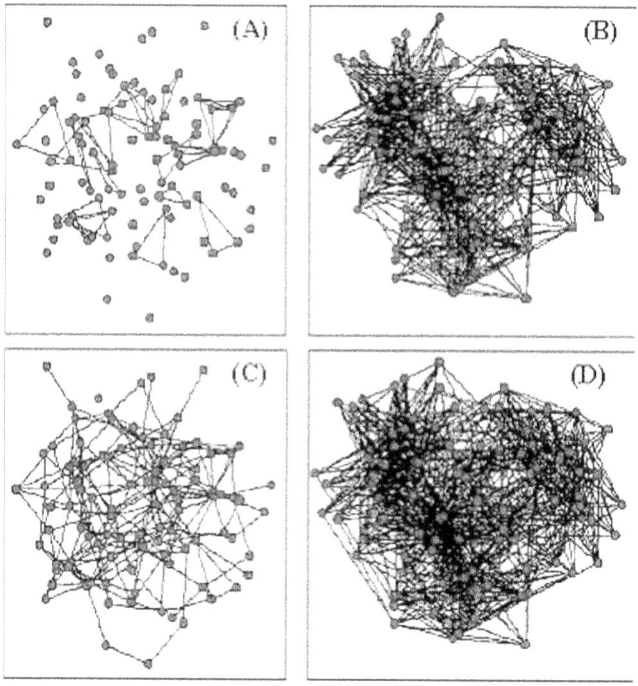

Figure 5. Formation of giant cluster: (A) Professional networks at t=0; (B) Professional networks at t=10; (C) Professional and friendship networks at t=0; (D) Professional and friendship networks at t=10

revealed in Figure 4(A-C) in case of professional network only. Noticeable, however, is the strong coalescent effect of the friendship relationships that drive the 61 initial clusters for professional networks down to approximately 3.5 clusters when friendship and professional networks are considered jointly (Figure 4 D). We also note that, in Figure 4 (F) average shortest path length between nodes falls over the time period. The evolution of network in this case, starts with a few giant clusters with long paths (Figure 5 C). As the connectivity increases through time the shortest path length, on average, becomes smaller. At the end of 68 time steps, the degrees of separation between any two agents are marginally smaller than in case of professional networks only (Figure 4 F).

5.3. Evolution of Uptake

To analyze the differences in the speed of diffusion depicted in Section 1.4, we look at how Tetracycline uptake evolves under three scenarios. We, therefore, define an *uptake cluster* as a set of agents who belong to the same contact group and each has already adopted Tetracycline. In this context the uptake cluster can be thought of as a *'cluster'* commonly encountered in percolation studies (Stauffer 1979). Figure 6 shows how the uptake of Tetracycline evolves through time.

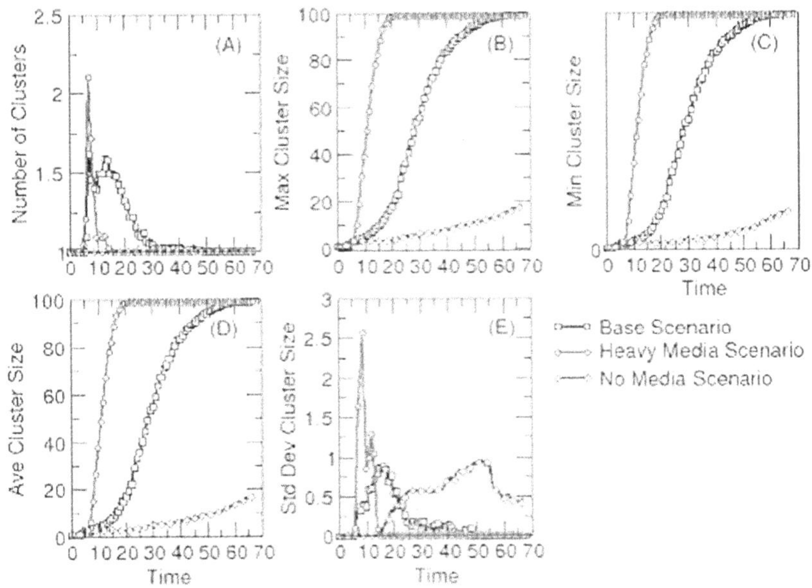

Figure 6. Uptake clusters for three scenarios: A) Number of clusters within the system; (B) Maximum cluster size; (C) Minimum cluster size; (D) Average cluster size; and (E) Standard deviation of cluster size

In the base scenario, starting from one seed (innovator) the average number of uptake clusters increases up to 1.6 at time step 15 (Figure 6 A). Then, it decreases to one giant uptake cluster (time step 30) as the size of the existing clusters increases gradually before merging. We also observe that the number of uptake clusters explodes rapidly under the heavy media scenario (Figure 6 A). Compared to the base scenario, much faster dynamics is displayed with the giant uptake cluster forming at time step 15 with heavy media exposure. At time step 20, 100% of the doctors have adopted the drug and the size of the uptake cluster equals the size of the social network. The integration scenario displays a very slow process of diffusion. On average, only one uptake cluster forms during the 68 time steps, and its maximum size barely reaches 20 by the end of the simulation (Figure 6 B). Thus, despite a very efficient deployment of the giant group, the diffusion struggles to spread across the whole network without external influence.

6. Integration and Diffusion

In this section we investigate: integration into which network matters more, for whom and at what stage. The previous sections analyzed the diffusion process for all doctors under different degrees of external influence. Given the existence of overlapping networks, this section attempts to disentangle the impacts of professional and friendship networks. For this purpose, we implement a normalized version of the model with uniform distribution of agents in terms of their professional integration. In this case, 101 doctors are initialized as: 33 in 33 private practices, 33 in 11 centers and 35 in 7 clinics. To distinguish between impacts of social and professional relationships, we select the most connected and least connected groups for comparison: i. both socially and professionally integrated, ii. integrated professionally, but socially isolated, iii. isolated professionally, but socially integrated, and iv. both professionally and socially isolated.

Given three types of office partnership and four types of friendship status, a range of definitions for professional or social integration or isolation is comprehensible. First, we define professionally integrated doctors as the ones with maximum number of colleague or have four colleagues in a clinic. Socially integrated doctors have maximum number of friends or three friends as specified in *Gammanym*. Isolates are the ones practicing alone or having no friends. The diffusion curves (Figure 7A) for the above groups of doctors are derived after averaging over 100 simulations for the benchmark scenario (with one initial adopter and one detailman).

Another set of adoption curves is derived by relaxing the definition of integration to friendship networks. The group of socially integrated doctors now includes doctors with two or more friends; and the socially isolated doctors are the ones with one or no friends (Figure 7B).

Normalized adoption curves, in both cases indicate that although underlying networks evolve in predictable ways, as has been analyzed in Section 5, the uptake is a function of initial starting conditions. Initial conditions, however, are determined by level of social or professional integration. Diffusion proceeds exactly the same way, for all groups of doctors, up to the formation of the giant cluster and after that, integration to different networks generates different impacts.

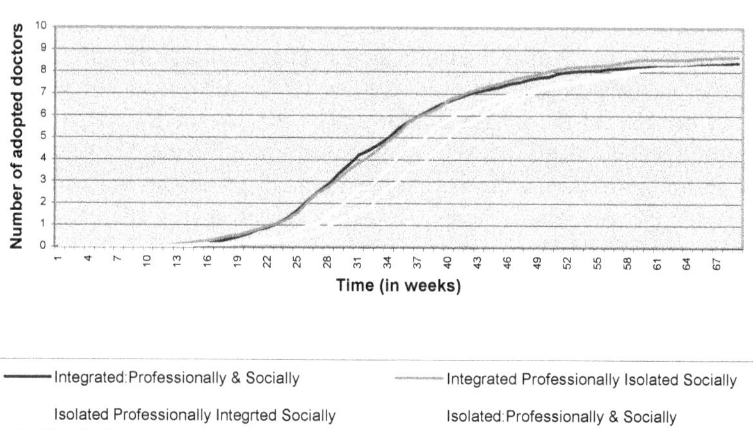

Figure 7A. Impact of integration on diffusion

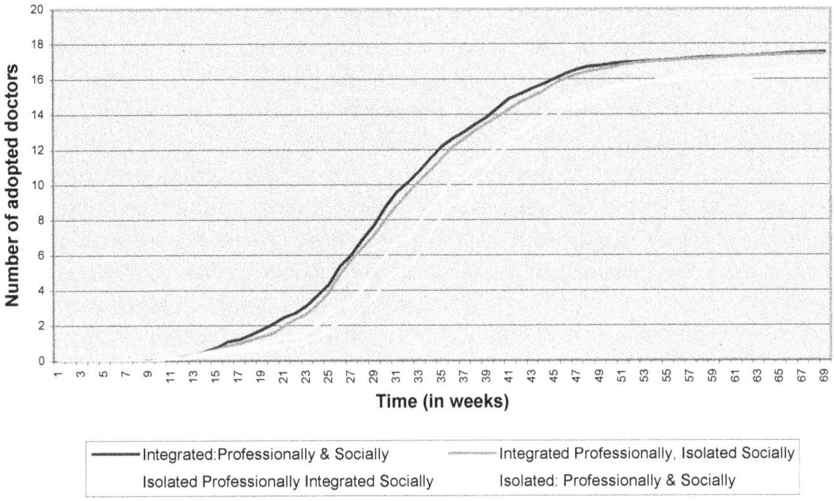

Figure 7B. Impact of integration on diffusion

Our results show that integration is crucial for initial uptake as the number of doctors, who are both socially and professionally integrated, is higher than for the rest of the groups. Though the speed of diffusion is not significantly different from the rest of the groups, diffusion among the least connected doctors, both socially and professionally isolated, is the slowest at the initial stage (Figure 7(A,B)). It is important to note that the simulation results for the two groups of doctors, where their integration status in two different networks is mutually reinforcing each other, replicate the findings in the original study.

Our results are rather interesting for two groups of doctors with contrasting integration status: integrated professionally, but socially isolated; and isolated professionally, but socially integrated. The degree of professional integration has a stronger impact. The similarity of adoption rates for group one (doctors with four colleagues and two or more friends) and group two (doctors with four colleagues, but with one or no friends) implies that, even if doctors are socially isolated, professional interactions enable them to adopt quickly (Figure 7(A,B)). Nevertheless, integration to a friendship network is more significant for doctors who are professionally isolated. Given the differences in the adoption rates between group three and group four doctors, we can imply that, to some extent, integration to friendship networks can make up for professional isolation. In other words, even if a doctor is practicing alone, social connections can work as a source of information.

7. Concluding Remarks

We develop an agent-based model, called *Gammanym*, to analyze the diffusion process. This is inspired by the classic *Medical Innovation* study on the adoption of Tetracycline in the Midwestern US in the 1950s by Coleman et al. (1966). Due to the limited availability of proper techniques/methods during the 1950s, the original study focused on interpersonal influence for pairs of individuals. In our study, we overcome this limitation of addressing complexity and dynamics of actual adoption using agent-based modeling. Our integrated modeling with *Gammanym* reveals original results within the existing literature of diffusion research, and also complements the extant work on *Medical Innovation.*

Our approach of integrated modeling enabled us to look at network properties such as connectivity and degree distribution, among others, in terms of the interaction matrices generated by the agent-based model. On the basis of these properties we determine that the interaction networks depicted in *Gammanym* are random graphs. Complexity of the diffusion process is explained by analyzing evolution of networks or dynamics on the networks. We find that initially the system consists of a number of disconnected components and quickly evolves, after 7–10 time steps, to form a single connected component. The analysis of network topology also indicates that underlying networks evolve in predictable ways, and the uptake is a function of the initial starting conditions.

Analysis of the evolution of uptake or adoption of Tetracycline enables us to disentangle the extent of different factors affecting adoption. Despite stressing the complementarity between network theory and diffusion research, a large body of diffusion literature has so far failed to examine the dynamic structures of the interpersonal networks and their evolutions over the diffusion process. Our model shows that although the media does not influence the network structure, it does have a major impact in accelerating the diffusion process. Under heavy media exposure undertaken by the pharmaceutical company to increase sales of Tetracycline, the average size of uptake group with agents who have adopted the new drug rises faster than otherwise. Moreover, all the agents adopt the new drug within 25 time steps, much earlier than with a baseline scenario of much less media exposure. We validate the previous evidence of a dominant media influence by comparing the speed of diffusion for three scenarios: baseline, heavy media and integration. The dominant role of media also suggests there is a trade-off between the efforts expended in promoting a new drug, and the speed of the diffusion process.

The comparison of the cumulative diffusion curves of *Gammanym* with those of *Medical Innovation,* is largely restricted because of unavailability of the original dataset. Nevertheless, our exploration with normalized adoption curves for groups of doctors, based on different degrees of social and professional integration, reveals that after the formation of the joint cluster, the initial uptake is dependent on the level of integration. Despite the predictability of evolution of networks, as suggested by network topology, this signifies the importance of social or professional ties for adoption decisions, as was outlined in the original study. Overall, our results provide a step towards understanding the complexity and dynamics of the diffusion process in the presence of professional and social networks.

References

Appold S J, 2004, "Research Parks and the Location of Industrial Research Laboratories: An Analysis of the Effectiveness of a Policy Intervention." *Research Policy, 33*. pp.225–243.

Balconi M, Breschi S and Lissoni F, 2004, "Networks of Inventors and the Role of Academia: An Exploration of Italian Patent Data." *Research Policy,33*. pp.127–145.

Barabási a L, 2002, *Linked: The New Science of Networks*, Cambridge, Massachusetts.

Becker W and Dietz J, 2004, "R&D Cooperation and Innovation Activities of Firms: Evidence for the German Manufacturing Industry." *Research Policy,33*. pp.209–223.

Burt R S, 1987, "Social Contagion and Innovation: Cohesion Versus Structural Equivalence." *American Journal of Sociology,92*. pp.1287–1335.

Doutriax J, 2003, "University-Industry Linkages and the Development of Knowledge Clusters in Canada." *Local Economy,18(1)*. pp.63–79.

Erdös P and Rényi A, 1959, "On Random Graphs." *Publcationes. Mathematicae Universatatis Debreceniessis,6*. pp.290–297.

Fritsch M and Franke G, 2004, "Innovation, Regional Knowledge Spillovers and R&D Cooperation." *Research Policy,33*. pp.245–255.

Grafton R Q, Knowles S and Owen D P, 2004, "Total Factor Productivity, Per Capita Income and Social Divergence." *Economic Record,80(250)*. pp.302–313.

Howells J R L, 2002, "Tacit Knowledge, Innovation and Economic Geography." *Urban Studies,39(5–6)*. pp.871–884.

Mahajan V and Peterson R A, 1985. *Models of Innovation Diffusion*, Newbury Park, CA: Sage.

Rogers E M, 1995, *Diffusion of Innovations*, New York: The Free Press.

Ryan B and Gross N C, 1943, "The Diffusion of Hybrid Seed Corn in Two Iowa Communities." *Rural Sociology, 8(1)*. pp.15–24.

Stauffer D, 1979, "Percolation." *Physics Reports, 54*. pp.1–74.

Strang D and Tuma N B, 1993, "Spatial and Temporal Heterogeneity in Diffusion." *The American Journal of Sociology,99(3)*. pp.614–639.

Valente T W, 1995, *Network Models of the Diffusion of Innovations*, Cresskil, New Jersy: Hampton Press, Inc.

Van Den Bulte C and Lilien G., 1999, "A Two-Stage Model of Innovation Adoption with Partial Observability: Model Development and Application", Working Paper: 1–44.*Institute for the Study of Business Markets, The Pennsylvania State University.*

Van Den Bulte C and Lilien G, 2001, "*Medical Innovation* Revisited: Social Contagion Versus Marketing Effort." *American Journal of Sociology, 106(5)*. pp.1409–35.

Watts D J, 2003, *Six Degrees: The Science of a Connected Age*, New York: W. W. Norton Company.

Watts D J and Strogatz S H, 1998, "Collective Dynamics of 'Small-World' Networks." *Nature, 393*: 440–442.

PART IV

Methodological Issues in Research on Complex Decision Making

Chapter 14

Measuring Change in Mental Models of Complex Dynamic Systems

James K. Doyle and Michael J. Radzicki
Department of Social Science and Policy Studies
Worcester Polytechnic Institute
doyle@wpi.edu
mjradz@wpi.edu

W. Scott Trees
Department of Economics
Siena College
trees@siena.edu

As scientists who are interested in studying people's mental models, we must develop appropriate experimental methods and discard our hopes of finding neat, elegant mental models, but instead learn to understand the messy, sloppy, incomplete, and indistinct structures that people actually have.

Donald A. Norman [1983, p. 14]

1. Introduction

A complex dynamic system can be thought of as a collection of interrelated variables whose structure determines the behavior of the system over time. The more variables that are involved and the more highly interconnected they are, the greater the complexity of the system. In addition to tight coupling of variables and a tendency toward self-organization, the features that give rise to dynamic complexity include the existence of feedback mechanisms, nonlinear relationships between variables, irreversible processes, adaptive processes, and time delays [Sterman 2000].

Human failure to control the behavior of such dynamically complex systems in desired ways has been demonstrated by case studies and experiments on human subjects in a wide variety of domains, including urban planning [Forrester 1969; Dörner, 1996], inventory control [Sterman 1989], resource management [Moxnes 1998], medical care

[Kleinmutz and Thomas 1987], and forest fire control [Brehmer 1989]. In all of these domains experimental subjects have been found to have great difficulty anticipating side effects and predicting the long-term impacts of their decisions, often resulting in severe negative consequences for the human actors in the system. Another common finding in these studies is the inadequacy of subjects' internal mental representations of the systems they are trying to control, often referred to as "mental models."

It should be noted here that the term "mental models" has a long and varied history and has been described differently by various authors. It has also been used by different academic disciplines in various ways, sometimes to indicate the entire range of mental representations and cognitive processes, and sometimes to indicate a specific subset of mental phenomena (e.g., images, assumptions, generalizations, perceptions). The present work adopts a definition of "mental models of dynamic systems" suggested by Doyle and Ford [1998] for use in research on complex dynamic systems:

> "A mental model of a dynamic system is a relatively enduring and accessible, but limited, internal conceptual representation of an external system whose structure maintains the perceived structure of that system."

Thus, for the purposes of this chapter the term mental model indicates a particular type of mental representation that is an attempt to mentally replicate the structure and relationships of a complex dynamic system. It is proposed that such mental models are developed naturally through experience with systems and that they play an important role in guiding dynamic decision making [Doyle et al. 2002]. Further, it is assumed that people have conscious access to these mental models and can articulate them through a knowledge elicitation process. However, it is important to distinguish between the output of such an elicitation process and the mental models themselves, which exist only in people's minds and cannot be directly observed. The phrase "measuring change in mental models" in the title of this chapter was carefully chosen to recognize the difficulty in externalizing mental models with any degree of certainty. You can never be sure that the external representation, whether in verbal or diagrammatic form, is a complete and accurate representation of the internal mental model on which it is based. In fact, the external representation obtained is likely to be highly dependent on the particular knowledge elicitation procedure that was employed, and the process of measurement is subject to error. But it is assumed that a reasonably accurate measurement of *change* in a mental model can be obtained so long as identical procedures are used each time knowledge is elicited and the procedures are valid and unbiased.

Although different researchers may disagree about how to define mental models, there is substantial agreement on their characteristic shortcomings. Many researchers conclude that mental models are often oversimplified and ill-defined, typically fail to adequately incorporate important features (e.g., feedback mechanisms, nonlinear relationships, and time delays) of the real system, and are subject to various forms of bias and error [Doyle and Ford 1998].

Acknowledging these shortcomings of mental models and the important role they play in dynamic decision making, the system dynamics computer modeling tradition founded by Jay Forrester [1961] (see Sterman [2000] for an overview of the field) adopts the goal of improving mental models as one of its primary aims. It is proposed that the act of building a system dynamics simulation model (or learning by interacting with a model built by someone else) can help decision makers overcome cognitive limitations as well as various external barriers to learning in the real world (e.g., time delays, irreversible actions), allowing mental models to become more dynamically complex. A wide variety of system dynamics and systems thinking interventions have been developed and implemented over the years, many with the explicitly stated goal of changing the mental models of participants to make them more complete, complex, and dynamic in order to improve their ability to manage a complex system [e.g., Vennix 1990, 1996; Cavaleri and Sterman 1997; Huz et al., 1997].

Of course, in order to judge the effectiveness of these interventions in promoting learning, participants' mental models must be elicited, organized, represented externally, and compared before and after the intervention, whether formally or informally. Toward this and other ends system dynamics and systems thinking researchers typically apply one or more of a wide variety of formal techniques for organizing and representing mental model information, including system flow diagrams [Forrester 1961; Morecroft 1982], causal loop diagrams [Richardson and Pugh 1981], various forms of influence diagrams [Axelrod 1976; Coyle 1977; Eden and Jones 1984], hexagons [Hodgson 1992], and social fabric matrices [Gill 1996]. Researchers have also developed several distinct sets of procedures and methods for eliciting or mapping mental models, typically during facilitated group sessions, including the Strategic Options Workshop [Eden and Huxham 1988], the Strategic Forum [Richmond 1987], the corporate system modeling policy session [Roos and Hall 1980], and the group model building approach described by Vennix [1996].

However, these established techniques were not originally designed primarily to measure mental models but to facilitate change and improvement in mental models. In fact, the very features that make them valuable for changing mental models (the introduction of new, systematic ways of thinking about mental models, the assistance and direction provided by the facilitator, and the consensus achieved during group processes) simultaneously make them unsuitable for measuring that change in an accurate and unbiased way. For example, the introduction of new and unfamiliar ways of thinking about mental models may cause participants to change their mental models during the elicitation procedure, masking their pre-intervention content and structure. And, the involvement of the facilitator and other group members during elicitation procedures introduces a host of potential ways in which the mental models of others can interfere with the elicitation of the mental model of any particular individual.

These criticisms by no means imply that systems interventions based on existing methodologies for eliciting and representing mental models are not effective in promoting learning; it simply means that their effectiveness cannot be demonstrated beyond a reasonable doubt by current practice. The resulting inability to judge the relative effectiveness of different interventions with a high degree of confidence likely inhibits the ability of individual researchers to learn from experience and the ability of the research field as a whole to learn by comparing the experiences of different research teams. (Indeed, the very existence and use of so many different methods and procedures for eliciting, representing, and changing mental models of systems suggests that the difficult work of documenting their comparative advantages and disadvantages has yet to be done.)

The goal of accurate, unbiased measurement of changes in mental models of complex dynamic systems can only be fulfilled by a program of controlled, rigorous experimental research designed to supplement and support more realistic field studies [Doyle 1997]. Such an effort will require the development and testing of new methods and procedures that emphasize accuracy of measurement of mental models rather than methods and procedures that emphasize facilitation and improvement of mental models and that can be adapted to test multiple hypotheses related to the relative effectiveness of alternate intervention protocols. The purpose of the present chapter is to define the necessary features of *any* methodology that aims to rigorously measure change in mental models of complex dynamic systems (see also Doyle et al. [2002]); to describe the development and implementation of one specific new methodology designed to fulfill these criteria; and to present the results of an exploratory application of the method to measuring changes in mental models due to a system dynamics intervention based upon the simulation game of the economic long wave, or Kondratiev cycle, developed by Sterman and Meadows [1985].

2. Experimental Design and Procedure for Measuring Change in Mental Models

The most appropriate and accurate techniques for measuring change in mental models of complex dynamic systems have yet to be established by the research literature [Vennix 1990], and there is a demonstrated need for research programs that will "make more precise and less artful the process of eliciting and mapping knowledge" [Richardson et al. 1989, p. 355]. Thus, it is not yet possible to prescribe the use of specific measurement instruments or protocols, and researchers should be encouraged to conduct exploratory work that tests alternate measurement techniques drawn from different literatures and research traditions. However, the more general requirements for the design and conduct of rigorous research on mental models and learning are well known in the psychology and education literatures and are largely agreed upon. Although there is room for researchers to exercise choice in how to operationalize these requirements, we believe that any method for measuring change in mental models that aims to be rigorous must strive to achieve at least the following nine goals:

1. *Attain a high degree of experimental control.* In designing any study of human cognition or behavior, choices must be made that affect the degree to which the study emphasizes experimental control (the ability to hold variables other than the one under examination constant) and external validity (the extent to which the observed results also apply to realistic settings outside the context of the study). Usually, but not always, a methodological choice that increases external validity decreases experimental control, and vice versa. For example, a study examining the effect of a system dynamics intervention that emphasizes realism might want to engage managers in a thorough, perhaps months-long examination of an important, real problem that affects the future of the company, in a setting that incorporates the incentives for performance, time pressures,

and accountability of real business settings. In such a study, however, one would have great difficulty controlling important variables in the face of the other priorities of the company and the unpredictability of external events and would not be able to rule out alternate possible explanations for results. In a study where the emphasis is on experimental control, one might instead choose to engage a convenient sample of subjects (e.g., undergraduate students) randomly assigned to experimental conditions in a simplified, brief examination (perhaps lasting a day or a week) of a problem in a somewhat artificial setting devoid of the complications of realistic intervening variables. In this study, at the expense of raising questions about the applicability of its findings to realistic settings, the researcher would be in a much better position to unambiguously determine the causes of observed changes in thought and behavior. Of course, both types of study are valuable, important, and necessary. But rigorous studies that emphasize experimental control are particularly lacking in the systems thinking field and are unavoidable if questions about the ability of systems thinking interventions to change mental models are to be answered with a high degree of confidence.

2. *Separate measurement and improvement.* Any study that intends to assess the cognitive effects of a system dynamics intervention must attempt to both measure and improve mental models. However, it is important that these goals be separated; for example, if the first technique participants use to express their mental models is one that is thought to increase the degree of organization in mental models or to encourage completeness, then the first mental models elicited will not represent a true pre-intervention benchmark. Measurement and improvement of mental models should occur through distinct and separate procedures that take place during different experimental sessions.

3. *Collect data from individuals in isolation.* Group sessions coordinated by a facilitator are an important part of many systems interventions. However, there are several problems that make it difficult to accurately elicit the mental models of individuals in such settings. First, group discussions tend to be dominated by a few individuals and participants may fail to share ideas and opinions due to the effects of social loafing [Latané et al. 1979], evaluation anxiety [Guerin 1986], or cognitive distractions [Baron 1986]. Second, in public forums people often comply with the views of others, while keeping their private opinions to themselves, in order to obtain rewards or avoid punishments [Kelman 1958]. Third, facilitators can inadvertently give participants clues about what ideas they believe to be better than others or lead discussions in a direction they favor rather than the direction the participants would choose on their own, resulting in what psychologists call "experimenter bias" [Rosenthal 1966]. To avoid these problems, measurement procedures should be conducted in a setting that ensures confidentiality and effectively isolates participants from the influence of each other and any facilitators.

4. *Collect detailed data from the memory of each individual.* Mental models reside in the minds of individuals, and it is not possible to unerringly infer the contents of individual mental models without a detailed examination of the memory of each individual participant. For example, although an individual during an intervention may

express agreement with statements made by other participants, or may indicate acceptance of a mental model representation developed by a group, it is possible that the relevant ideas are only held in memory in a temporary state. If so, the ideas may be forgotten or may be replaced by prior or subsequent information rather than become incorporated into mental models held in long-term memory. To control for this possibility, each individual participant should be asked to generate completely from memory their full mental model, in all of its messy, often fragmented detail, both before and after an intervention.

5. *Measure change rather than perceived change.* It is often tempting for researchers evaluating systems interventions to assess mental changes simply by having participants look inward and describe the effect the intervention has had on their mental models. However, there are serious problems with accepting such evidence at face value, including the possibility that participants may simply report what they think the researcher wants to hear, a phenomenon which psychologists call "subject bias" [Orne 1962]. Over-reliance on self-evaluation should be avoided: changes in mental models should instead be inferred by the researcher from a comparison of controlled pre- and post-intervention measures.

6. *Obtain quantitative measures of characteristics of mental models.* Efforts to improve systems thinking often fail to define the specific changes in mental models they hope to bring about. When this occurs, researchers are forced to judge the magnitude of observed changes in subjective, unsystematic, and possibly idiosyncratic ways. In addition, when dependent variables are defined post hoc, bias may result from focusing only on those measures that confirm expectations. To avoid these problems, researchers should explicitly define a priori such characteristics of mental models as detail complexity and dynamic complexity [see Senge 1990] and precisely how they will be quantified.

7. *Employ a naturalistic task and response format.* Research on human cognition suggests that memory and decision making are largely constructive processes. Which information is recalled from memory depends to a substantial degree on precisely how memory is measured [Roediger et al. 1989], whether the task being performed during retrieval is similar to the task performed during learning [Moscovitch and Craik 1976], whether external aids are used [Intons-Peterson and Newsome 1992], and subtle characteristics of the surrounding environment [Tulving 1983]. Similarly, the mental models and processes used in decision making are often highly variable depending on task characteristics, goals, response modes, and even seemingly inconsequential changes in the way questions are ordered, worded, or framed [Kahneman and Tversky 1984; Hogarth 1982; Payne et al. 1992]. This implies that researchers that elicit mental models with a new or unfamiliar task run the risk of measuring different mental models than the ones participants use when free to follow their more typical habits of thinking, deciding, and problem solving. The degree to which an elicitation task encourages participants to think systematically or to exhaustively examine all of the relationships between relevant variables is also

important and should be appropriate for the level of expertise of the participants. If, for example, the task encourages more effortful thinking than participants normally engage in, the study may end up measuring new, transient mental representations constructed during the elicitation procedure rather than the more durable mental models participants formed before the intervention. To avoid these problems, elicitation procedures designed for measurement rather than improvement should use tasks and response modes that approximate as closely as possible the way participants naturally go about representing and conveying their knowledge.

8. *Avoid bias*. Ideally, the elicitation procedures should be designed without making any a priori assumptions about the form or content of subjects' mental models. For example, asking how variable A is related to variable B may induce subjects to add variables and relationships that didn't exist in their mental model prior to answering the question. Asking subjects to identify feedback loops assumes that there are feedback loops in their mental models, which may or may not be true. Although the systems in which people operate are often complex and dynamic, their mental models of these systems may or may not be. The elicitation procedure should allow for the possibility that subjects' mental models, particularly novices, are relatively simple and static.

9. *Obtain sufficient statistical power*. The paradigm of controlled research on human subjects requires that sufficient numbers of participants be studied to allow hypotheses to be tested for statistical significance. Given that the magnitude of the effect of system dynamics interventions on various characteristics of mental models is not yet known, data should be collected independently from enough participants to allow the detection of even moderate to small changes in mental models.

3. Prior Research

The only prior research program conducted within the system dynamics community that meets all nine of the identified criteria for rigorous research on measuring change in mental models was described by Vennix [1990]; see also Verburg [1994]. In an ambitious, well-conceived, and thoroughly documented study, Vennix conducted a controlled experiment testing the effect of an intervention based upon a computer simulation of the Dutch social security system on several quantifiable features of mental models. One hundred and six college students participated in one of two sequences of experimental sessions involving a 40-hr. commitment over a 7-week period. Pre- and post-intervention mental models were elicited by asking subjects to prepare individually a two-page written policy note addressing a problem involving social security and the economy. These policy notes were subsequently coded into "cognitive maps," or directed graphs, following the procedures developed by Axelrod [1976]. Results showed that the intervention resulted in the following statistically reliable changes in mental models: an increase in the number of relationships that were quantified; an increase in the proportion of computer model concepts included (subjects' models, however, remained quite simple compared to the complexity of the computer model with which they interacted); an increase in the number of relationships between concepts; and an increase in the number of mentions of time delays. The intervention had no statistically reliable effect on the average length of paths or the number of feedback loops in the cognitive maps.

The present work applies the same general experimental approach to address a subset of the questions about the effect of a simulation-based intervention on mental models explored by Vennix [1990], and therefore can serve as a conceptual replication that may corroborate or qualify some of the conclusions of that work. However, to facilitate comparisons between the two studies, it is worth noting the following important differences:

1. Vennix [1990] tested the effects on mental models of interaction with an econometrics-based simulation. The present study tests the effects of interaction with a system dynamics-based model, which might be expected to better promote feedback thinking.
2. The intervention we tested is much briefer and simpler than the one tested by Vennix. While this limits the potential impact of the intervention and decreases external validity, it allows a higher degree of experimental control than longer, more complicated interventions. For example, unlike Vennix [1990], in the present work all data collection procedures were supervised and the time investment of subjects was kept constant. One of the goals of the present research is to develop a more practical, less time-consuming, yet rigorous methodology for measuring changes in mental models that will allow research results to accumulate more quickly. Toward this end we are interested in determining if statistical reliable changes in mental models can be obtained by a relatively brief intervention.
3. Both studies rely on a detailed content analysis of written documents that subjects are asked to produce in response to a set of questions. However, the present study borrowed its elicitation and coding procedures not from Axelrod [1976] but from a research tradition in cognitive psychology that holds that knowledge and beliefs are organized in memory in narrative or story-like structures that are variously termed narrative models [Bower and Morrow 1990], scripts [Schank and Abelson 1977], schemas [Fiske 1993], or, simply, stories [Schank 1990].[1] Research by Pennington [1981] and Pennington and Hastie [1986, 1988] has confirmed that these story-like structures are spontaneously constructed and used to guide decision making when judgments are based on large amounts of interrelated information or experience that must be reviewed and organized. Thus the present study attempts to achieve naturalism by asking participants to convey their mental models the way they are typically conveyed in conversation: by creating a causal explanation or scenario that explains the available evidence.

[1] The assumption that the mental models of the participants in the present study will be based upon the specific, concrete events and relationships characteristic of stories rather than more abstract concepts and variables is consistent with the results of several studies that have reported the mental models of novices to be more representational and less abstract than those of experts [Larkin 1983; Chi et al. 1988].

4. In the Vennix [1990] experiment, subjects conveyed their mental models by writing a two-page essay that presumably took several hours, during which time subjects could refer to an introductory text on the topic at hand. Although this task is realistic and has the advantage of encouraging thoroughness, it is not clear whether the mental models derived from it represent durable models held in long-term memory. For example, a subject could include in the essay ideas and concepts he or she has read just minutes before but has not really learned and incorporated into mental models. The present study takes a different approach, giving subjects much less time to write a briefer essay without access to any reference materials, in order to ensure that the ideas being conveyed are coming from long-term memory.
5. In the Vennix [1990] experiment, subjects were asked to read a 25-page introductory text before pre-intervention mental models were measured. This allows for a much stricter test of the effectiveness of the intervention, as pre/post differences will reflect changes in mental models over and above any changes caused by reading the text. However, this also means that the more naive mental models subjects held before reading the text were not measured. Such naive mental models can be important, as several empirical studies have found that they tend to persist and influence behavior even after instruction in "correct" models [e.g., DiSessa 1982; Clement 1983; McCloskey 1983]. In the present study, we chose to compare post-intervention mental models with the naive mental models, based on life experience, that subjects held before engaging in any activities related to the experiment.

4. Method

4.1. Subjects

Sixty-four undergraduates enrolled in an introductory social science course at Worcester Polytechnic Institute participated in the experiment in order to fulfill a course requirement. Forty-six of the 64 students completed the experiment; the other 18 were dropped from the study due to failure to attend one or more of the experimental sessions or to participation in pretests of the experiment. The students were almost exclusively science and engineering majors taking the course to fulfill a breadth requirement and they had little or no prior exposure to economics, management, or system dynamics at the college level. Forty-six percent of the students were female. Thirty-eight percent of the students were seniors; 38% were juniors, 17% were sophomores, and 4% were freshmen. The students were assured that their responses would be completely confidential and that, although they were required to participate, their performance in the experiment would not affect their course grade in any way.

4.2. Design

Subjects participated in a system dynamics intervention structured around their experience playing STRATEGEM-2, a simulation game of the Kondratiev cycle, or economic long wave, developed by Sterman and Meadows [1985] and employed in experiments on dynamic decision making by Sterman [1989]. The purpose of the game is to help students and managers learn about and gain confidence in a simplified behavioral theory of the causes of long-term cycles of overexpansion and depression in the economy [Sterman 1985]. According to the theory, which focuses on the capital-producing sector of the economy, management investment decisions lead to overexpansion due to time delays in production and the reinforcing nature of capital self-ordering. This overexpansion leads to a subsequent depression as excess capital slowly depreciates over time. The goal of the intervention was to encourage participants to develop mental models that are more sophisticated in terms of both detail complexity and dynamic complexity [see Senge 1990] and that include important elements of the expert model, for example, the recognition that (1) the causes of the long wave are largely internal to the economic system rather than external; (2) it is the structural characteristics of the system (capital self-ordering, time delays) that largely determine the behavior of the system; and (3) periods of economic expansion and depression are causally connected.

Mental models of the causes of the economic long wave were measured by administering identical survey instruments before and after the intervention. [2]

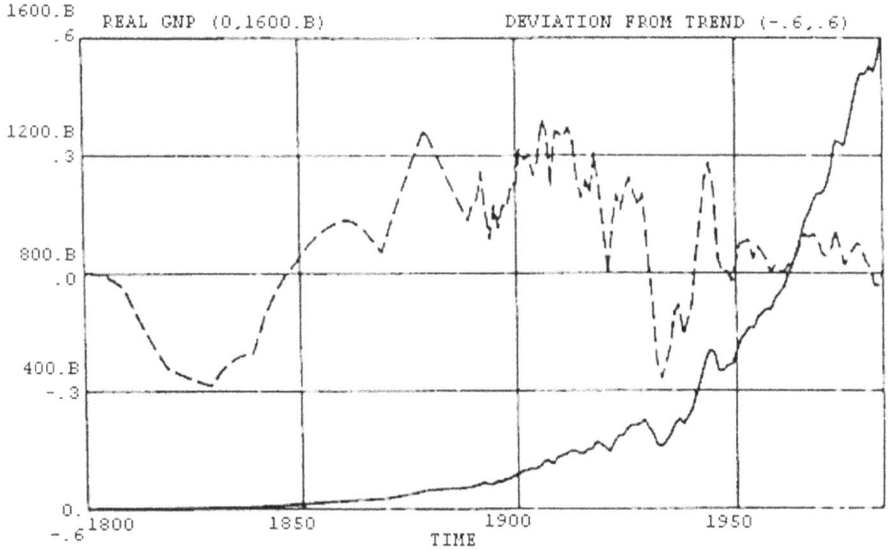

Figure 1. Percentage deviation from the exponential growth trend of real GNP (adjusted for inflation) for the U. S. economy from 1800 to 1984 (1972 dollars). Reprinted from Sterman [1986]

[2] It should be noted that, as in Vennix [1990] and most studies of systems thinking interventions, the present study is limited by lack of access to a traditional control group. That is, there was no group of subjects studied concurrently who did not participate in the intervention. Therefore the possibility that any measured changes in mental models are due to events external to the experiment, although remote, cannot be completely ruled out.

However, two different experimental conditions were created: only half of the subjects were assigned, at random, to receive the preliminary survey in order to control for the possibility of pretest effects, that is, the possibility that the act of taking a pretest can itself change mental models and behavior independently of the effects of any intervention. For example, those subjects who take a pretest are made more cognizant of the fact that they are being studied and tested and may therefore exert more effort than others, increasing the effectiveness of the intervention. A more likely possibility is that subjects who express their mental models during a pretest may feel compelled to defend them later on and dismiss new ideas, decreasing the effectiveness of the intervention. This issue was chosen for experimental study due to its implications for future research. If significant pretest effects are found and confirmed, for example, then costly controls for pretest effects may have to become a standard part of methodologies designed to document change in mental models.

The mental models surveys began by introducing subjects to the concept of gross national product (GNP) and presenting reference mode data: a graph displaying deviations in the trend of real U. S. GNP from 1800 to 1984 (see Fig. 1). After receiving instruction in how to interpret the graph, subjects were told that some researchers had identified in these and other economic data a cyclical pattern of expansion (e.g., in the 1850s, 1900s, and 1940s) and depression (or recession) (e.g., in the 1830s, 1880s, and 1930s) that recurs about every 50 years. Subjects were then given the following prompt to elicit from them a causal narrative or "story" that would explain the data:

> Explain, in a paragraph or more, your best theory of the causes of the 50-year cyclic pattern in the GNP data. What do you think caused these changes in GNP? Use the space below to "tell the story" behind the pattern in the data, including important events, factors, and variables and the relationships between them. Rather than simply labeling the depressions and expansions, try to explain them by thinking back to the ultimate basic factors in the economy or society that caused them.

Subjects were given two additional prompts to elicit further information. They were asked, "What do you think caused the period of depression between the years 1929 and 1933?" and "What do you think caused the period of expansion between the years 1933 and 1944?" The prompts were kept simple and open-ended to avoid providing clues to subjects about which events or variables might be relevant and to discourage subjects from reaching beyond their knowledge to "guess" at how variables might be related. Within the allotted time, subjects could decide for themselves how much or how little to write in answer to the prompts. Each prompt was followed by a question asking subjects to indicate, on a 1 to 7 scale, how confident they were that their explanation was correct.

4.3. Procedure

The experiment was conducted during 5 separate class sessions spread over a two-week period. On the first day, half of the subjects were randomly selected to complete the mental models pretest. During this time the remaining subjects completed a different survey that was unrelated to the present experiment. Subjects were allotted 25 minutes to complete the surveys individually under strict supervision.

On the second day, subjects received verbal and written instructions, based on the recommendations of Sterman and Meadows [1985], covering the purpose and operation of the simulation game, which was implemented in *Powersim*. The instructions included definitions of all economic terms, a detailed presentation and explanation of the structure of the game (see Fig. 2), and explicit definitions of the player's goal and how performance would be measured.

Subjects, in small groups of 3 to 10, participated in a 1-hr. session in a microcomputer laboratory on the third day. As in Sterman [1989], the game began in equilibrium and there was a simple step-function change in exogenous orders from the consumer goods sector from 450 to 500 units; the object of the game was to respond to this external shock and return the system to equilibrium. Working individually, subjects completed one play of the simulation game (36 periods) and submitted their printed data. During the game subjects had access to the structure diagram; graphs showing changes over time in their orders, production capacity, and desired production; and information on changes over time in all variables in tabulated form. Two monitors were present at all times to answer questions about the structure and operation of the game The monitors also closely supervised the session to ensure that there was no communication between subjects and that subjects turned in the

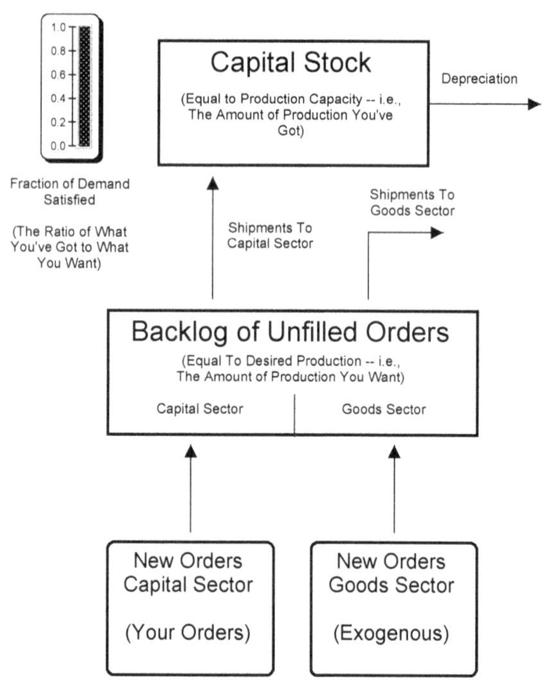

Figure 2. Diagram of the structure of STRATEGEM-2 presented to subjects. Adapted from Sterman and Meadows [1985]

results from their first play of the game.[3] As previously reported by Sterman [1989], subjects' performance in the simulation game was quite poor compared with the optimum possible performance. The mean score was 1806 (SD = 1085),[4] compared to a mean of 591 (SD = 1,176) reported by Sterman [1989] and an optimal score of 19.[5] Eighty-one percent of the subjects generated an oscillatory wave pattern in production capacity, despite the simple step-function change in exogenous orders; only 17% of subjects were able to reestablish equilibrium before the game was over. The game thus achieved its goal for the great majority of subjects, namely, to illustrate that the long-wave pattern could result solely from decisions made by managers of the capital sector of the economy.

The fourth day consisted of a 50-minute debriefing session led by a facilitator that followed, with some modifications, the procedure outlined in Sterman and Meadows [1985]. Subjects were shown typical examples of results from their own experimental sessions and were engaged in a group discussion concerning the thoughts and feelings they experienced while playing the game. The facilitator emphasized that poor scores on the game were not due to factors outside the players' control, since the structure and rules of the game and the state of the system were fully known. Subjects were asked to guess the pattern of orders from the consumer goods sector, and most suggested that there was probably a cyclic pattern in the exogenous orders. They were then shown the simple step-function that the simulation employed in order to emphasize the point that it was their own decisions that produced the cyclic wave pattern. At this point, the topic of the economic long wave was introduced and related to the simulation game, and subjects were shown examples of long-wave patterns in several different types of economic data. Summary results from the mental models pretest were presented to stimulate discussion about the group's pre-intervention mental models concerning long-term patterns in the economy. The facilitator presented data that contradicted some of the assumptions of subjects' pre-intervention mental models and asked the group to discuss the data. The facilitator then presented and explained, via causal loop diagrams, the simplified expert model of the economic long wave described by Sterman [1985], and led the group in a discussion of it. Finally, the facilitator closed the session by presenting an argument that the causes of the wave patterns in the simulation game were also a plausible explanation of similar patterns in the real economy.

On the fifth day the mental models post-test was administered to all of the subjects. This session was scheduled several days after the debriefing session in order to reduce recency effects (i.e., to reduce the chance of eliciting transient, unstable mental models). The post-test was administered in the same setting and employing the same procedures and time limits as the pretest.

---

[3] Pretests of the experiment determined that this level of supervision was, in fact, necessary. As first reported by Sterman [1989], in these pretests several subjects were so highly motivated to perform well that they attempted various forms of cheating.

[4] Four clear outliers (scores in excess of 20,000) were removed from this analysis.

[5] The difference in mean scores between the present study and the Sterman [1989] study is likely due to the differing degrees of subject matter expertise held by participants in the two studies.

4.4. Data Analysis

4.4.1. Content Analysis

The methodology employed in this study creates large amounts of verbal data which are often messy, incomplete, redundant, and idiosyncratic and which must be reduced, organized, and coded in a consistent and unbiased way. There are many different existing techniques for coding such data into diagrams or "cognitive maps" composed of concepts (or "nodes") and the relationships between them ("links"). The present study adopted a simplified version of a "causal chain analysis" coding scheme developed by Pennington [1981] based on Schank's [1972, 1975] conceptual dependency theory of causal relationships in natural language. In this type of analysis the nodes in a cognitive map of a mental model are not represented as abstract variables but as the concrete events that comprise stories or narratives. The nodes are organized temporally and connected with unidirectional, unsigned links to indicate that one event causes, enables, results in, or can lead to a second event. [6]

The coding process began by having an expert coder read through the entire data set to create a comprehensive list of all of the events described by subjects, resulting in a list of 103 events.[7] The experimental data were then coded while blind to experimental condition. The resulting lists of linked events were compiled into causal scenario diagrams to facilitate the coding of quantitative variables.

4.4.2. Quantitative Variables

Several quantitative variables were created based on the characteristics of the causal scenario diagrams, following the recommendations of Vennix [1990] when applicable. The content of mental models was quantified by calculating the percentage of subjects in each experimental condition who included each event in their narrative. Pre/post differences in these percentages were then subjected to a chi-square analysis. Since one of the goals of most systems interventions is to move the participants toward a "shared understanding" or "shared mental model," a measure of the degree to which mental model content was shared by the participants was created by computing the average percentage of subjects who included the most often-mentioned events in their narratives. This measure indicates the extent to which subjects are converging on a small number of important events versus holding competing mental models that include widely divergent events.

According to Senge [1990], mental models can exhibit two different kinds of complexity: detail complexity and dynamic complexity. Detail complexity is simply the amount of content, for example, the number of nodes and links. In contrast, dynamic complexity indicates the presence of feedback thinking and an understanding of other important system dynamics concepts (e.g., that the causes of events are often

---

[6] Unlike Pennington [1981], in order to simplify the analysis no distinctions were made between different types of events or links in the coding process.

[7] This approach, in which the coding categories are created from the experimental data itself rather than an independent source, is exploratory rather than confirmatory [see Carley and Palmquist 1992]. Such an approach is necessary when, as in the present case, the full set of concepts used by subjects cannot be predicted a priori.

remote in time and space from their effects). In this study detail complexity was assessed by counting the number of events and links in subjects' causal scenario diagrams. In addition, the number of links per event was calculated to control for the fact that an increase in the number of events can increase the number of links without increasing the extent to which the diagram is interconnected. As an additional measure of detail complexity, the average length of causal paths extending back from the primary to-be-explained events (changes in GNP and the occurrence of economic expansions or depressions) was calculated. Since subjects varied widely in the number of events they included, the average causal path length was divided by the maximum possible length for each subject (the number of events in the diagram – 1).

Three variables relevant to dynamic complexity were created. First, the number of feedback loops contained in each diagram was counted as an indicator of the degree of "feedback thinking." This number was divided by the number of events in the diagram and also by the average length of the feedback loops (since as the length of a feedback loop increases the chance that additional loops will be created by adding small variations to the single loop increases). Second, the percentage of subjects who described a causal link between economic depressions and expansions was noted for each experimental condition. Third, as a measure of the "remoteness in time and space" of initiating causes of events, the maximum causal path length extending back from the primary to-be-explained events, divided by the maximum possible length, was calculated for each subject.

For all of the continuous variables, it is possible to ask if an increase in the variable is due to subjects simply writing longer narratives rather than including more information in the same number of words. To control for this possibility, all of the continuous variables were divided by the number of words in the narratives from which they were drawn. [8]

5. Results and Discussion

5.1. General Characteristics of the Causal Scenario Diagrams

As suggested by Norman [1983], the causal scenario diagrams reported in this study, both pre and post, indicate that the subjects indeed held mental models that were "messy, sloppy, incomplete, and indistinct." The diagrams were relatively modest in both size (containing 11 events and 9 links, on average) and complexity (the mean average path length back from the primary to-be-explained events was 1.5, and the mean maximum path length was 3). In addition, evidence of feedback thinking was rare overall. For example, the average number of feedback loops in each diagram was only 0.5. And, most subjects did not even consider the possibility that expansions and depressions could act as causal agents, as is evident from the high number of links pointing toward these events compared with almost no links pointing away from them.

[8] Actually, post-test narratives proved to be reliably shorter, on average, than pretest narratives ($t = -2.66$, $p < .05$, mean number of words 158 vs. 180).

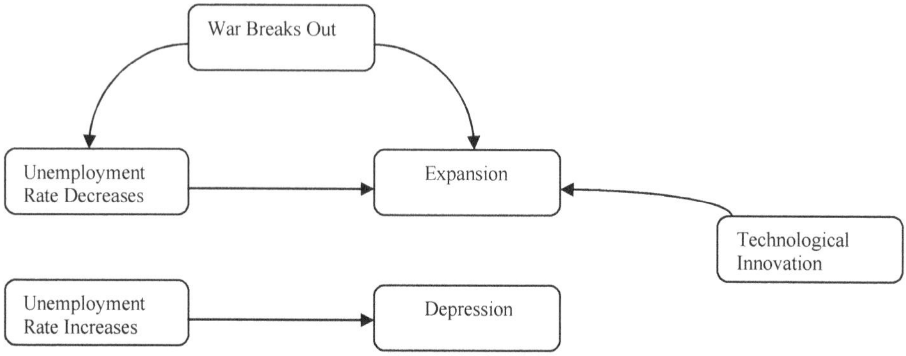

Figure 3. Pretest composite causal scenario diagram containing events and links mentioned by at least one-third of subjects

The diagrams were also highly variable across subjects in both size (the smallest contained only 4 events, while the largest contained 20) and content (subjects described a total of 103 unique codable events). One way of reducing this bewildering variety so that similarities between subjects can be more easily perceived is to create from the set of individual diagrams a "composite" diagram that includes only the events and links mentioned by a substantial number of subjects. This is done in Figs. 3 and 4 for pretest and posttest data, respectively.[9] The most obvious feature of these diagrams, both pre and post, is how greatly simplified they are compared with expert explanations of economic systems and the long wave. For example, the detailed chain of events by which technological innovation leads to economic growth is reduced in both diagrams to a single link. The diagrams are also very nonspecific: when events from the expert theory of the long wave are included, they are not precisely correct. For example, when subjects mention "management decision errors," they do not typically specify that they take place in the capital sector of the economy – they just know that some manager, somewhere ordered too much of something.

Even though the composite diagrams are highly sanitized, they still show evidence of mental sloppiness. For example, in more than one case both simple and more complicated explanations of the same causal chain exist simultaneously (e.g., "war causes economic expansion" and "war reduces unemployment, which causes expansion"). And, on occasion, as in Fig. 4, "dead ends" are included, that is, events or chains of events are included if they are thought to be relevant, even if it is not known how they relate to the rest of the events. In summary, these diagrams, as should be

[9] The causal scenario diagrams in Figs. 3 and 4 are closely related to but distinct from causal loop diagrams. The main difference is that increases and decreases in variables are treated as separate "events" and therefore the "sign" of relationships is incorporated into the nodes rather than the links, as they are in causal loop diagrams. Causal scenario diagrams can easily be converted into causal loop diagrams if desired. However, the assumption that subjects in this experiment are thinking at the level of abstraction represented by causal loop diagrams is not supported by the data.

expected given that the subjects typically had no formal training in the domains relevant to the intervention, reflect the exceedingly simple, occasionally confused thoughts of complete novices.

5.2. Pre/Post Differences in Causal Scenario Diagrams

5.2.1. Content of Mental Models

Tables 1 and 2 list the events contained in at least 19% of the pre- and post-test causal scenario diagrams, respectively. Table 2 also displays the χ^2 statistic and associated significance level for the pre/post difference in percentage of mentions for each post-test event. The most significant pre/post difference is the marked increase (from 5 to 43%) in the percentage of subjects mentioning the occurrence of a "poor management decision," a key element of the expert theory of the long wave. Two additional events important to the expert theory, "overproduction of goods (actually, capital)" and "time delays occur," also show a substantial increase after the intervention, although they are only marginally significant statistically. Thus there is reliable evidence that, post- intervention, subjects are incorporating at least some of the expert concepts into their mental models.

However, this does not mean that these new events are replacing the events in the pretest models. Instead, subjects seem to have integrated the new events from the expert theory into their existing naïve mental models. This is apparent, for example, in the fact that there is no statistically reliable post-test change in the percentage of subjects mentioning many of the most common pretest events (e.g., "war breaks out," unemployment rate increases"). In fact, for two events unrelated to the expert theory of the long wave, "desire to innovate increases" and "technological innovation occurs," there is a statistically significant and a marginally significant increase, respectively. This suggests that during the intervention subjects are not only learning some elements of the expert theory but are also learning about and accepting elements of the naïve mental models of their fellow subjects.

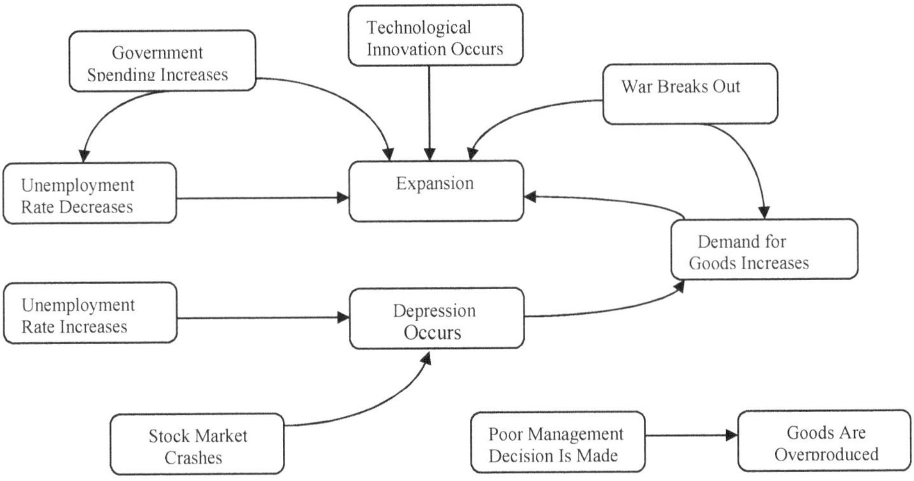

Figure 4. Posttest composite causal scenario diagram containing events and links mentioned by at least one-third of subjects

Table 1. Most Often Mentioned Events in Pre-Test Causal Scenarios of the Economic Long Wave (N = 21)

Event	Percentage of Subjects
War breaks out	81
Technological innovation occurs	52
Unemployment rate decreases	43
Unemployment rate increases	33
Consumer spending increases	29
Consumer spending decreases	29
Government spending increases	24
Demand for goods increases	24
Consumer morale increases	24
Consumer morale decreases	19
Stock market crashes	19
Savings are depleted	19
Amount of trade increases	19
Incomes decline	19
War ends	19

Overall, the method seems to be capturing a transitional state as subjects move from novices toward a level of somewhat more expertise. The pretest causal scenario diagrams suggest mental models in which the primary causes of the long wave are thought to be external to the economy (e.g., war, technological innovation). The post-test diagrams include these elements essentially unchanged while also incorporating aspects of the expert model. This pattern of results likely represents a problem endemic to the enterprise of trying to change mental models: old mental models are not easily gotten rid of, even after new mental models have been learned and accepted.

Although some aspects of the content of the diagrams changed due to the intervention, this did not result in subjects converging on a "shared mental model:" there was no reliable pre/post difference in the average percentage of subjects who mentioned either the top 5 (χ^2 (df = 1) = .38, n.s.) or top 15 events (χ^2 (df = 1) = .30, n.s.). This is not particularly surprising since the intervention did not incorporate group consensus building elements.

5.2.2. Complexity of Mental Models

Detail Complexity. Post-intervention causal scenario diagrams contained more events ($t = 2.92$, $p < .01$) and links ($t = 3.74$, $p < .002$) than pre-intervention diagrams. It should be noted, however, that this increase in detail complexity, although statistically reliable, was relatively modest (the mean number of events per 100 words of text increased from 6.2 to 8.1 and the mean number of links per 100 words increased from 4.7 to 6.9). The pre/post difference in the number of links per event, however, was only marginally significant (means .49 vs. .59, $t = 1.72$, $p = .10$). This suggests that, although subjects' mental models increased in size due to the intervention, they did not become substantially more intricate

Table 2. Most Often Mentioned Events in Post-Test Causal Scenarios of the Economic Long Wave (N = 25)

Event	Percentage of Subjects	χ^2 (df = 1) test for pre/post difference
Technological innovation occurs	76	3.60 (p < .10)
War breaks out	72	.52 (n.s.)
Demand for goods increases	52	1.62 (n.s.)
Poor management decision is made	43	18.0 (p < .001)
Unemployment rate increases	43	.40 (n.s.)
Unemployment rate decreases	38	.10 (n.s.)
Government spending increases	38	1.02 (n.s.)
Goods are overproduced	38	3.08 (p < .10)
Stock market crashes	38	1.88 (n.s.)
Desire to innovate increases	29	4.28 (p < .05)
Time delays occur	24	3.10 (p < .10)
Capital depreciates	19	2.04 (n.s.)
Consumer spending increases	19	.53 (n.s.)
Consumer spending decreases	19	.53 (n.s.)

or interconnected. This conclusion is supported by the finding that there was no reliable pre/post difference in the average path length back from the primary to-be-explained events, divided by the maximum possible length ($t = -.49$, n. s.).

Dynamic Complexity. Post-intervention causal scenario diagrams contained significantly more feedback loops than pre-intervention diagrams ($t = 2.45$, $p < .05$, mean number of loops/number of events/average length of loops/100 words .013 vs. .045). This result is particularly noteworthy because subjects received no training in describing, identifying, or constructing feedback loops. In addition, subjects were not asked to think diagrammatically; instead, the loops were implicit in the verbal narratives they were asked to write. Further evidence of an increase in feedback thinking is apparent in a statistically reliable increase in the number of subjects who described a causal link or chain connecting economic expansions and depressions, a key element of the expert theory of the long wave (χ^2 (df = 1) = 7.48, $p < .01$). However, there was no reliable evidence of a change due to the intervention in the second component of dynamic complexity, the "remoteness in time and space" of initiating causes of events: the pre/post difference in the maximum causal path length extending back from the primary to-be-explained events (divided by the maximum possible length) was not statistically significant ($t = .70$, n.s.).

5.2.3. Confidence

Few interventions attempt to measure the degree of confidence participants have in their mental models. However, people are remarkable explanatory creatures and are often quite willing to construct plausible-sounding explanations on the spot that they don't necessarily hold much confidence in. This leaves open the possibility that any measured changes in mental models due to an intervention may be illusory, since confidence may not have increased. To rule out this possible interpretation of results, the present study included both pre- and post-measures of participants' confidence in their mental models. After each prompt for verbal data, subjects were asked to indicate on a 1 to 7 scale, where 1 was not at all confident and 7 was extremely confident, how confident they were that their explanation was correct. The results indicate that subjects were reliably more confident in their explanations of the long wave after the intervention than they were prior to the intervention ($t = 2.55$, $p < .05$, means 3.5 vs. 4.1), although the average level of confidence remained near the middle of the scale.

5.2.4. Pretest Effects

Chi-square analyses were conducted to determine if the 25 subjects who did not participate in the mental models pretest were more or less likely to include in their narratives the top 15 events listed in Table 2. For thirteen of these events the percentage of subjects did not reliably differ between the two groups, and were often quite similar. However, there were two statistically reliable differences: subjects who took the pretest were significantly more likely to include the events "desire to innovate increases" (χ^2 (df = 1) = 5.3, $p < .05$, percentages 29 versus 4) and "technological innovation occurs" (χ^2 (df = 1) = 17.0, $p < .001$, 76 versus 16) than subjects who took only the post-test. In addition, analysis of variance tests were conducted to determine the effects of participation in the mental models pretest on the quantitative variables related to detail complexity, dynamic complexity, and confidence described above. Two marginally significant effects were found: both the average ($t = 1.89$, $p < .07$, mean avg. path length/maximum possible length/100 words of text .14 vs. .24) and maximum ($t = 1.65$, $p = .11$, mean maximum path length/maximum possible length/100 words of text .34 vs. .22) length of paths extending back from the primary to-be-explained events were longer for subjects who did not participate in the mental models pretest.

Thus, while the pretest had no effect on the majority of the variables related to change in mental models, there were a small number of significant or marginally significant effects. Apparently, the mere act of answering the pretest survey made subjects much more likely to include at least two of the popular pretest variables on the post-test and also somewhat more likely to write narratives that were similar to their pretest narratives in terms of detail and dynamic complexity. This tendency for post-test models to be more similar to pretest models than they would have been if no pretest had been conducted represents another example of how the goals of measurement and promoting learning and change can be in conflict. While pretests are unavoidable for rigorous assessments to be conducted they may at the same time reduce the effectiveness of the interventions they are designed to assess.

5.2.5. Moderating Variables

Several moderating variables relating to subjects' experience playing the Kondratiev game and to subjects' general background characteristics were examined as possible predictors of how much their mental models changed due to the intervention. These variables included subjects' Kondratiev game scores (log-transformed since they varied over more than 2 orders of magnitude) as well as the timeshape of production capacity generated during the game. [10] In addition, subjects filled out a post-intervention survey in which they were asked to indicate, on a 1 to 7 scale, how much they enjoyed playing the game, how hard they tried to get a good score, and how carefully they thought about each decision before submitting it, as well as to report the number of economics classes taken prior to the experiment and self-rated computer skill. Finally, the average exam score in the class in which the experiment was conducted was included as a proxy variable for general academic ability.

These variables were included together in ordinary least squares (for continuous dependent variables) and logit (for categorical dependent variables) regression models. [11] The dependent variables included all of the quantitative variables related to detail and dynamic complexity described above as well as the content-related variables relevant to the expert theory of the long wave. Several of the variables were significant or marginally significant predictors of pre/post differences in the number of events and links in subjects' causal scenario diagrams. The pre/post difference in number of events and links decreased as enjoyment of the game increased (events $t = -1.9$, $p < .10$; links $t = -2.3$, $p < .05$), as computer skill increased (events $t = -2.3$, $p < .05$; links $t = -3.2$, $p < .01$), and as exam score increased (events $t = -2.5$, $p < .05$; links $t = -3.2$, $p < .01$). Thus the causal scenarios of those students with more computer skill and who like computer games more were less likely to change in size as a result of the intervention, perhaps because these students had less interest or skill in writing the verbal narratives through which mental models were assessed. The models of those students with higher academic ability were also less likely to change in size, perhaps because these students had larger models to begin with. Exam scores were, however, related positively to increases in the average and maximum causal path length of links in the causal scenario diagrams (average $t = 2.4$, $p < .05$; maximum $t = 2.0$, $p < .10$). How hard students tried during the game was also positively related to increases in average and maximum path length (average $t = 2.8$, $p < .05$; maximum $t = 3.4$, $p < .01$), perhaps because effort in playing the game is correlated with effort during the debriefing sessions and post-test. None of the predictors were significantly related to changes in the number of feedback loops or to changes in the percentage of subjects including expert concepts in their diagrams. Finally, Kondratiev game score and timeshape did not reliably predict any of the measures of change in mental models, after controlling for the other predictors. This may mean that what happens during game play is less important for fostering change in mental models than what happens during the subsequent debriefing session.

---

[10] The timeshape variable was examined because it can be unrelated to the game score, it may be a better indicator of performance than the game score (i.e., did the subject bring production capacity and desired production back into equilibrium or not?), and it is a good indicator of whether subjects experienced the simulated long-wave the game was designed to induce.

[11] In these models the statistical tests are for partial regression coefficients: the test asks whether the given variable reliably explains a portion of the variation in the dependent variable after controlling for all the other variables included in the model. With covariation among the predictor variables, this can produce conservative conclusions about the importance of a variable.

6. Conclusion

Measuring change in the mental models of the participants in systems thinking and system dynamics interventions is unavoidable if the relative effectiveness of different interventions in promoting learning about complex dynamic systems is to be assessed. However, few efforts have been made to design and implement rigorous methods that emphasize measurement of mental models rather than improvement of mental models. As a step toward encouraging such efforts, this chapter has described the general features of rigorous methods designed to measure change in mental models of complex dynamic systems and provided a detailed example of the design, implementation, and analysis of one such method.

The experimental results produced by this process have in the present case been generally encouraging about the effectiveness of system dynamics interventions in promoting learning. Although the intervention tested was quite modest, involving simply a single play of a management flight simulator and related preparatory and debriefing sessions comprising about 5 hours of time spread out over a two week period, several statistically significant changes in participants' mental models resulted, including: an increase in the size of subjects' mental models, a change in the content of mental models toward the expert model of the problem posed by the intervention, and an increase in the degree of feedback thinking contained in subjects' models. [12] Certainly it would be expected that longer interventions or interventions that focus on building systems thinking or system dynamics modeling skills would elicit even stronger changes in mental models.

However, the present results must be interpreted with a great deal of caution: the paradigm of controlled laboratory experimentation does not allow the luxury of drawing grand conclusions from a single experiment, but instead relies on systematic replication that inches toward truth one small step at a time. In fact, the experiment reported herein has several important limitations that can only be addressed (and should be addressed) by future research. For example, the study demonstrated positive effects due to the intervention, but does not address the possibility that some other type of intervention, unrelated to system dynamics, might be even more effective. The study also does not address what aspects of the intervention are important for producing the observed results: for example, were the changes in mental models due primarily to subjects' experience with the management flight simulator or to the information presented during the debriefing session? In addition, the limited time frame covered by

---

[12] It is worth noting that these results are similar in several ways to the results reported by Vennix [1990], despite the important differences between the interventions and methods applied by the two studies. For example, both studies reported an increase in the number of expert concepts included in post-test cognitive maps, an increase in the number of links between concepts, an increase in the number of mentions of time delays, and no change in the average lengths of causal paths. One major difference between the results of the two studies is that the present work reported a statistically reliable increase in feedback thinking, whereas Vennix [1990] did not. This is most likely because the present work studied a system dynamics-based intervention, while the Vennix [1990] study examined an econometrics-based intervention.

the intervention did not offer the opportunity to assess the stability of the measured changes in mental models: it is entirely possible that the gains achieved by the intervention disappeared or at least decayed in the weeks and months after the intervention. The time constraints on the experiment also did not allow for the study of the correlation between the measured mental models and decision behavior as represented, for example, in post-intervention plays of the management flight simulator: as described in Doyle [1997], it cannot simply be assumed that this correlation is strong and positive.

Given these limitations on interpreting the experimental results, the main contribution of this research lies not in resolving questions related to the effectiveness of systems thinking interventions, but in demonstrating how they can most appropriately be studied. The described method proved to be both practicable and capable of capturing and quantifying even subtle changes in mental models due to the intervention, and it can be adapted in a straightforward manner to resolve the above-stated questions as well as a wide variety of other questions related to the effectiveness of systems interventions in promoting learning. Finally, it is hoped that the present work, which documents the messiness and sloppiness characteristic of mental models, as well as the problems faced by those who would attempt to measure and change them (including the persistence in memory of old mental models in the face of new information and the existence of pretest effects) will lead to an increased appreciation of the high degree of difficulty inherent in studying mental models of complex dynamic systems.

References

Axelrod, R. (ed.), 1976, *The Structure of Decision: The Cognitive Maps of Political Elites*, Princeton Univ. Press. (Princeton, NJ).

Baron, R. S., 1986, Distraction-conflict theory: Progress and problems, in *Advances in Experimental Social Psychology*, edited by L. Berkowitz, Academic Press (Orlando, FL).

Bower, G., & Morrow, D., 1990, Mental models in narrative comprehension, *Science*, **247**(5), 44–48.

Brehmer, B., 1989, Feedback delays and control in complex dynamic systems, in *Computer Based Management of Complex Systems*, edited by P. Milling and E. Zahn, Springer-Verlag (Berlin).

Carley, K., & Palmquist, M., 1992, Extracting, representing, and analyzing mental models, *Social Forces*, **70**(3), 601–636.

Cavaleri, S., and Sterman, J. D. 1997, Towards evaluation of systems thinking interventions: A case study. *System Dynamics Review*, **13**(2), 171–186.

Chi, M. T. H., Glaser, R., & Farr, M. J. (eds.), 1988, *The Nature of Expertise*, Erlbaum (Hillsdale, NJ).

Clement, J., 1983, A conceptual model discussed by Galileo and used intuitively by physics students. In *Mental Models*, edited by D. Gentner and A. L. Stevens, Erlbaum (Hillsdale, NJ), pp. 325–340.

Coyle, R. G., 1977, *Management System Dynamics*, Wiley (London).

DiSessa, A. A., 1982, Unlearning Aristotelian physics: A study of knowledge-based learning, *Cognitive Science*, **6**, 37–75.

Dörner, D., 1996, *The Logic of Failure: Recognizing and Avoiding Error in Complex Situations*, Addison-Wesley (Reading, MA).

Doyle, J. K., 1997, The cognitive psychology of systems thinking, *System Dynamics Review*, **13**(3), 253–265.

Doyle, J. K., & Ford, D. N., 1998, Mental models concepts for system dynamics research, *System Dynamics Review*, **14**(1), 3–29.

Doyle, J. K., Ford, D. N., Radzicki, M. J., & Trees, W. S., 2002, Mental models of dynamic systems, in System Dynamics and Integrated Modeling, edited by Y. Barlas, from *Encyclopedia of Life Support Systems (EOLSS)*, developed under the auspices of the UNESCO, EOLSS Publishers, Oxford, UK (http://www.eolss.net)

Eden, C., & Huxham, C., 1988, Action-oriented management, *Journal of the Operational Research Society*, **39**(10), 889–899.

Eden, C., & Jones, S., 1984, Using repertory grids for problem construction, *Journal of the Operational Research Society*, **35**(9), 779–790.

Fiske, S. T., 1993, Social cognition and social perception, in *Annual Review of Psychology*, **44**, edited by L. Porter & M. R. Rosenzweig , pp. 155–194.

Forrester, J. W., 1961, *Industrial Dynamics*, MIT Press (Cambridge, MA).

Forrester, J. W., 1969, *Urban Dynamics*, Pegasus Communications (Waltham, MA).

Gill, R., 1996, An integrated social fabric matrix/system dynamics approach to policy analysis, *System Dynamics Review*, **12**(3), 167–181.

Goodman, M. R., 1974, *Study Notes in System Dynamics*, MIT Press (Cambridge, MA).

Guerin, B., 1986, Mere presence effects on human beings: A review, *Journal of Personality and Social Psychology*, **22**, 38 77.

Hodgson, A. M., 1992, Hexagons for systems thinking, *European Journal of Operational Research*, **59**, 220–230.

Hogarth, R. M. (ed.), 1982, *Question Framing and Response Consistency*, Jossey-Bass (San Francisco).

Huz, S., Andersen, D. F., Richardson, G. P., and Boothroyd, R. (1997). A framework for evaluating systems thinking interventions: An experimental approach to mental health systems change. *System Dynamics Review*, **13**(2), 149–169.

Intons-Peterson, M. J., & Newsome, G. L., III. , 1992, External memory aids: Effects and effectiveness, in *Memory Improvement: Implications for Memory Theory*, edited by D. Herrmann, H. Weingartner, A. Searleman, & C. McEvoy, Springer-Verlag (New York), pp. 101–121.

Kahneman, D. & Tversky, A., 1984, Choices, values, and frames, *American Psychologist*, **39**, 341–350.

Kelman, H. C., 1958, Compliance, identification, and internalization: Three processes of attitude change, *Journal of Conflict Resolution*, **2**, 51–60.

Kleinmutz, D., and Thomas, J., 1987, The value of action and inference in dynamic decision making, *Organizational Behavior and Human Decision Processes*, **39**(3), 341–364.

Larkin, J. H., 1983, The role of problem representation in physics, in *Mental Models*, edited by D. Gentner & A. L. Stevens, Erlbaum (Hillsdale, NJ). pp. 75–98.

Latane', B., Williams, K., & Harkins, S., 1979, Many hands make light the work: The causes and consequences of social loafing, *Journal of Personality and Social Psychology*, **37**, 822–832.

McCloskey, M., 1983, Naive theories of motion, in *Mental Models*, edited by D. Gentner and A. L. Stevens, Erlbaum (Hillsdale, NJ), pp. 299–324.

Morecroft, J. D. W., 1982, A critical review of diagramming tools for conceptualizing feedback structure, *Dynamica*, **8**, 20–29.

Moscovitch, M., & Craik, F. I. M., 1976, Depth of processing, retrieval cues, and uniqueness of encoding as factors in recall, *Journal of Verbal Learning and Verbal Behavior*, **15**, 447–458.

Moxnes, E., 1998, Not only the tragedy of the commons: Misperceptions of biofeedback. *Management Science*, **44**(9), 1234–1248.

Norman, D. A., 1983, Some observations on mental models, in *Mental Models*, edited by D. Gentner and A. L. Stevens, Erlbaum (Hillsdale, NJ), pp. 7–14.

Orne, M., 1962, On the social psychology of the psychology experiment, *American Psychologist*, **17**, 776–783.

Payne, J. W., Bettman, J. R., & Johnson, E. J., 1992, Behavioral decision research: A constructive processing perspective, *Annual Review of Psychology*, **43**, 87–131.

Pennington, N., 1981, Causal Reasoning and Decision Making: The Case of Juror Decisions, Ph. D. Thesis, Harvard University, Cambridge, MA.

Pennington, N., & Hastie, R., 1986, Evidence evaluation in complex decision making, *Journal of Personality and Social Psychology*, **51**, 242–258.

Pennington, N., & Hastie, R., 1988, Explanation-based decision making: Effects of memory structure on judgment, *Journal of Experimental Psychology: Learning, Memory, and Cognition*, **14**(3), 521–533.

Richardson, G. P., & Pugh, A., III, 1981, *Introduction to System Dynamics Modeling with DYNAMO*, MIT Press (Cambridge, MA).

Richardson, G. P., Vennix, J. A. M., Andersen, D. F., Rohrbaugh, J., & Wallace, W. A., 1989, Eliciting group knowledge for model building, in *Computer-Based Management of Complex Systems*, edited by P. M. Milling & E. O. Zahn, Springer-Verlag (Berlin).

Richmond, B., 1987, *The Strategic Forum: From Vision to Strategy to Operating Policies and Back Again*, High Performance Systems (Hanover, NH).

Roediger, H. L., III, Weldon, M. S., & Challis, B. H., 1989, Explaining dissociations between implicit and explicit measures of retention: A processing account, in *Varieties of Memory and Consciousness*, edited by H. L. Roediger, III & F. I. M. Craik, Erlbaum (Hillsdale, NJ), pp. 3–41.

Roos, L. L., & Hall, R. I., 1980, Influence diagrams and organizational power, *Administrative Science Quarterly*, **25**(1), 57–71.

Rosenthal, R., 1966, *Experimenter Effects in Behavioral Research*, Appleton-Century-Crofts (New York).

Schank, R. C., 1972, Conceptual dependency: A theory of natural language understanding, *Cognitive Psychology*, **3**, 552–631.

Schank, R. C., 1975, The structure of episodes in memory, in *Representation and Understanding: Studies in Cognitive Science*, edited by D. G. Bobrow and A. Collins, Academic Press (New York).

Schank, R. C., & Abelson, R. P., 1977, *Scripts, Plans, Goals, and Understanding: An Inquiry into Human Knowledge Structures*, Erlbaum (Hillsdale, NJ).

Schank, R. C., 1990, *Tell Me a Story: A New Look at Real and Artificial Memory*, Macmillan (New York).

Senge, P., 1990, *The Fifth Discipline: The Art and Practice of the Learning Organization*, Doubleday (New York).

Sterman, J. D., 1985, A behavioral model of the economic long wave, *Journal of Economic Behavior and Organization*, **6**(1), 17–53.

Sterman, J. D., 1986, The economic long wave: Theory and evidence, *System Dynamics Review*, **2**(2), 87–125.

Sterman, J. D., 1989, Misperceptions of feedback in dynamic decision making, *Organizational Behavior and Human Decision Processes*, **43**, 301–335.

Sterman, J. D., 2000, *Business Dynamics: Systems Thinking and Modeling for a Complex World*, McGraw-Hill (Boston).

Sterman, J. D., & Meadows, D., 1985, Strategem-2: A microcomputer simulation game of the Kondratiev cycle, *Simulation and Games*, **16**(2), 174–202.

Tulving, E., 1983, *Elements of Episodic Memory*, Oxford Univ. Press (New York).

Vennix, J. A. M., 1990, *Mental Models and Computer Models: Design and Evaluation of a Computer-Based Learning Environment for Policy-Making*. CIP-Gegevens Koninklijke Bibliotheek (Den Haag).

Vennix, J. A. M., 1996, *Group Model Building: Facilitating Team Learning Using System Dynamics*, Wiley (New York).

Verburg, L., 1994, *Participative Policy Modeling Applied to the Health Care Insurance Industry*. CIP-Gegevens Koninklijke Bibliotheek (Den Haag).

Chapter 15

A Comprehensive Model of Goal Dynamics in Organizations: Setting, Evaluation and Revision[1]

Yaman Barlas
Bogazici University, Department of Industrial Engineering
34342 Bebek, Istanbul, Turkey
ybarlas@boun.edu.tr

Hakan Yasarcan
Accelerated Learning Laboratory
Australian School of Business (Incorporating the AGSM)
The University of New South Wales
Sydney, NSW 2052, Australia
h.yasarcan@unsw.edu.au

1. Introduction

Goal setting plays crucial role in decision making in organizations as well as in individuals. Most improvement activities consist of the following cycle: set a goal, measure and evaluate the current performance (against the set goal), take actions (e.g. training) to improve performance, evaluate and revise the goal itself if necessary, again measure and evaluate the performance against the current goal, and so on … [Forrester 1975; Senge 1990; Lant 1992; Sterman 2000]. So goals constitute a base for the decisions and the managerial actions. In an organization, the performance level is evaluated against a goal and, further, the effectiveness of the goal itself can and must be periodically evaluated.

Among various research methods to analyze the dynamics of goals in organizations (and individuals), an important one is simulation modeling – more specifically *system dynamics* modeling that is particularly suitable to model qualitative, intangible and 'soft' variables involved in human and social systems [Forrester 1961; Forrester 1994; Morecroft and Sterman 1994; Sterman 2000; Spector et al 2001]. *System dynamics* is designed specifically to model, analyze and improve dynamic socio-economic and managerial systems, using a feedback perspective. Dynamic strategic management

[1] Supported by Bogazici University Research Grant no. 02R102.

problems are modeled using mathematical equations and computer software and dynamic behavior of model variables are obtained by using computer simulation [Forrester 1961; Ford 1999; Sterman 2000; Barlas 2002]. The span of applications of the system dynamics field includes: *corporate planning and policy design, public management and policy, micro and macro economic dynamics, educational problems, biological and medical modeling, energy and the environment*, and more [Forrester 1961; Roberts 1981; Senge 1990; Morecroft & Sterman 1994; Ford 1999; Sterman 2000]. Since these problems are typically managerial-policy oriented, structures that deal with goal dynamics play an important role in most system dynamics models. It is therefore no surprise that modeling of goal dynamics is an explicit research topic in system dynamics.

A fundamental notion and building block used in most policy models, is a 'goal seeking' structure that represents how a certain condition (or state) is managed so as to reach a given 'goal' (see Figure 1 and 2 below). For instance, the state may be the inventory level or delay in customer service, and the goals would be a set (optimal) inventory level or a targeted service delay respectively. [For numerous examples see Sterman 2000 and Forrester 1961]. Management would then take the necessary actions (*Improvement* in Figure 1) so as to bring the states in question, closer to their set goal levels. The most typical heuristic used to formulate *Improvement* is:

where *SAT*, *State adjustment time*, is some time constant. The dynamics of this simplest goal seeking structure is depicted in Figure 2: *State* approaches and reaches the *Goal* gradually, in a negative exponential fashion. In more sophisticated and realistic goal-seeking models, the goal is not fixed; it varies up and down depending on current conditions, called a 'floating goal' structure [Sterman 2000; Senge 1990]. In this case, if the performance of the system is persistently poor (in approaching the originally set goal), then the system implicitly or deliberately lowers the goal (*eroding* goal). If, on the other hand, the system exhibits a surprisingly good performance, then the goal may be pushed further up (*evolving goal*). Another key component of a general goal setting structure is *expectation formation*: Goals are set and then adjusted in part as a function of future expectations [Forrester 1961; Sterman 1987]. This may involve the expectations of management, expectations of participants, or typically both. Formulation of expectations is a rich research and modeling topic with its roots in theories of cognitive research in policy making [Spector and Davidsen 2000] and

$$Improvement = (Goal - State)/SAT$$

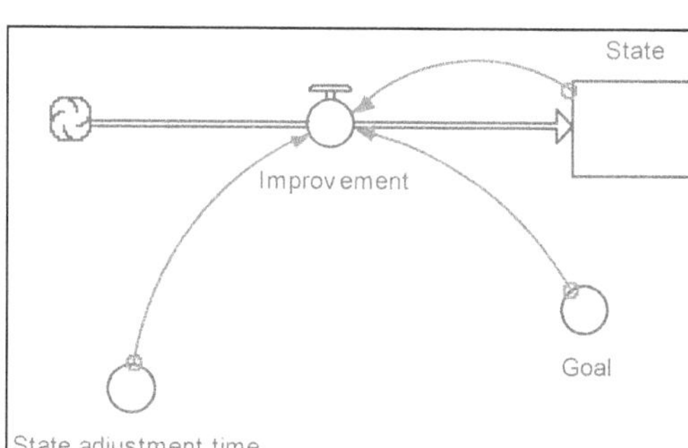

Figure 1. The simplest goal seeking structure (stock-flow diagram)

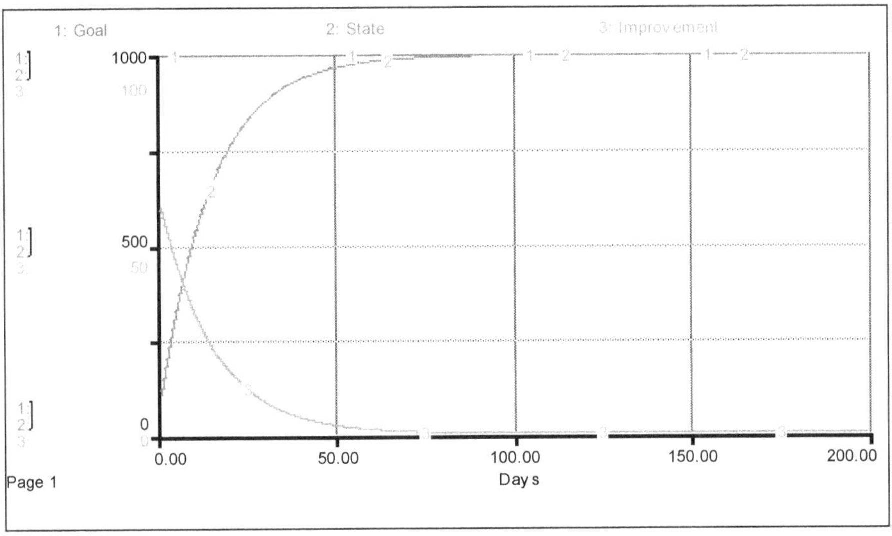

Figure 2. The simple goal seeking behavior generated by the model of Figure 1

rationality; ranging from rational expectations to *satisficing* and bounded rationality [Simon 1957; Morecroft 1983]. So, formulation of goal setting, evaluation and seeking is a deep and important research topic in system dynamics modeling.

In this chapter we present a comprehensive goal dynamics model involving different types of explicitly stated and implicit goals, expectation formation and potential goal erosion as well as positive goal evolution dynamics. We further include a host of factors not considered before, such as organizational capacity limits on performance improvement rate, performance decay when there is no effort, time constraints, pressures and, motivation and frustration effects. We show that the system can exhibit a variety of subtle problematic dynamics in such a structure. The modeling setting assumed in the chapter is an organization in which a new 'performance' goal is set and a new program (e.g., a training activity) is started to achieve the goal. In a service company, this may be a new training program set 'to increase the customer satisfaction from 60% to 80% in one year' as measured by customer surveys. Or in a public project, this may be a new educational program in a poor neighborhood set 'to increase the functional literacy rate from 80% to 90% in three years' as measured by periodic tests. So there is a goal and there is also a time horizon set to reach the goal. Note also that the environment described above implies that the *Goal* is always approached from below and higher *State* levels mean always better for the system. (As opposed to inventory management, where there is some 'optimal' target inventory level so that the management increases the inventory if it is too low compared to the set goal and lowers the inventory if it is too high).

The chapter starts by elaborating on the simplest model of constant goal seeking dynamics, by including a *constraint* on improvement *capacity* and a nominal *decay* rate out of the state variable (Figure 3). We then gradually add a series of more realistic and complex goal-related structures to the initial model. The purpose of

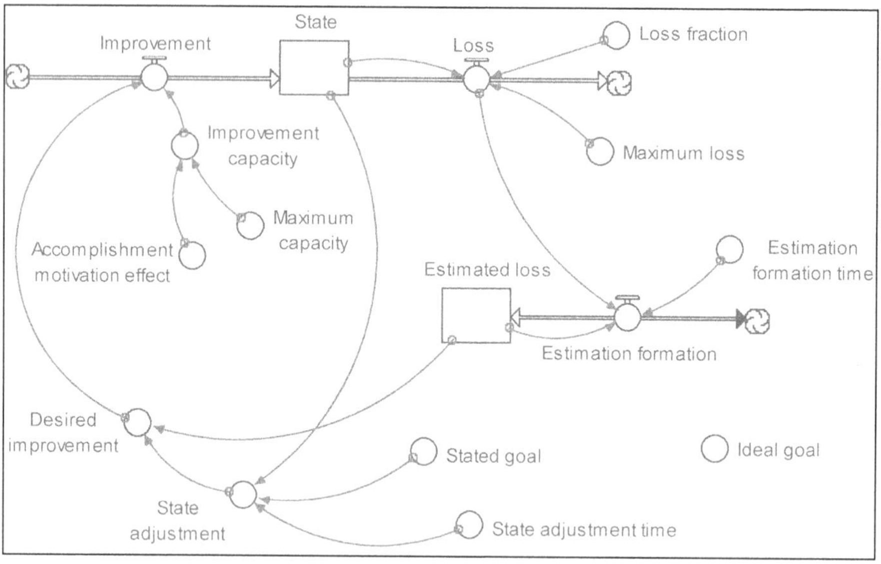

Figure 3. Simple goal seeking model with improvement capacity limit and loss flow

each addition is to introduce and discuss a new aspect of goal dynamics in increasingly realistic settings. In the first enhancement, we introduce how the implicit goal in an organization may unconsciously erode as result of strong past performance traditions. Next, we discuss under what conditions recovery is possible after an initial phase of goal erosion. In the following enhancement, we include the effects of *time constraints* on performance and goal dynamics. Many improvement programs have explicitly stated time horizons and the pressure (frustration) caused by an approaching time limit may be critical in the performance of the program. We also show that in such an environment, the implicit *short-term* self-evaluation horizons of the participants may be very critical in determining the success of the improvement program. Finally we propose and test an adaptive goal management policy designed to assure satisfactory goal achievement, by taking into consideration the potential sources of failures discovered in the preceding simulation experiments. We conclude with some observations on implementation issues and further research suggestions.

2. The Simple Goal Seeking Structure and its Dynamic Behavior

The simplest goal-seeking structure consists of a single fixed goal, a condition (called *State* in Figure 1) that is managed and a management action (*Improvement* in Figure 1). Once the goal is set, it is not challenged by the internal dynamics of the system or by any external factor. As mentioned before, assuming a typical stock adjustment heuristic (i.e. *Improvement* = (*Goal–State*)/*SAT*), this structure exhibits a pure negative exponential goal-seeking behavior shown in Figure 2.

In our enhanced version of this simplest goal-seeking structure, we introduce two new factors to make it more general and realistic:

i. If there is no improvement effort, *State* experiences a natural decay (*Loss* flow) and,
ii. *Improvement* rate is (naturally) limited by some *Maximum capacity*.

The outflow, *Loss* is assumed to be simply a fraction (*loss fraction*) of the actual *State*. This implies in essence that due to rapidly changing hi-tech organizational setting, the performance level tends to decay over time, if no improvement effort is undertaken. Proportional formulation means that as *State* goes up, so does the *Loss* rate. This is realistic to some extent, but to prevent exaggeratedly high *Loss* rates, we place a limit on *Loss* as well, called *Maximum loss*. (We choose *Maximum loss* to be less than *Improvement capacity*, otherwise it would be impossible to ever fulfill the *goal*).

Maximum capacity is constant in our models, as capacity management is beyond the scope of this paper. The effective *Improvement capacity*, on the other hand is variable (*Accomplishment_motivation_effect* × *Maximum_capacity*), but in this first model *Accomplishment motivation effect* is equal to one, so it has no role yet. (It will have an important role later in the enhanced versions of the model). In any case, *Improvement* is thus formulated as:

$$Improvement = MIN(Desired\_improvement,\ Improvement\_capacity)$$

The *Desired improvement* formulation is the standard 'anchor-and-adjust' formulation [Sterman 1989; Sterman 2000; Barlas and Ozevin 2001; Yasarcan 2003]. It is given by:

$$Desired\_improvement = Estimated\_loss + State\_adjustment$$

$$State\_adjustment = (Stated\_goal - State)/State\_adjustment\_time$$

So the *Desired improvement* decision uses *Estimated loss* as an anchor and then adjusts the decision around it, depending on the discrepancy between the goal and the current state. Since it is not possible to know the *Loss* immediately and exactly, it must first be estimated by the decision maker or system participants. So, *Estimated loss* is the output stock (see Figure 3) of an expectation formation structure, using simple exponential smoothing formulation:

$$Estimation\_formation = (Loss - Estimated\_loss)/Estimation\_formation\_time$$

Estimation formation time represents the delay in learning the actual value of performance *Loss*. All stock, flow and converter variables are shown in Figure 3.

Note that there are two different goal variables in Figure 3; *Stated goal* is set and declared by the management, and *Ideal goal* is defined as the best (highest) possible goal for the system. In this first simple model, *Stated goal* is assumed to be equal to the *Ideal goal*. (In all simulation experiments, initial *State* is taken as 100 and *Ideal goal* as 1000. In the simpler models and experiments, *Stated performance goal* is equal to *Ideal goal*, but especially in more complex models, *Stated goal* will not be necessarily equal to the *Ideal goal*, and in the final model it will not even be constant).

Dynamic behavior generated by simulating the model of Figure 3 can be seen in Figures 4 and 5. The behavior is a variant of standard 'goal-seeking behavior' seen in Figure 2. *State* (line 3 in Figure 4) gradually seeks the *Stated goal.* Since the *Improvement* rate (line 3 in Figure 5) is above the *Loss* rate (line 4 in Figure 5), the

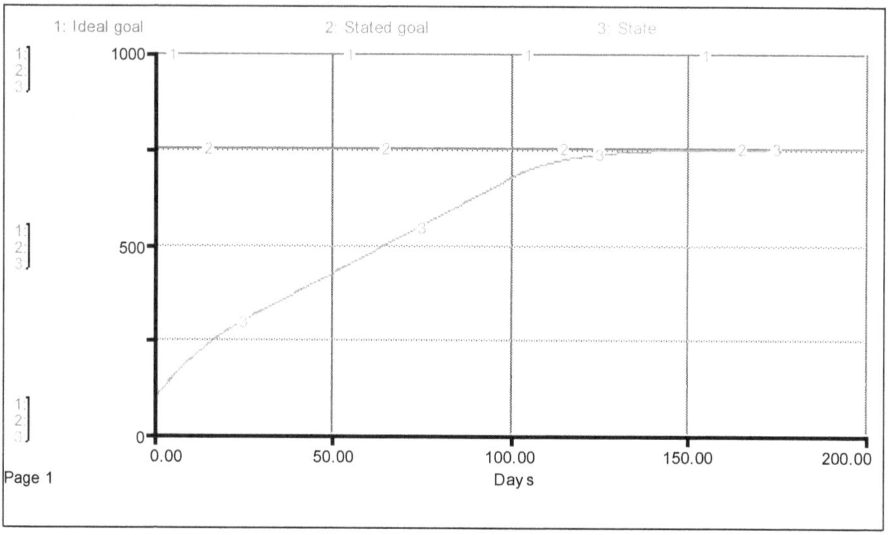

Figure 4. Goal seeking behavior generated by the model of Figure 3 (*Stated_goal* = 750)

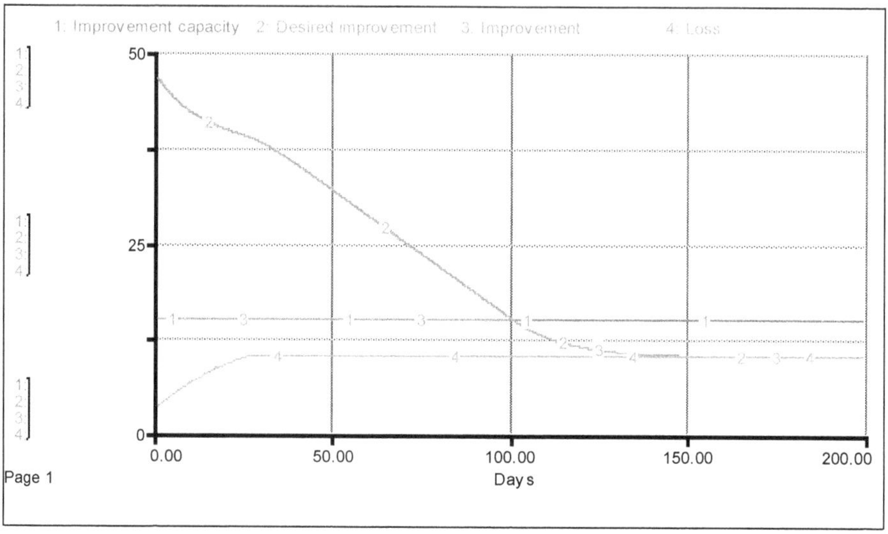

Figure 5. Dynamics of flows related to the run in Figure 4

State level keeps increasing until it reaches the *Stated goal*, at which point the improvement is lowered down to the *Loss* rate, since the goal is reached. Also observe (in Figure 5) that when the *Desired improvement* (line 2) is above the maximum improvement capacity (line 1), the actual improvement rate stays constant at *Maximum capacity*. When *Desired improvement* is below the *Maximum capacity*, then the actual improvement becomes equal to the *Desired improvement*.

This model is still too simple, being basically of introductory pedagogical value. The model can explain simple goal seeking dynamics like heating of a room or a water tank filling up to a desired level after flushing. In order to represent goal dynamics of human systems and organizations, we incorporate a series of realistic enhancements in the following sections.

3. Traditional Performance, Implicit Goal and Erosion

Goal erosion may occur if there is an endogenously created, undeclared, *Implicit goal* that system seeks, instead of the explicitly set goal (*Stated goal*). The model shown in Figure 6, is more realistic and complex version of the simple goal seeking model, involving the structures related to *Implicit goal* and eroding goal dynamics. Observe two important new variables in Figure 6: *Traditional performance* and *Implicit goal*. *Traditional performance* [Forrester 1975] represents an implicit, unconscious habit formation in the system. The human element in the system gradually forms a belief (a self image) about his/her own performance as time passes, and this learned performance (*Traditional performance*) may start to have even more effect than the *Stated goal* [Forrester 1975; Senge 1990; Sterman 2000]. The accumulation of the individual beliefs creates a belief within the system that the system can realistically perform around this past performance. To represent this mechanism, we assume that the system creates its own internal goal called *Implicit goal*, and seeks this new goal instead of the managerially *Stated goal*.

Observe that in the model of Figure 6, there are only two new variables compared to Figure 3. The first one is *Traditional performance* which is essentially a historical (moving) average of past performance (*State*). This is formulated by simple exponential smoothing (with a tradition formation time of 30 days), just as it was done to formulate *Estimated loss*, above. The second one is *Implicit goal*, which is assumed to gradually tend to the *Traditional performance*, starting with an initial value of *Stated goal*. (There is a new parameter in Figure 6 called *Weight of stated goal* that should be ignored at this point, as it has no role in this version of the model; this parameter will have a role in the following section). So in a nutshell, *Implicit goal* is also an exponentially delayed function of *traditional performance:*

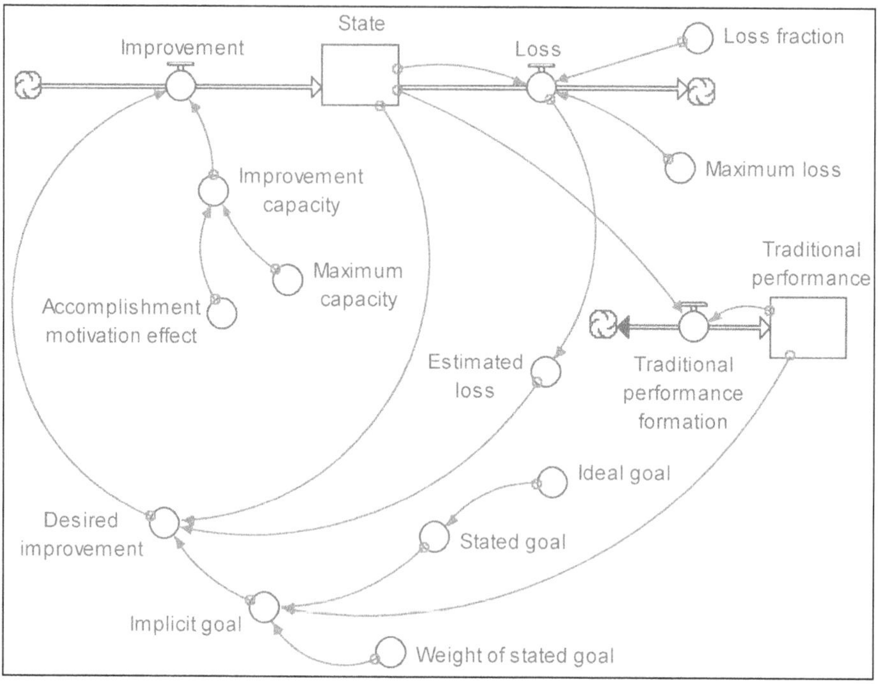

Figure 6. A basic model of eroding goal dynamics

Implicit_goal = SMTH3(Traditional_performance, 10, Stated_goal)

The third-order exponential smoothing function (SMTH3) used above means that the output (*Implicit goal)* does not immediately react to a change in the input (*Traditional performance);* there is a period of initial inertia. *Implicit goal* does erode to *Traditional performance*, but the erosion should not start immediately and should not react to any temporary change in traditional performance. Finally, the goal used in the *Desired improvement* equation is now *Implicit goal* (instead of *Stated goal*).

We assume that in a new improvement program, in the beginning there is no concept of past performance, so the human participants in the system completely accept the *Stated goal* as their goal (i.e. *Implicit_goal = Stated_goal)*. But as time passes, *Traditional performance* starts to have bigger effect and the *Implicit goal* starts approaching the *Traditional performance* instead of staying at the managerially *Stated goal*, a phenomenon often called *'goal erosion*'. Note that together with this goal erosion, the *State* starts to pursue the *Implicit goal*, not the *Stated goal*, so the result of the improvement program is a failure. The dynamic behaviors of goal erosion can be seen in Figure 7.

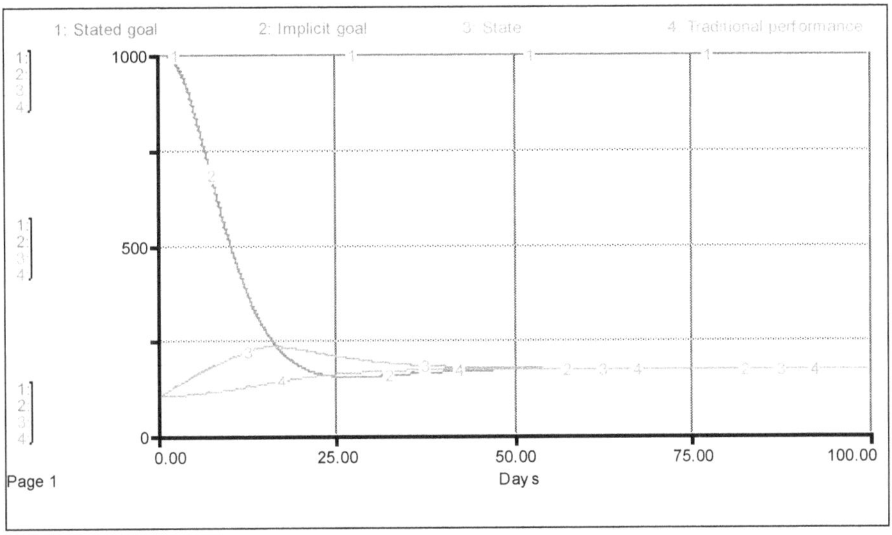

Figure 7. Strong erosion in goal, caused by *Traditional performance* bias

Goal erosion can be severe or mild, depending on some environmental factors. In the scenario represented in Figure 7, we assume that the *Traditional performance formation time* is relatively long (there is a strong past tradition), so the *Implicit goal* erodes toward *Traditional performance* and may erode to the point of even crossing below the current *State* level, since the highly delayed *Traditional performance* determines the *Implicit performance goal* (see Figure 7). Erosion can be extreme if there is a very strong past tradition. Conversely, if the tradition formation time is short, then goal erosion is milder and also simpler: since with a short formation time, *Traditional performance* is almost equal to current *State*, the *Implicit goal* would be effectively seeking the *State* (and the *State* naturally seeking the *Implicit goal*).

4. Goal Erosion and Recovery

In the model shown in Figure 6, if the parameter *Weight of stated goal* is given a value between 0 and 1, then the equation of *Implicit goal* becomes:

$$Implicit\_goal = SMTH3[Weight\_of\_stated\_goal \times \mathrm{Stated\_goal} + (1 - Weight\_of\_stated\_goal) \times Traditional\_performance,\ 10]$$

The above equation states that *Implicit goal* is now basically a weighted average of the S*tated goal* and *Traditional performance*. This weighted average is then passed through a third-order smoothing to give it a realistic inertia, just as in the previous model. (But in the previous version, *Weight of stated goal* was set to 0, so that *Implicit goal* was simply exponentially eroding to *Traditional performance*, starting with an initial S*tated goal*). In the current version of the model, participants are affected both by the external

Stated goal and by their *Traditional performance,* so that their *Implicit goal* is somewhere in between the two extremes [Forrester 1961; Sterman 2000; Barlas and Yasarcan 2006]. *Weight of stated goal* determines how much the system believes in *Stated goal.* The resulting dynamics with *Weight_of_stated_goal* = 0.5 are shown in Figure 8. After a significant initial erosion, *Implicit goal* and hence *State* gradually recover towards *Stated goal.* Observe that since *Traditional performance* is an exponential average of past *State* values, the former moves - fast or slow- towards the *State* after some delay. On the other hand, the other component of the weighted average, *Stated goal* is fixed. The net result is that all variables, including *Traditional performance* eventually recover and gradually approach *Stated goal* over time (Figure 8). Thus, although this model is more realistic than the previous version in some sense, it suffers from the following fundamental weakness: the formulation implies that, even though a system may suffer from initial goal and performance erosion, in time it will always recover (fast or slow) and attain the *Stated goal.* The weakness is that there is no notion of 'time horizon' and potential frustration (or motivation) to be experienced by the system participants having evaluated their performances against the time constraints. In reality, *Implicit performance goal* may continue to erode if there is a belief in the system that the set goal (*Stated goal*) is too high or impossible to reach in some given time horizon. These concepts are introduced in the following enhancements.

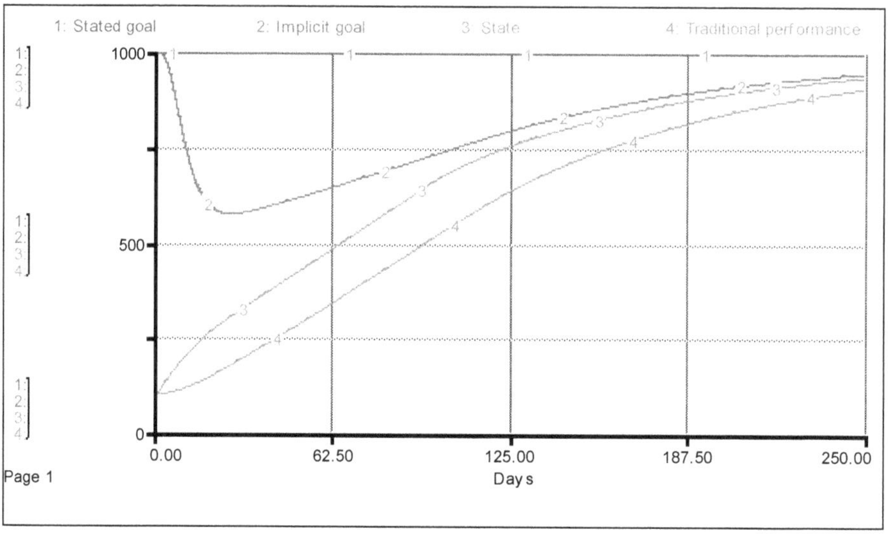

Figure 8. Behavior of goal erosion and recovery model (*Weight_of_stated_goal* = 0.5)

5. Time Horizon Effects: Frustration, Motivation and Possible Recovery

The two essential enhancements in this model are the concepts of project *Time horizon* and accomplishment motivation (or frustration) that results from the participants' assessment of their performance against this time horizon. The project time horizon is taken as 200 days in the following simulation runs. *Time horizon* is a stock variable representing how many days left to the end of the project, starting at the initial horizon value, and depleting day by day (see Figure 9). One subtle feature of this *Time horizon* stock is that it does not quite deplete to 0; it stops when it reaches what we call '*Short term horizon*' taken as 12 days. The idea is that if the goal is reachable in just another 12 days, these twelve days are always allowed. (The *Short term horizon* will have an active role in the next enhancements and will be discussed in the following section). So the depletion rate of *Time horizon* is given by:

$$Time\_horizon\_depletion\_rate = \text{IF } Time\_horizon > Short\_term\_horizon \text{ THEN 1 ELSE 0}$$

Accomplishment motivation effect is a factor that represents the participants' belief that the stated performance goal is achievable within the *Time horizon.* To formulate this, we represent *Accomplishment motivation effect* as a decreasing function of *Remaining work and time ratio* as shown in Figure 10. *Remaining work and time ratio* is an estimate of how many days would be needed to close the gap between the current *State* and *Stated goal*, relative to the remaining *Time horizon*:

$$Remaining\_work\_and\_time\_ratio = [(Stated\_goal - Perceived\_performance)/5]/Time\_horizon$$

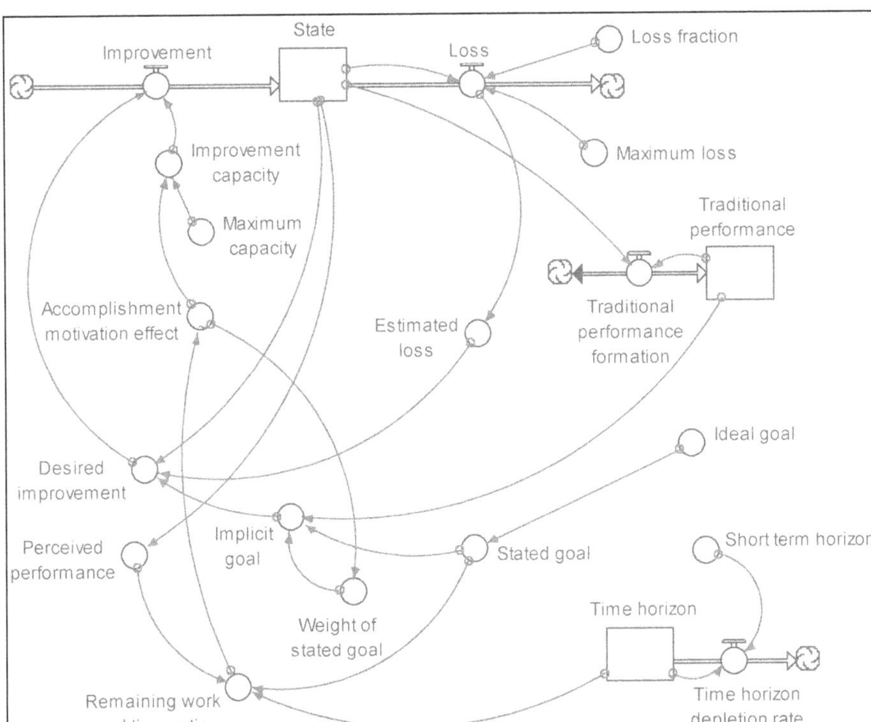

Figure 9. A model of goal erosion and possible recovery with a time horizon

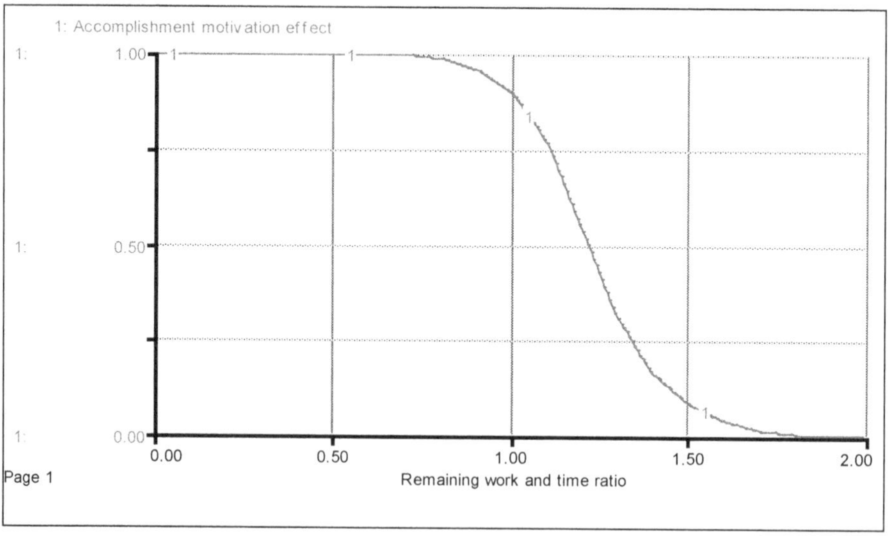

Figure 10. *Accomplishment motivation effect* is a function of *Remaining_work_and_time_ratio*

In the above formulation, the discrepancy between *Stated goal* and *Perceived performance* is first divided by the maximum rate at which *State* can be improved (which is *Maximum_capacity – Maximum_loss* = 15–10 = 5), yielding how many days would be needed at least to close the gap. (*Perceived performance* is just an exponentially smoothed average of *State*). The result is then divided by remaining *Time horizon* to provide a normalized ratio. (This ratio is also smoothed in the model, so that motivation does not change too fast, without any inertia – see the equations list in the appendix). Figure 10 states that motivation is full (equal to one) when the above ratio is less than or equal to 0.7, it starts dropping afterwards and when the ratio becomes about 2, it denotes complete frustration (equal to zero),.

Accomplishment motivation effect plays two different roles in the model. First, this motivation increases the effective improvement capacity of the participants:

$$Improvement\_capacity = Accomplishment\_motivation\_effect \times Maximum\_capacity$$

where $Maximum\_capacity = 15$

Thus, the more motivated the participants are, the closer becomes the actual *Improvement capacity* to *Maximum capacity*. (In the earlier versions of the model, *Accomplishment motivation effect* was set to 1, so it had no effect). This first role of *Accomplishment motivation effect* is represented by the positive feedback loop shown in Figure 11). The second role of *Accomplishment motivation effect* is to influence *Weight of stated goal* as follows:

$$Weight\_of\_stated\_goal = Accomplishment\_motivation \times Reference\_weight$$

where $Reference\_weight = 1.0$

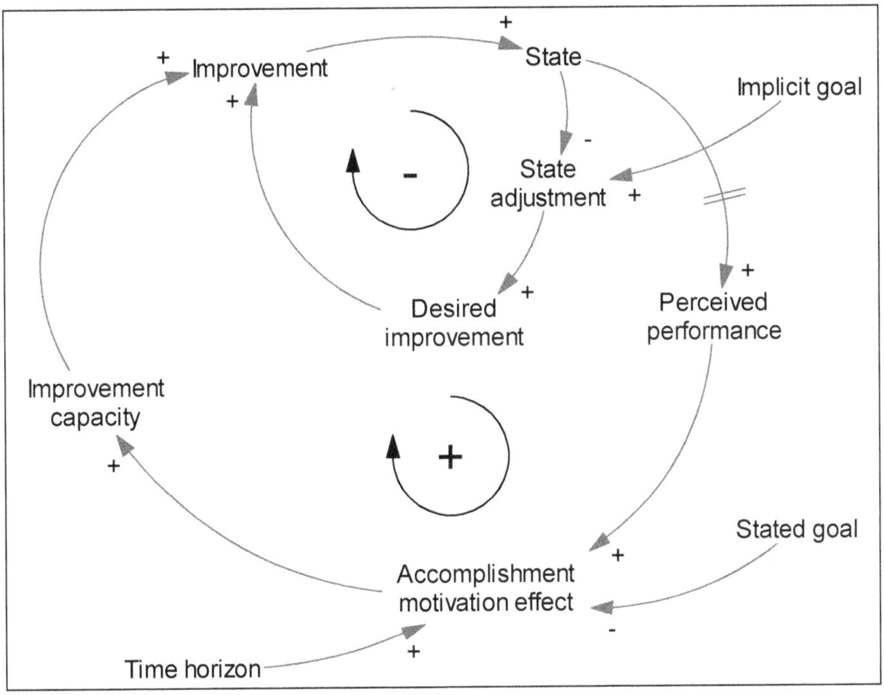

Figure 11. The *State* seeks the *Implicit goal*, via *Improvement* (the basic inner goal-seeking loop). Simultaneously, the *Perceived performance* relative to *Time horizon* determines the *Accomplishment motivation effect* which in turn affects the *Improvement* (the outer reinforcing loop).

Thus, *Weight of stated goal* is now a variable, depending on the motivations of the participants. When participants have full motivation (1.0), then *Weight of stated goal* becomes 1.0 as well, so that in forming their *Implicit goal, participants give* 100% weight to S*tated goal* and no weight at all to *Traditional performance.* At the other extreme, with zero motivation, all weight is given to *Traditional performance* and zero weight to S*tated goal* (meaning complete goal erosion). In between these two extremes, some non-zero weights are given both to S*tated goal* and to *Traditional performance*, depending on the level of motivation (see Figure 10).

Two typical dynamics generated by this model are depicted in Figures 12 and 13. In both dynamics, there is an initial phase of strong erosion in *Implicit goal*, because the *Stated goal* level is too high compared to the initial *State* (hence *Traditional performance*). After this initial erosion, the next phase is one of recovery: Implicit goal starts moving up gradually, and thus pulling up the *State* and *Traditional performance.* Finally, there is a third phase in the dynamics that is interesting: In Figure 12, all three variables, after having improved significantly, reverse their patterns and a final phase of erosion begins, continuing all the way to the end. The mechanism behind this final erosion is related to the time horizon of the project and the negative effects of de-motivation resulting from the impossibility of reaching the goal, given the time constraint. This is observed in Figure 12, where *Stated goal* is set to 1000 and *Time horizon* at 200. In Figure 13, on the other hand, *Stated goal* is set at a lower value of 750 and the second phase of erosion never takes place.

The reason why all variables (*Implicit goal, State* and *Traditional goal*) continue to improve is that *Stated goal* is now set at a more 'realistic' value, relative to the given *Time horizon*. So the gap between *Stated goal* and *Perceived performance* relative to the remaining time horizon never becomes so high as to cause a hopeless situation for the participants. At the end, *State* can sustain a value of 750, which is below the *Ideal goal* of 1000, but much better than the 'giving up' dynamics of Figure 12.

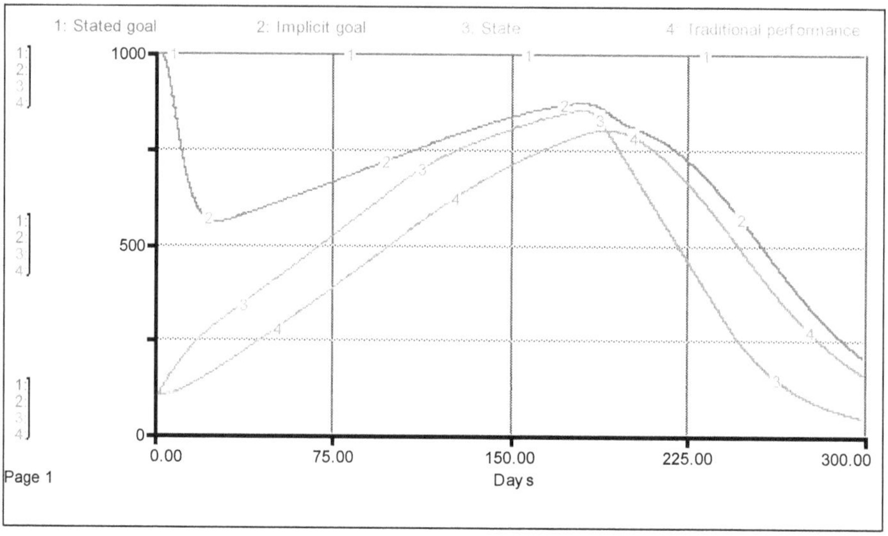

Figure 12. Erosion dynamics, when *Stated goal* is too high for the given *Time horizon* (*Stated_goal* = 1000, *Time_horizon*$_0$ = 200)

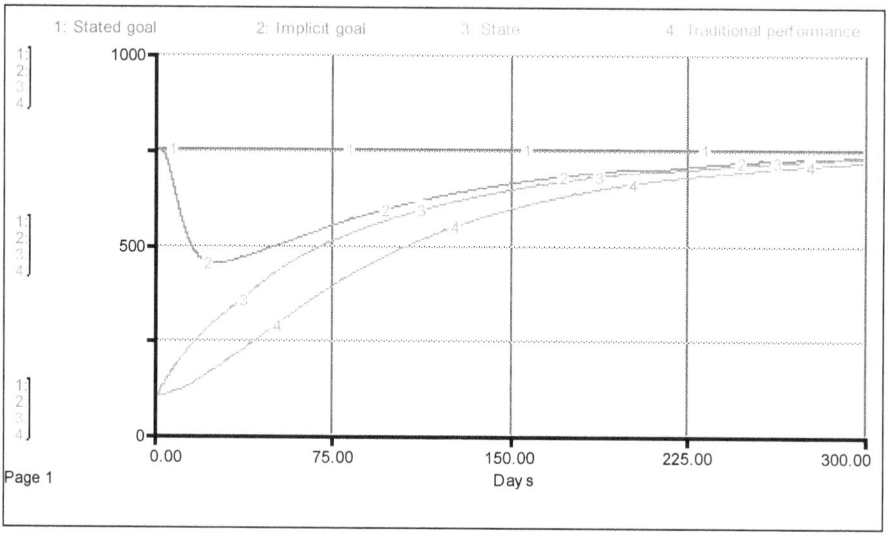

Figure 13. Erosion-then-recovery, when *Stated goal* is low enough for the given *Time horizon* (*Stated_goal* = 750, *Time_horizon*$_0$ = 200)

6. Role of Short-term Horizon in Potential Recovery

A second component of motivation may have to do with the participants' own intrinsic *Short term horizon*, (assumed to be 12 days). The assumption is that participants judge their own performance over a 12-day horizon and if they are satisfied (high *sort term motivation*), then they give more weight to the *Stated goal*, otherwise they lower this weight (meaning that the weight of their *Traditional performance* increases).

The new equation for *Weight of stated goal* becomes:

$$Weight\_of\_stated\_goal = Accomplishment\_motivation \times Short\_term\_motivation \times Reference\_weight$$

where $Reference\_weight = 1.0$

The formulation of *Short term motivation effect* is very similar to that of long term *Accomplishment motivation effect* described in the previous section: The motivation is a decreasing (from 1 down to 0) function of *Short term work and time ratio*, defined just like *Remaining work and time ratio* defined above, except that the ratio is divided by *Short term horizon* (12 days constant) instead of *Time*

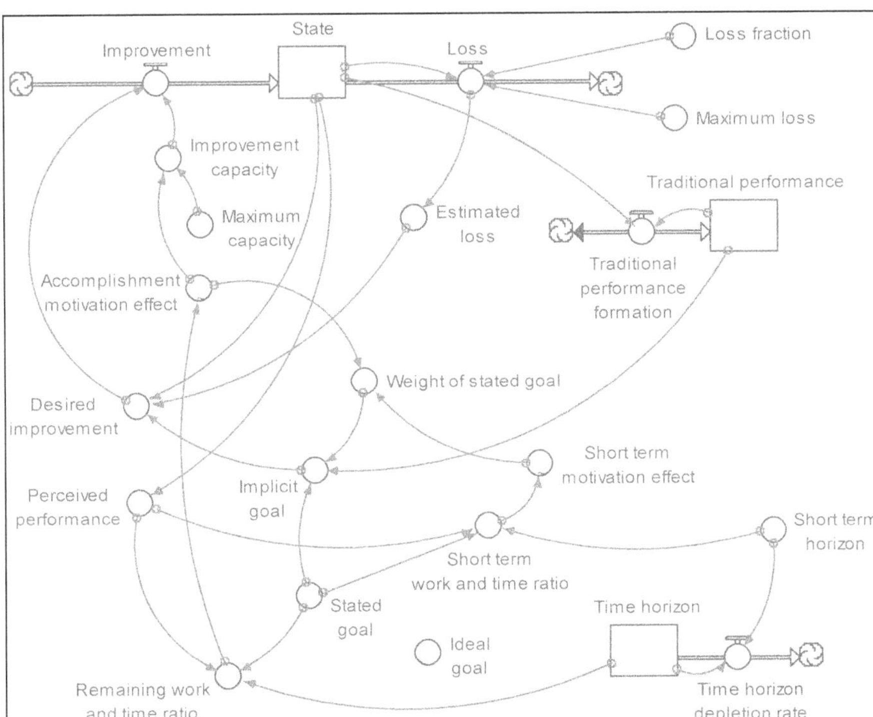

Figure 14. *Time horizon, Short term horizon* and their corresponding motivation effects

horizon of the project. (See Appendix for complete equations). One implication is that if participants perceive that only 12 days of full effort can take them to the goal, then they never give up, even if one day is left to the project *Time horizon*. Thus, the weight of *Stated goal* (hence potential erosion of *Implicit goal)* depends more subtly on the dynamics of both the short and long term accomplishment motivations of the participants. These long term and short term motivation interactions and main loops are shown in figure 15.

As will be seen below, where the *Stated goal* is set turns out to be critically important in this new model. The dynamics for the different values of *Stated goal* are plotted in Figures 16, 17 and 18. If the *Stated goal* is too high (for instance 600) relative to the initial *State* (100) and *Time horizon* (200), there develops a disbelief in the system that the stated goal is ever reachable. This disbelief results in de-motivation, which further causes the *Weight of stated goal* to reduce to near zero and the *Implicit goal* erodes (Figure 16). After this initial eroding goal behavior, motivations never become high enough to ignite a performance improvement, but they can at least sustain the performance at some level for some duration of time. Finally, the system participants recognize that the remaining time to accomplish the *Stated goal* is impossibly too short, which results in 'giving-up' behavior and improvement activity eventually dies down (Figure 16).

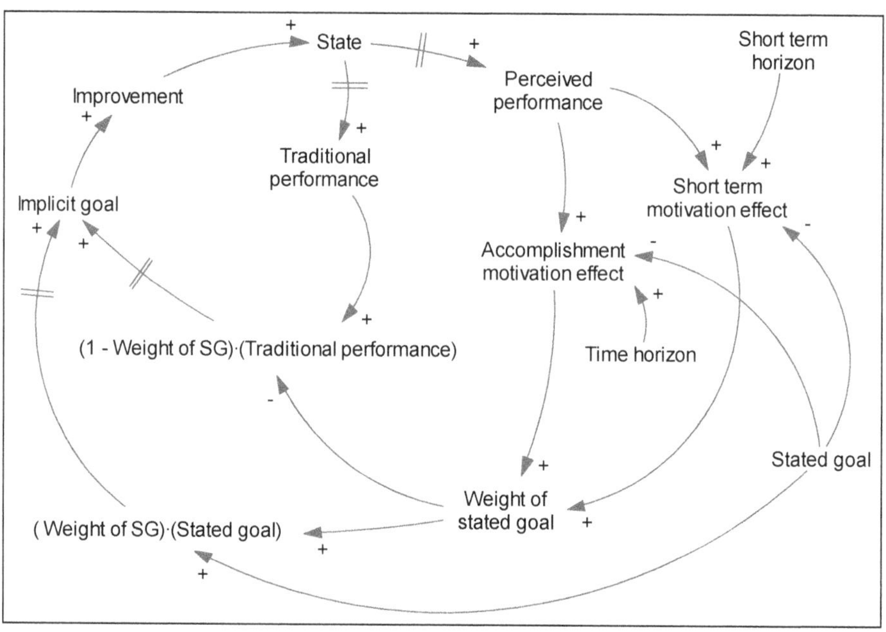

Figure 15. A weighted average of *Traditional performance* and *Stated goal* determines *Implicit goal* after some delay. The weight depends on the short term and long term accomplishment motivations, which in turn depend on perceived remaining performance gaps (relative to the time horizons)

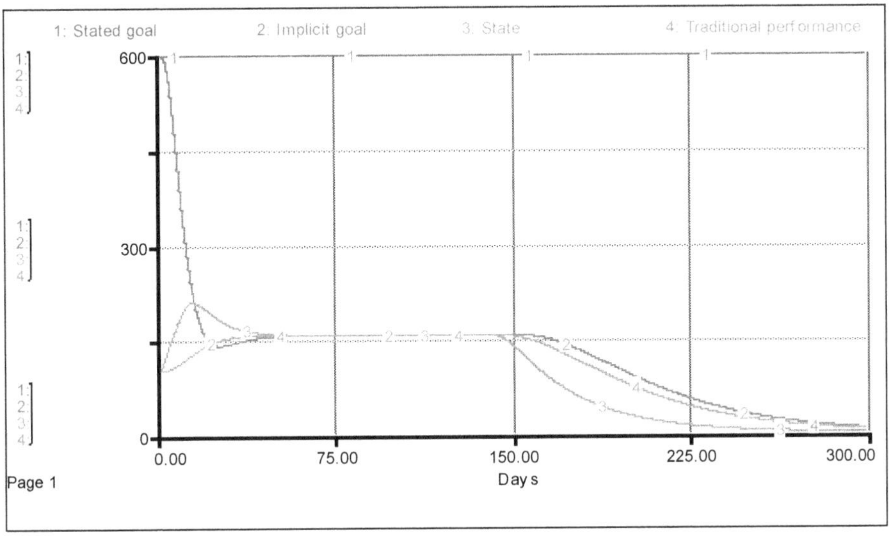

Figure 16. After an initial erosion in *Implicit goal*, long term stagnating *State* eventually results in giving-up behavior (*Stated_goal* = 600; *Time_horizon*$_0$ = 200)

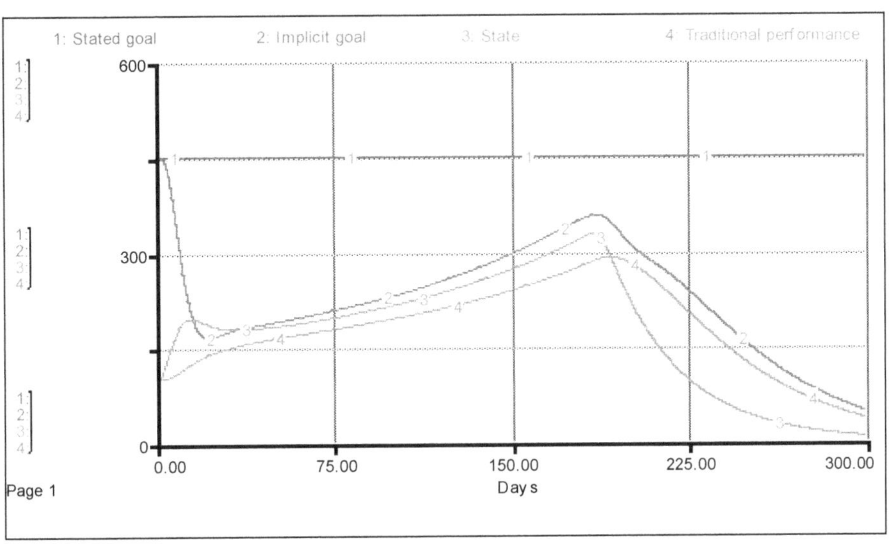

Figure 17. Initial erosion, then recovery and finally giving-up behavior due to time limit (*Stated_goal* = 450; *Time_horizon*$_0$ = 200)

The next simulation experiment starts with a lower *Stated goal.* In Figure 17, firstly we observe an eroding goal behavior in the short term, due to the initial gap between *Stated goal* and *Traditional performance.* But because *Stated goal* is not too high (450), this time the *Weight of stated goal* does not become too small, which allows a goal recovery phase between days 50 and 175. But, still the *Stated goal* is not low enough to prevent a giving-up behavior in longer term due to the time limit. So, in Figure 17, three different stages of goal dynamics can be observed: initial goal erosion, goal (and state) recovery, and finally giving-up behavior.

Finally in Figure 18, *Stated goal* is low enough (400) to create a success: the recovery is sustained and the goal is reached in the given *Time horizon.* The three dynamics (Figures 16, 17 and 18) show that if *Stated goal* is low enough relative to initial *State* and *Time horizon* so as to create enough motivation, the *State* will improve towards the goal and achieve it.

But, how can a manager know the 'correct level' of the *Stated goal*? Furthermore, what if this level of performance is too low (conservative) compared to the true potential of the system participants? Our comprehensive model and simulation experiments so far illustrate these issues. As a result, we conclude that solutions to the above problems necessitate use of a "dynamic and adaptive goal setting" strategies by the management, as will be addressed in the next section.

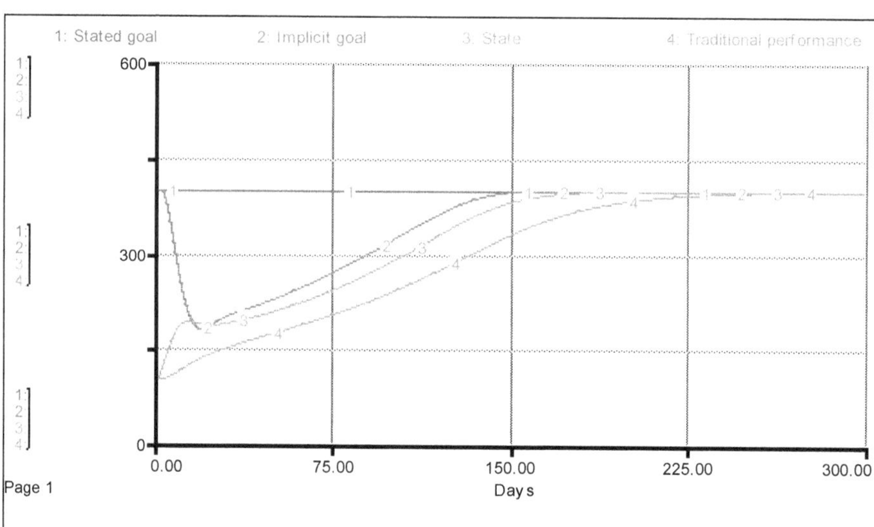

Figure 18. If *Stated goal* is low enough, sustainable recovery is possible (*Stated_goal* = 400; *Time_horizon*$_0$ = 200)

7. Adaptive Goal Setting Policy for Consistent Improvement

J. W. Forrester (Forrester 1975) states: "…The goal setting is then followed by the design of actions which intuition suggests will reach the goal. Several traps lie within this procedure. First, there is no way of determining that the goal is possible. Second, there is no way of determining that the goal has not been set too low and that the system might be able to perform far better. Third, there is no way to be sure that the planned actions will move the system toward the goal."

In order to address these uncertainties, *Stated goal* should be set and managed dynamically and adaptively. The management must continuously monitor the *State* and must evaluate its level and its trend. The *Stated goal* should be then set realistically within the bounds of a "reachable region", which is a function of the *State level* and its trend (net improvement rate). If *State* is improving, *Stated goal* must also be gradually moved up to guide and motivate the improvement activities. On the other hand if the *State* is stagnating, this means that *Stated goal* is unrealistically high, so it must be lowered till there is a sign of sufficient improvement in the *State* level. The structure in Figure 19 is designed to implement and test such an adaptive management strategy.

In the related formulations, we assume that top management does not know *Implicit goal*, *Weight of stated goal*, the *motivation effects*, *Short time horizon*, *Capacity* and *Loss flow*. Management can only perceive the system performance (*State*) over time, so the *Stated goal* decisions must be based on this information. If *State* is not improving enough, *Stated goal* should be lowered, and if *State* is improving then *Stated goal* must

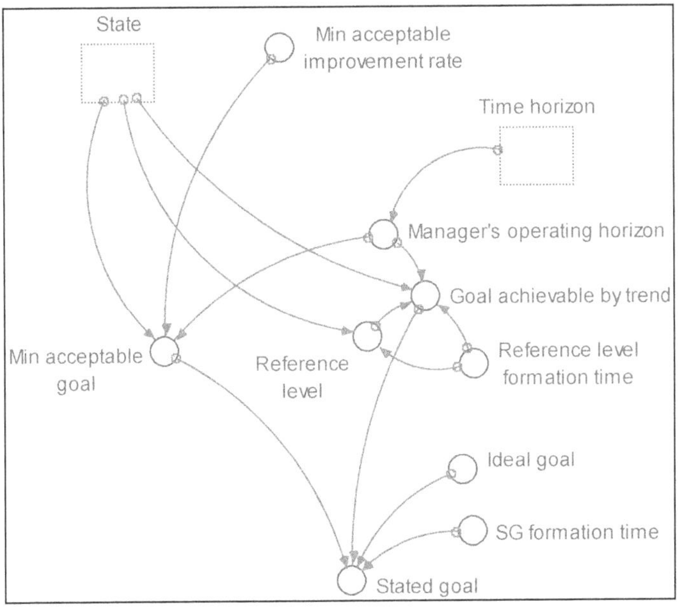

Figure 19. Proposed adaptive dynamic *Stated goal* setting policy structure

also be gradually moved up. Beyond this, exactly how much *Stated goal* should be moved up or down, depends on several factors: *Stated goal* can not be bigger than *Ideal goal* and it should not be lower than some minimum acceptable goal determined by the top management. If the level determined by the *trend* in *State* is in acceptable region, then *Stated goal* must be equal to this level:

$$Stated\_goal = MIN(Ideal\_goal, MAX(Goal\_achievable\_by\_trend, Min\_acceptable\_goal))$$

Ideal goal is a given constant (1000) as discussed earlier. The second variable, *Goal achievable by trend* is a managerial estimate of the current improvement trend and what level can be attained at this rate, in some time horizon:

$$Goal\_achievable\_by\_trend = State + ((State - Reference\_level)/Reference\_level\_formation\_time) \times Manager's\_operating\_horizon$$

$$Reference\_level = SMTH3(State, Reference\_level\_formation\_time)$$

$$Manager's\_operating\_horizon = MIN(90, Time\_horizon)$$

In the above formulation, *Reference level* is an average of past *State*, used in estimating the trend, by dividing the improvement by *Reference level formation time. Manager's operating horizon* is used in extrapolating the trend into the future. This horizon is equal to the time horizon of the project as the project advances. But in early phases, the managerial horizon is set to a smaller value (90), because it is assumed that extrapolating the current trend farther into the future would be too uncertain. Finally, the third component of Stated goal equation is some minimum acceptable goal determined by the top management. This *Min acceptable goal* is determined by adding some minimum acceptable improvement rate (in *Manager's operating horizon)* on top of the current *State*. It is assumed that a given management has some minimum acceptable improvement rate, below which is simply unacceptable:

$$Min\_acceptable\_goal = State + Min\_acceptable\_improvement\_rate \times Manager's\_operating\_horizon$$

$$Min\_acceptable\_improvement\_rate = 0.5$$

In the above formulation, managerial constants like *Manager's operating horizon, Min acceptable improvement rate*, and *Reference level formation time* are set at some reasonable values that serve our research purpose. In an actual study, these constants must be well estimated by data analysis and interviews and also tested by sensitivity analysis.

The adaptive goal setting policy structure is integrated in the full model (of Figure 14) and simulation experiments are run under different scenarios. In all scenarios, it is assumed that management starts with a rather high initial *Stated goal* (900), to demonstrate the fact that the starting stated goal is not important anymore, because *Stated goal* is a variable in the adaptive goal setting policy. In figure 20, the provided

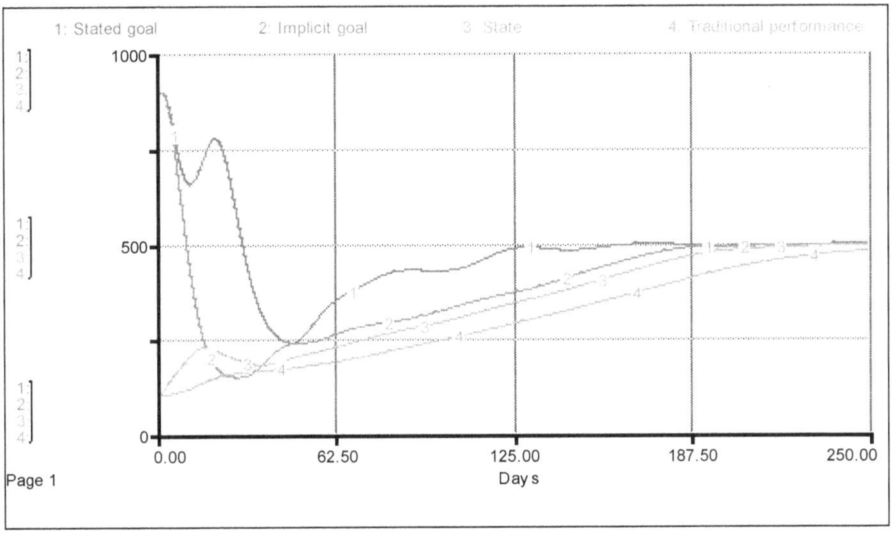

Figure 20. A satisfactory result with the dynamic goal management policy, even when the initial *Stated goal* is high (900) and *Time horizon*$_0$ is short (200)

Time horizon is quite short (200), so the expected behavior, based on previous experiments is one of strong erosion and give-up behavior towards the end (for instance, Figures 16 and 12). But the dynamics in Figure 20 display an obvious improvement: Initially, *Stated goal* is deliberately lowered (with some oscillations) by top management so as to ignite participant motivation, and then later it is moved up gradually and adaptively, pulling together with it the *Implicit goal* and *State*. At the end, although reaching the ideal goal of 1000 was impossible in the given *Time horizon*, *State* has improved to a reasonable level (500) within the time limits, without displaying any giving-up behavior.

In the second experiment, a longer *Time horizon* (350) is provided. The main dynamic characteristics are the same as those observed in the previous run: *Stated goal* is deliberately lowered initially by top management, and then later, it is moved up gradually and adaptively, pulling together the *Implicit goal* and *State* (Figure 21). At the end, since the *Time horizon* is longer, *State* is improved to a higher level (750), compared to the previous run (although still lower than the ideal goal of 1000). From these two runs, the important contribution of the adaptive goal setting policy is obvious: The performance consistently improves and eventually settles down at a level without any giving-up behavior at the end, which apparently is a strong improvement given *Time horizon*.

Finally, when *Time horizon* is sufficiently long relative to the *Ideal goal* of the project, the last simulation experiment demonstrates that *Ideal goal (1000)* can be reached. In Figure 22, *Time horizon* is taken as 500 and we observe that with sufficient time, the *Ideal goal* is eventually attained within the program *Time horizon*, implying maximum organizational success.

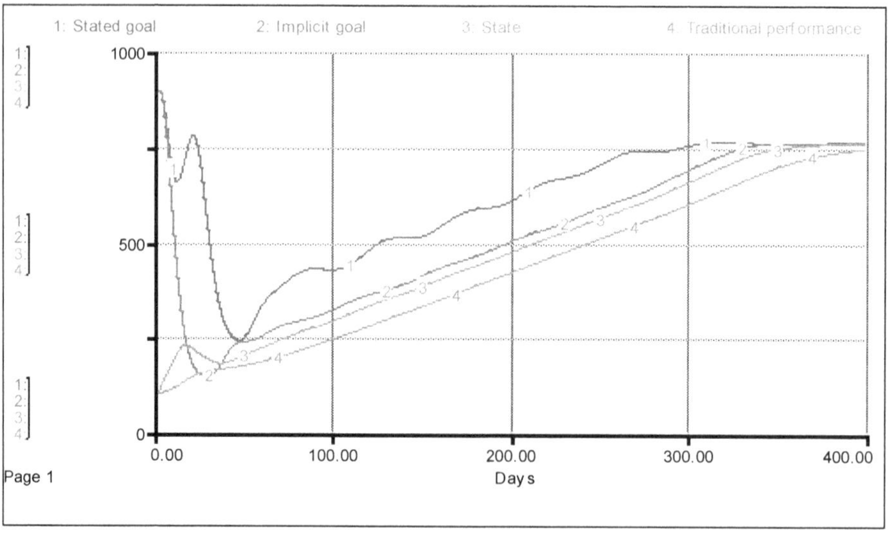

Figure 21. A better result with dynamic goal management policy, when more time is provided (initial *Stated_goal* = 900 & $Time\_horizon_0 = 350$)

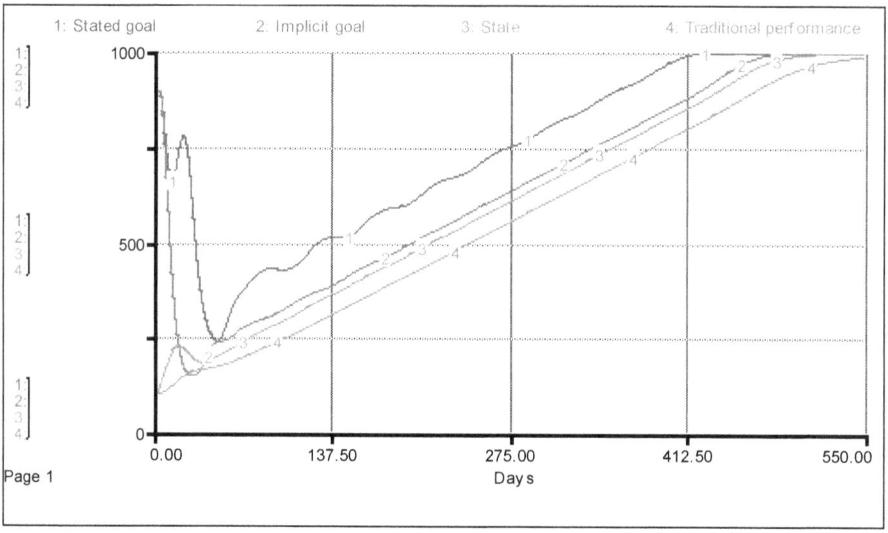

Figure 22. A close optimal result*: Ideal goal* is attained via dynamic goal management, when *Time horizon* is long enough (initial *Stated_goal* = 900 & $Time\_horizon_0 = 500$)

8. Conclusions

In the simplest computer/simulation models of goal-seeking in organizations, there is a constant goal and the model describes the dynamic difficulties involved in reaching that given goal. In more sophisticated models, the goal itself is variable: it can erode as a result of various phenomena such as frustration due to persistent failure or it can evolve further as a result of confidence caused by success. There exist some models of limited and linear goal erosion dynamics in the literature. We extend the existing models to obtain a comprehensive model of goal dynamics, involving different types of explicitly stated and implicit goals, expectation formation and potential goal erosion as well as positive goal evolution dynamics. The model constitutes a general theory of goal dynamics in organizations; involving a host of factors not considered before, such as organizational capacity limits on performance improvement rate, performance decay when there is no improvement effort, time constraints, pressures, and motivation and frustration effects. We show that the system can exhibit a variety of subtle problematic dynamics in such a structure.

We build a series of more and more realistic and complex goal-related structures. The purpose of each enhancement is to introduce and discuss a new aspect of goal dynamics in increasingly realistic settings. In the first enhancement, we introduce how the implicit goal in an organization may unconsciously erode as result of strong past performance traditions. Next, we discuss under what conditions recovery is possible after an initial phase of goal erosion. In the following enhancement, we include the effects of time constraints on performance and goal dynamics. We also show that in such an environment, the implicit short-term self-evaluation horizons of the participants may be very critical in determining the success of the improvement program. Finally we propose and test an adaptive goal management policy that is designed to assure satisfactory goal achievement, by taking into consideration the potential sources of failures discovered in the preceding simulation experiments.

Our theoretical model and management strategies can be implemented to specific improvement program settings, by proper adaptation of the model structures and calibration of parameters. Several managerial parameters in our models are set at some reasonable values that serve our research purpose. In further applied research and actual studies, these constants must be well estimated by data analysis and interviews and also tested by sensitivity analysis. Our models can also be turned into interactive simulation games, microworlds and larger learning laboratories so as to provide a platform for organizational learning programs. More generally, our models may provide useful starting points for different research projects on goal setting, performance measurement, evaluation and improvement.

References

Barlas, Y., 2002, System dynamics: systemic feedback modeling for policy analysis. In *Knowledge for sustainable development: an insight into the encyclopedia of life support systems* (pp. 1131–1175). UNESCO Publishing – Eolss Publishers (Paris, France; Oxford, UK).

Barlas, Y., & Ozevin, M. G., 2004, Analysis of stock management gaming experiments and alternative ordering formulations. *Systems Research & Behavioral Science,* Vol. 21, Issue 4, 439–470.

Barlas, Y and H. Yasarcan. 2006. Goal setting, evaluation, learning and revision: A dynamic Modeling approach. *Evaluation and Program Planning*. 29(1). pp 79–87.

Ericsson, K. A., 2001, Attaining excellence through deliberate practice: Insights from the study of expert performance. In M. Ferrari (ed.), *The pursuit of excellence in education* (pp. 21–55). Erlbaum (Mahwah, NJ).

Ford, A., 1999, *Modeling the Environment*. Island Press (Washington, DC).

Forrester, J. W., 1961, *Industrial dynamics*. MIT Press (Massachusetts).

Forrester, J. W., 1975,. Planning and goal creation: excerpts. In *Collected papers of Jay Forrester* (pp. 167–174). Wright-Allen Press (Massachusetts).

Forrester, J. W., 1994, Policies, decisions, and information sources for modeling. In J. D. W. Morecroft & J. D. Sterman (eds.), *Modeling for learning organizations* (pp 51–84). Productivity Press (Portland, Ore.).

Lant, T. K., 1992, Aspiration level adaptation: An empirical exploration, *Management science*, 38 (5), 623–644.

Lant, T. K., & Mezias, S. J., 1992, An organizational learning model of convergence and reorientation. *Organization science, 3*, 47–71. (Reprinted in Cohen, M. D. & Sproull, L.S. (eds.) (1995). *Organizational learning*. Sage Publications).

Morecroft, J. D. W., 1983, System dynamics: Portraying bounded rationality. *OMEGA The international journal of management science, 11*(2), 131–142.

Morecroft, J. D. W., & Sterman, J. D. (eds.), 1994, *Modeling for learning organizations*. Productivity Press (Portland, Ore.).

Newcomer, K., & Neill, S. (Eds.), 2001, Assessing program performance: Measurement and evaluation. Special issue of *Evaluation and program planning, 24*(1).

Roberts, E. W., 1981, *Managerial applications of system dynamics*. MIT Press (Cambridge, Massachusetts).

Schalock, R. L., & Bonham, G. S., 2003, Measuring outcomes and managing for results. *Evaluation and program planning, 26*(3), 229–235.

Senge, P. M., 1990, *The fifth discipline: The art and practice of the learning organization*. Currency and Doubleday (New York).

Simon, H., 1957, *Administrative behavior: A study of decision making processes in administrative organizations*. Macmillan (New York).

Spector, J. M., & Davidsen, P. I., 2000, Cognitive complexity in decision making and policy formulation: A system dynamics perspective. In R. Sanchez & A. Henee, *Selected papers from the 4th international conference on competency based management*.

Spector, J. M., Christensen, D. L., Sioutine, A. V., & McCormack, D., 2001, Models and simulations for learning in complex domains: using causal loop diagrams for assessment and evaluation. *Computers in Human Behavior, 17,* 517–545.

Sterman, J. D., 1987, Expectation formation in behavioral simulation models. *Behavioral science, 32,* 190–211.

Sterman, J. D., 1989, Modeling managerial behavior: Misperceptions of feedback in a dynamic decision making experiment. *Management science, 35*(3), 321–339.

Sterman, J. D., 2000, *Business dynamics: systems thinking and modeling for a complex world*. McGraw-Hill (New York).

Yasarcan, H., 2003, *Feedback, delays and non-linearities in decision structures*. Doctoral Dissertation, Department of Industrial Engineering, Bogazici University.

Yasarcan, H., & Barlas Y., 2005, A generalized stock control formulation for stock management problems involving composite delays and secondary stocks. *System dynamics review, 21*(1), 33–68.

Appendix

List of Final Model Equations

State(t) = State(t–dt) + (Improvement - Loss) * dt
INIT State = 100
INFLOWS:
Improvement = MIN(Desired_improvement, Improvement_capacity)
OUTFLOWS:
Loss = MIN(Loss_fraction * State, Maximum_loss)

Time_horizon(t) = Time_horizon(t–dt) + (Time_horizon_depletion_rate) * dt
INIT Time_horizon = 200
OUTFLOWS:
Time_horizon_depletion_rate =
IF Time_horizon>Short_term_horizon THEN 1 ELSE 0

Traditional_performance(t) = Traditional_performance(t - dt)
+ (Traditional_performance_formation) * dt
INIT Traditional_performance = State
INFLOWS:
Traditional_performance_formation = (State – Traditional_performance)/30

Desired_improvement = Estimated_loss + (Implicit_goal – State)/15

Estimated_loss = SMTH1(Loss,5)

Goal_achievable_by_trend = State + ((State – Reference_level)/
Reference_level_formation_time) * Manager's_operating_horizon

Ideal_goal = 1000

Implicit_goal = SMTH3(Weight_of_stated_goal * Stated_goal
+ (1–Weight_of_stated_goal) * Traditional_performance, 10, Stated_goal)

Improvement_capacity = Accomplishment_motivation_effect * Maximum_capacity

Loss_fraction = 1/30

Manager's_operating_horizon = MIN(90, Time_horizon)

Maximum_capacity = 15

Maximum_loss = 10

Min_acceptable_goal = State
+Min_acceptable_improvement_rate *
Manager's_operating_horizon

Min_acceptable_improvement_rate = 0.5

Perceived_performance = SMTH3(State,5)

Reference_level = SMTH3(State, Reference_level_formation_time)

Reference_level_formation_time = 10

Reference_weight = 1

Remaining_work_and_time_ratio = SMTH3(((Stated_goal
– Perceived_performance)/5)/Time_horizon, 10)

SG_formation_time = 10

Short_term_horizon = 12

Short_term_work_and_time_ratio = ((Stated_goal - Perceived_performance)/5)
/Short_term_horizon

Stated_goal = SMTH3(MIN(Ideal_goal, MAX(Goal_achievable_by_trend,
Min_acceptable_goal)), SG_formation_time, 900)

Weight_of_stated_goal = Accomplishment_motivation_effect *
Short_term_motivation_effect * Reference_weight

Accomplishment_motivation_effect = GRAPH(Remaining_work_and_time_ratio)
(0.7, 1.00), (0.8, 0.99), (0.9, 0.96), (1, 0.9), (1.10, 0.77), (1.20, 0.54), (1.30, 0.31), (1.40, 0.16), (1.50, 0.08), (1.60, 0.035), (1.70, 0.01), (1.80, 0.00)

Short_term_motivation_effect = GRAPH(Short_term_work_and_time_ratio)
(0.00, 1.00), (0.2, 0.998), (0.4, 0.995), (0.6, 0.985), (0.8, 0.967), (1.00, 0.933), (1.20, 0.887), (1.40, 0.836), (1.60, 0.78), (1.80, 0.72), (2.00, 0.66), (2.20, 0.599), (2.40, 0.54), (2.60, 0.482), (2.80, 0.427), (3.00, 0.376), (3.20, 0.328), (3.40, 0.284), (3.60, 0.244), (3.80, 0.209), (4.00, 0.177), (4.20, 0.149), (4.40, 0.124), (4.60, 0.102), (4.80, 0.084), (5.00, 0.068), (5.20, 0.055), (5.40, 0.043), (5.60, 0.034), (5.80, 0.026), (6.00, 0.019), (6.20, 0.014), (6.40, 0.009), (6.60, 0.005), (6.80, 0.002), (7.00, 0.00)

PART V

Furture Directions on Complex Decision Making

Chapter 16

Future Directions on Complex Decision Making Using Modeling and Simulation Decision Support

Hassan Qudrat-Ullah
School of Administrative Studies
York University, Canada
hassanq@yorku.ca

1. Introduction

The increasing complexity of the environments in which we live and work requires us to continue the search for ways, methods, and techniques to improve our ability to make better decisions. Although fragmented research exists on recent developments in the fields of decision support, dynamic decision making, theories on learning and managing complexities, and system dynamics and agent-based modeling approaches. We strongly believe this research deserves greater attention and provides impetus for an integrated volume in the service of complex decision making. This book is a modest attempt to publish the contributions from the scholars who have dedicated their research efforts to align our best understanding of decision making in complex, dynamic environments.

Methodology

In our call for chapters on specific topics we went through various e-mail lists of professional bodies. We also posted the call for chapters at the message boards of a few international conferences in the fields. Personal invitations were sent to target authors as well. We received twenty-one chapters in total. All the chapters received from the contributors went through a double-blind review process. The reports from the independent reviewers were sent to the authors to include changes on the basis of comments made by the reviewers. Only fifteen chapters made it to the final stage of acceptance. The final versions of the chapters have been edited and included in this volume.

Research categories

The chapters thus complied are classified under five main categories following the structure of the book. The first category examines the nature and the dimensions of decision making and learning in dynamic tasks. The second category deals with the range of tools, methods, and technologies that support decision making in complex, dynamic environments including Method of Dynamic Handling of Uncertainty, System Dynamics, System Dynamics and Multiple Objective Optimization based hybrid approach, Object Oriented Programming based simulation methods, Data Envelopment Analysis, Group Model Building, and Agent-based Modeling that can help people to make better decisions in complex systems. The third category provides the empirical evidence to the application of both system dynamics and agent-based modeling approaches to variety of complex decision making situations covering both individual and group decision making perspectives. The fourth category deals with two key methodological issues: (i) understanding of "goal dynamics" and (ii) measurement of change in the mental models of the decision makers in complex decision making environments. The current category summarizes and synthesizes all the previous chapters and discusses the emerging trends and challenges in using decision aiding tools and techniques for improved decision making in complex, dynamic environments in the coming years.

2. Decision Making and Learning in Complex, Dynamic Tasks

The two chapters in the first part of the book shed light on the concept of dynamic tasks in general, and more specifically on the role of decisional aids, both in the form of technology and human facilitation, in developing expertise in dynamic tasks. The debate centers on the notion of the misperception of feedback (MOF) hypotheses (Sterman, 1989)—people perform poorly in dynamic tasks because of their inability to identify endogenous positive feedback gains enlarging apparently tiny decision errors and side effects. On the other hand, researchers have shown that various decisional aids, e.g., computer simulation based interactive learning environments (ILEs) show promise to overcome the misperception of feedback. Also it is difficult to measure learning and development of expertise in dynamic, complex problems.

In Chapter 1, Karakul and Qudrat-Ullah review research on dynamic decision making and learning in ILEs. The review is based on published articles in the area in English-language journals, books, conference proceedings and so on. The authors review thirty-nine experimental studies based on the key criteria- the underlying task in each study has to be a dynamic task. For each study, types of predictor factors, dynamic task structure, decisions required, measurements of performance, as well as, a short summary of major results are provided. The authors find that decision making and learning in dynamic task is contingent on four major factors of an ILE–decision-maker, decision task, decision-making environment, and human facilitation. Their research indicates that the decisional aids in the form of human facilitation and human-computer interaction design principles are crucial to optimal performance in dynamic tasks. An overwhelming majority of these studies support the MOF hypothesis. Only three studies examine how the decisional aids can help overcome the MOF hypothesis. This uncovers the fact that much remains to be done in the area. This chapter highlights some gaps in future directions of research on decision making and learning in complex, dynamic tasks. Specifically, the authors present a conceptual model suitable for future experimental studies aimed at improving people' performance in dynamic task using the human-facilitation based decisional aids and HCI design principles for ILEs.

Spector (Chapter 2) characterizes the complex, ill-structured tasks as dynamic tasks as a result of changing circumstances, evolving systems and structural considerations, altered understanding and perceptions of situations and those involved, or any combination of these. As a consequence of this characterization, he argues that expertise in dynamic tasks is not directly measurable by a particular solution or test score or knowledge item. Instead, within the context of learning in and about complex domains, people are required to develop expert-like flexible processes and techniques for approaching a variety of problem situations. He then suggests the DEEP methodology as a promising solution to the problem of measuring expertise in ill-structured problem solving situations. Finally, the author draws some conclusions that help further development of the methodology for measuring expertise in dynamic tasks.

It appears from the above analysis that although the authors of the first two chapters emphasize the need for (i) the human-facilitation and HCI principles based decisional aids, and (ii) a flexible process-oriented approach to the measurement of expertise in dynamic tasks, a pessimistic view of people's decision making and learning in dynamic tasks is present. However, it is strongly argued that people's performance in dynamic tasks can be achieved to the highest possible extent through a variety of logical and technological decision support systems. The next section aims at identifying a range of tools, methods, and technologies that support decision making in complex, dynamic environments.

3. Tools and Techniques for Decision Making in Complex Systems

Uncertainty is the fundamental characteristics of any complex decision making environment. Lenny (Chapter 3) addresses the question of how to deal with uncertainty in MRP/ERP environments (a special type of complex decision making environments). He argues that many uncertainties in MRP/ERP environments are treated as 'controllable' elements, with a variety of buffering, dampening and smoothing techniques such as overtime, multi-skilled labour, subcontract and holding of safety stock for parts being used to cope. He argues that more dynamic handling of uncertainty is required to support responsive decision making on ways to deal with uncertainty.

He performed an empirical study using two performance measures: the '*K-factor*' and 'classification of enterprises'. Lenny concludes (i) that the use of these two performance measures, benchmarked against the levels of uncertainties and levels of late delivery to customer, provide a step forward to dynamically handle uncertainty and (ii) such evaluation supported a better decision making on the adoption of suitable buffering, dampening, and smoothing techniques on specific uncertainty. These results are encouraging yet the generalization of this approach--dynamic handling of uncertainty—to other complex decision making environments than MRP/ERP environments warrants a series of future empirical studies.

Researchers in the domain of dynamic decision making and learning in complex, dynamic tasks face the challenge to search for ways, methods, and techniques to improve our ability to make better decisions. Jim (Chapter 4), by presenting a synthesis of system dynamics and multiple objective optimization, demonstrates how this hybrid approach can be used to help decision makers apply a quantitative and qualitative approach in order to improve their ability to make better and more informed decisions in a complex decision making environment. He summarizes (i) dynamic complexity and conflicting objectives, (ii) the traditional approach to single objective optimization used in system dynamics modelling, and (iii) multiple objective optimization. He then presents an illustrative case study based on the Beer Game, which integrates multiple objective optimization with system dynamics, and provides a practical way for decision makers to fully explore policy alternatives in a complex decision making environment. Despite the limited generalization of these results based on single case study, the proposed hybrid approach holds a sound promise as a modelling methodology to improve decision making and learning in complex, dynamic systems.

In Chapter 5, Latorre et al. have proposed a simulation method for hydro basins, based on Object Oriented Programming paradigm. Hydro reservoir management is complex decision making environment. Authors provide a solution to hydro reservoir management issues with a three-phase algorithm, whose main purpose is to adapt the given strategies to the fulfillment of the minimum outflows and avoiding superfluous spillages. The three phases of the algorithm are (i) decide an initial reservoir management, (ii) modify the previous management to avoid spillage or to supply irrigation and ecological needs, (iii) and determine the hydro units' output with previous outflows.

Latorre et al. have implemented this simulation method in a computer program to represent a realistic case for a Spanish hydro basin. Different strategies including pre-calculated decision table, production of the incoming inflow, minimum operational level, and maximum operational level, representing all possible real alternatives to manage reservoirs with diverse characteristics, are analyzed. The results of this simulation model have been compared with those obtained by direct optimization. Their simulation model achieves a 95% of the total optimal hydro production at a much lower computational cost. Thus, authors have demonstrated convincingly the utility of this Object Oriented Programming based simulation methods in dealing with complex, dynamic issues.

Complex decision problems often involve a large number of input and output variables. The choice of decision options becomes very difficult. Alexander (Chapter 6) presents Data Envelopment Analysis (DEA), a nonparametric mathematical programming approach which allows for comparisons of decision options with similar objectives, canonically termed decision making units (DMUs), as a better solution methodology. He first examines the method's characteristics relative to decision analysis problems. According to Alexander, the main strengths of DEA are (i) its ability to handle the evaluation of alternatives with input and output vectors consisting of elements on different units and scales, which are common to complex decision problems, (ii) no assumptions on the form of the decision functions, common to parametric approaches, have to be made, and (iii) DEA is not susceptible to parametric modeling challenges in parameter estimation, such as for input variables showing the presence of autocorrelation of an error term with an output variable. The main disadvantage is the limited discriminatory ability of the approach when the dimensionality of inputs, outputs and decision options do not fall within the outlined bounds.

Alexander then outlines the formation of the efficiency frontier, the two base models Charnes, Cooper and Rhodes (CCR) and the Banker, Charnes and Cooper (BCC) model. In addition, extended Data Envelopment Analysis models on scale determination and (decision option) target choice determination are discussed. Data Envelopment Analysis is shown as a viable modeling methodology that carries few required model parameter choices while providing an effective means to analyze otherwise hard to model complex decision problems.

Effective management of any complex system requires the understanding of the structure that causes the (problematic) behavior in that system. Zagonel and Rohrbaugh (Chapter 7) present group model building (GMB), based on system dynamics methods and the structured involvement of client group in the process of model construction, as a model-building framework suitable for understanding the behavior of the system. They argue that GMB interventions are particularly useful in situations in which there is strong interpersonal disagreement in the client group regarding the problem and/or the policies that govern system behavior.

In this chapter, they report the process, content and outcomes of a GMB effort in which a complex issue in the public sector was modeled and analyzed. They demonstrate how GMB was used to inform welfare reform policy making and implementation, drawing upon the perspectives and knowledge of the client teams. This chapter addresses the policy aspects of the research by answering an important question using the model elicited, built, simulated, tested and extensively used throughout the intervention. Finally, Zagonel and Rohrbaugh draw some conclusions regarding the utility of GMB in dealing with complex problems. They conclude the chapter with a discussion of the limitations of the analysis.

In their provocative chapter on agent-based modeling, Schieritz and Milling (Chapter 8) argue that because of the different level of aggregation the system dynamics and the agent-based approach have the potential to complement each other: High-level system-dynamics structures can be identified and quantified using an agent-based model. They demonstrate this particular applicability of the agent-based approach to the problem of a "population dynamics"—a system of complex interactions between the population characteristics of *Fertility* and *Mortality*. First, they provide an overview of the agent-based modeling approach and its applicability to better understand the complex, dynamic systems e.g., a population system. The key assumptions are (i) the variables characterizing the problem on a macro level first have to be identified together with their causal structure, and (ii) the policies of the individual agents have to be identifiable. Then they run the simulation experiments with the agent-based model to develop a macro theory of the evolution of a population. Limitations of the methodology are also mentioned at the end of this chapter.

These six chapters in the second part of the book shed light on some state-of-the-art tools, methods, and technologies—method of dynamic handling of uncertainty

(Chapter 3), System Dynamics and Multiple Objective Optimization based hybrid approach (Chapter 4), Object Oriented Programming based simulation methods (Chapter 5), Data Envelopment Analysis (Chapter 6), Group Model Building (Chapter 7), and Agent-based Modeling (Chapter 8) that can help people to make better decisions in complex systems. Two of these modeling approaches, system dynamics and agent-based modeling, stand out as the most promising modeling approaches to aid decision making in complex systems. In the next section, we present some real world cases where these modeling approaches have successfully been applied.

4. System Dynamics and Agent-Based Modeling in Action

System dynamics and agent-based models have been developed and used in various domains to better understand the dynamics of underlying complex systems (for a good reference please see Chapter 1 and Chapter 13 of this volume). All these models have great insights into specific areas, helping to make better managerial as well as policy decisions. The case studies summarized below are some examples of the most recently developed model based empirical work by various researchers with the aim facilitating the achievement of better decision making and learning about complex systems in five crucial and important fields: health care management, Knowledge-based Economy, Tourism Management, High-tech Industry Retail Management, and Dynamics of the Diffusion Process for Medical Innovations.

Health Care Management

Wolstenholme et al. (Chapter 9) presents a classic real world application of system dynamics modeling in dealing with complex, dynamic system: health and social care system in the UK. Their work have been at the forefront of both national and local policy analysis and has achieving outstanding success in influencing behavior and decisions by its timeliness and relevance. The specific objective of the modeling project was to better understand the dynamics of "delayed hospital discharge" problem:

> *"Delayed hospital discharge is a key issue which first came onto the legislative agenda in late 2001. The 'reference mode' of behaviour over time for this situation was that of increasing numbers of patients occupying hospital beds, although they had been declared "medically fit". In March 2002, 4,258 people were "stuck" in hospital and some were staying a long time, pushing up the number of bed days and constituting significant lost capacity. The government's approach to this issue was to find out who was supposed to "get the patients out" of acute hospitals and threaten them with 'fines' if they did not improve performance. This organisation proved to be social services who are located within the local government sector and who are responsible for a small but significant number of older people needing ex-hospital ('post-acute') care packages. Such patients are assessed and packages organised by hospital social worker (Wolstenholme* et al., Chapter 9)".

They concstuct a system dynamics model consisting of five key sectors of the hopital discharge system including Primary Care, Medical Bed, Surgical Beds, Intermediate Care, Post-Acute Care. Their dynamic hypothese about the underlying issue is that complex behavior over time arises in patient pathways by the interaction of the proportions of patients flowing (converging and diverging), their lengths of stays in different parts of the pathway and the capacities of each pathway, together with the actions taken by the agencies owning each piece of the pathway as patient movements conspire in and against their favor. They argue that the resultant behavior of the whole system over time can be very counter intuitive, which is a major justification for the use of system dynamics based computer analysis to simplify the problem and improve understanding and communication.

Applying the developed system dynamics model for the delayed hospital discharge, Wolstenholme et al. draw some conclusions as:

- Common sense solutions can be misleading. For example, the obvious unilateral solution of adding more acute capacity can exacerbate the delayed discharge situation. More hospital admission will be made in the short term, but no additional discharges will be possible. Hence, the new capacity will simply fill up and then more early discharges and outliers will be needed.
- Fines may have unintended consequences. For example, this solution will also be worse if the money levied from social services is given to the acute sector to finance additional capacity. It will be worse still if it causes the post-acute sector to cut services. The effects of service cuts may also spill over into other areas of local government including housing and education.
- There are some interventions that can help:
- Increasing post acute capacity gives a win-win solution to both health and social care because it increases all acute and post acute sector performance measures. Further, counter intuitively, increasing medical capacity in hospital is more effective than increasing surgical capacity for reducing elective wait times. This again is because 'outliers' are avoided.
- Reducing assessment times and lengths of stay in all sectors is beneficial to all performance measures, as is reducing variation in flows, particularly loops like re-admission rates
- If fines are levied they need to be re-invested from a whole systems perspective. This means re-balancing resources across all the sectors (NOT just adding to hospital capacity).
- In general the model showed that keeping people out of hospital is more effective than trying to get them out faster. This is compounded by the fact that in-patients are more prone to infections so the longer patients are in hospital, the longer they will be in hospital
- Improving the quality of data is paramount to realising the benefits of all policies.

They found an interesting generalisation of the findings was that increasing stock variables where demand is rising (such as adding capacity) is an expensive and unsustainable solution. Whereas increasing rate variables, by reducing delays and lengths of stay, is cheaper and sustainable. Because of the success of the delayed hospital discharge model, authors produced another system dynamic model for the emergent local health and social care economy model.

Finally, authors provide some practical suggestions for having the best out of a system dynamics study including (i) well-defined issues, (ii) consistent and representative teams of management professionals from all agencies involved, well briefed in the principles of the approach, (iii) well-defined expectations and well-scoped models in time and space, (iv) mechanisms for managing the politics of the management team and keeping senior people in touch with the project, (vi) ways of maintaining momentum, and (vi) ways of reinforcing the benefits.

Knowledge-based Economy

Managing the transition from a traditional production-based economy to a knowledge-based economy is indeed a complex task for any nation, state, or a country. Dangerfield (Chapter 10) presents a case study where the objective of the underlying system dynamics modelling project is to provide the state of Sarawak (in Malaysia) with a tool to aid their future economic and social planning. On the use of a model as a policy support tool, Dangerfield asserts that it is impossible in a complex system of interacting influences to even start to plan sensibly without the aid of a model specifically designed to deal with such complex networks of influences.

Dangerfield developed a system dynamics model based on some key assumptions including (i) development of a knowledge economy would be constrained if the supply of science graduates and suitable sub-professional knowledge-workers are not forthcoming, (ii) initial staffing of R & D centers may be a problem but it is possible that, with sufficiently attractive remuneration packages, qualified Sarawak expatriates would be tempted to return, (iii) the proposed R & D centers can be seen as crucial catalysts in the stimulation of the quaternary sector and they will offer a primary supply of people with the necessary skill sets to enthuse the creation and development of knowledge-based industry and services, and (iv) the key the dynamic flows to be considered in the simulation model are:

- Skills and technology transfer
- Money and other financial resources
- Capital equipment
- Human resources

Based on the developed and validated system dynamics model, Dangerfield then created a micro world whereby the decision makers can play out various policy scenarios.

Finally, Dangerfield draws some conclusions about this system dynamics modeling project that (i) this contribution has described an empirical study conducted over 27 months and which has resulted in a system dynamic tool for macro-economic planning being made available in a developing country, (ii) as well as the availability of the tool, knowledge of the capability of system dynamics as a modelling methodology for addressing complex systems has taken root, reinforced by a technology and knowledge transfer element in the project, and (ii) it is anticipated that, as new policy issues arise, the Sarawak State government will be pre-disposed to their analysis using the system dynamics methodology.

Tourism Management

In the context of the Gastein Valley (Gasteinertal) in the county of Salzburg, one of the top tourist regions of Austria, Schwaninger (Chapter 11) makes a convincing case for the notion of "systemic decision", by showing how qualitative and quantitative evaluations were combined within a systemic framework, to arrive at a decision whose superior quality could not have been attained if a non-systemic approach had been used. In this chapter, Schwaninger presents (i) a description of the context of this case study, (ii) the conceptual approach taken, (iii) the steps of the decision process with decision analysis and synthesis, (iv) an account of the final decision, and (v) a reflective summary of this model-based case study.

On the use and justification for a systemic study to manage a complex system as the Gastein Valley, Schwaninger concludes that (i) the case study reported on here brings several features to the fore, features which characterize a systemic approach to studying complex and dynamic issues, (ii) the system under study exhibited exceptionally high complexity, (iii) the intervention and the ramifications of its effects as studied in this project made up a multidimensional web of relationships susceptible to multi-dimensional, holistic and integrative analysis where the dynamic patterns are elicited by the use of system dynamic model, (iv) analytic and synthetic reasoning stands in contrast to the monochromatic dominance of analysis and the lack of synthesis widely observable in studies of a comparable kind, (v) at any rate, the quantitative methods were an essential complement of the qualitative ones, and (vi) the system dynamics model showed that the results of conventional calculations underestimated the fatal consequences of the planned intervention. He argues that this was a case of applying systemic thinking to a complex situation and thereby generating superior results in comparison to what might have been achieved with traditional approaches.

Finally, he claims that this study was a vital ingredient in the decision-making process, and ultimately proved to supply a crucial argument for choosing the most beneficial, although most demanding, of the variants - *cum grano salis* - for the case in point: A valley was saved by means of a system study.

High-tech Industry Retail Management

Bivona and Ceresia (Chapter 12) contribute with an innovative application of system dynamics modeling to the high-tech industry retail management. They present an analysis based on a real case aimed at helping *Jeppy Ltd* management to build successful policies to enhance the relationship with its retail chain. Through several meetings, which included the management, a System Dynamics model was built to explore alternative scenarios and related strategies. In particular, the use of a System

Dynamics approach allowed the authors to identify an effective set of internal policies based on HRM practices (recruitment, training and goal setting policies), as well as external efforts aimed at increasing the awareness of potential customers about the company's product benefits.

In this chapter, the authors provide (i) an analysis of the main contributions of Distribution Channel Management, (ii) a review the role of Strategic Human Resource Management, and (iii) the case-study. In the case study, they discuss the evolution of the business, problem issues, feedback analysis of managerial growth policies, unperceived potential limits to growth and policy design to remove business limits to growth are discussed here. They report that problem issues and feedback analysis of the adopted company growth policies are the results of several meetings with the involvement of the management. Finally, they present an analysis of the main stock and flow structure of the System Dynamics model used to explore alternative scenarios and report the results of two simulation runs.

Based on a comprehensive analysis of the feedback structure of the case, the authors built an insightful system dynamics model. The developed system dynamics model shows the dynamics of company's key variables and helps the exploration of the effectiveness over time of the suggested strategies (e.g., retail stores' employees' recruitment, advanced training and goal setting policies, and investments in developing potential customers' awareness of UT concept). The system dynamics model covers three main sub-systems:

1. Company Customers related to "Trojan horse" and "UT" products sales
2. Company retail store
3. Human Resources Management

The authors draw two main conclusions: (i) in order to successfully design long-term policies to foster manufacturer-retailer relationships, manufacturers must not make decisions exclusively oriented to generate short term gains, and (ii) the System Dynamics methodology is effective in exploring and understanding the complex and dynamic phenomenon related to the manufacturer and retail chain relationship.

Dynamics of the Diffusion Process

Understanding the dynamics of the diffusion process is a complex task largely due to the complex interactions involved in it. Ratna et al. (Chapter 13) reanalyze *Medical Innovation* by Coleman, Katz and Menzel (1966), the classic study on diffusion of Tetracycline, which at that time was a newly introduced antibiotic. Based on the findings in *Medical Innovation*, the authors develop a diffusion model, based on agent-based modeling approach, called *Gammanym*. They also analyze the topology of networks generated in *Gammanym*, and its evolution to evaluate the network structure influencing the diffusion process.

In this chapter, the authors (i) describe the original study and the rationale for this study, (ii) present the modeling framework, modeling sequences and methods, (iii) analyze the simulation results under different scenarios, (iv) depict the structure of the social networks in their model, (v) analyze the evolution of the social networks,

(vi) explore the significance of types of network integration after normalizing adoption curves in terms of numbers for professionally integrated and isolated doctors, and (vii) discuss and review the implications of the simulation results.

Authors develop an agent-based model, called *Gammanym*, to analyze the diffusion. Their integrated modeling with *Gammanym* reveals original results within the existing literature of diffusion research, and also complements the extant work on *Medical Innovation.*

Their approach of integrated modeling enabled us to look at network properties such as connectivity and degree distribution, among others, in terms of the interaction matrices generated by the agent-based model. On the basis of these properties we determine that the interaction networks depicted in *Gammanym* are random graphs. Complexity of the diffusion process is explained by analyzing evolution of networks or dynamics on the networks. We find that initially the system consists of a number of disconnected components and quickly evolves, after 7–10 time steps, to form a single connected component. The analysis of network topology also indicates that underlying networks evolve in predictable ways, and the uptake is a function of the initial starting conditions.

Analysis of the evolution of uptake or adoption of Tetracycline enables the authors to disentangle the extent of different factors affecting adoption. They argue that despite stressing the complementarities between network theory and diffusion research, a large body of diffusion literature has so far failed to examine the dynamic structures of the interpersonal networks and their evolutions over the diffusion process. Their model shows that although the media does not influence the network structure, it does have a major impact in accelerating the diffusion process. Under heavy media exposure undertaken by the pharmaceutical company to increase sales of Tetracycline, the average size of uptake group with agents who have adopted the new drug rises faster than otherwise, they found. Based on their analysis with the agent-based model, they conclude (i) that all the agents adopt the new drug within 25 time steps, much earlier than with a baseline scenario of much less media exposure, (ii) the previous evidence of a dominant media influence by comparing the speed of diffusion for three scenarios: baseline, heavy media and integration is valid, (iii) the dominant role of media also suggests there is a trade-off between the efforts expended in promoting a new drug, and the speed of the diffusion process, and (iv) although the comparison of the cumulative diffusion curves of *Gammanym* with those of *Medical Innovation,* is largely restricted , the exploration with normalized adoption curves for groups of doctors, based on different degrees of social and professional integration, reveals that after the formation of the joint cluster, the initial uptake is dependent on the level of integration. Finally, they claim that their results provide a step towards understanding the complexity and dynamics of the diffusion process in the presence of professional and social networks.

5. Some Methodological Issues in Research on Complex Decision Making

Prior research on dynamic decision making and learning with model-based decision support systems has attempted, with some success, to improve people's decision making capabilities in complex, dynamic tasks.

For example, in a system dynamics simulation based learning environments (ILEs), decision makers can test their assumptions, practice exerting control over a business situation, learn from the "immediate" feedback of their decisions to achieve the desired performance goals (Qudrat-Ullah, 2006). ILEs allow the decision makers to discover the misconceptions about the dynamic task system and improve their mental models. On the one hand, the performance goals are dynamic in nature. On the other hand, the evaluation of mental models of decision maker is a very complex task. In order to advance the science of decision making in complex, dynamic tasks, we need better understanding of the role of goal dynamics as well a systematic and scientific approach to the measurement of mental models. The following two chapters (14 and 15) take this challenge with some success.

On understanding of goal dynamics, Barlas and Yasarcan (Chapter 14), present a theoretical model that constitutes a general theory of goal dynamics in organizations; involving a host of factors not considered before, such as organizational capacity limits on performance improvement rate, performance decay when there is no improvement effort, time constraints, pressures, and motivation and frustration effects.

They begin the chapter by elaborating on the simplest model of constant goal seeking dynamics, by including a *constraint* on improvement *capacity* and a nominal *decay* rate out of the state variable. They then gradually add a series of more realistic and complex goal-related structures to the initial model. In the first enhancement, they introduce how the implicit goal in an organization may unconsciously erode as result of strong past performance traditions. Next, they discuss under what conditions recovery is possible after an initial phase of goal erosion. In the following enhancement, they include the effects of *time constraints* on performance and goal dynamics. They also show that in such an environment, the implicit *short-term* self-evaluation horizons of the participants may be very critical in determining the success of the improvement program. Finally they propose and test an adaptive goal management policy designed to assure satisfactory goal achievement, by taking into consideration the potential sources of failures discovered in the preceding simulation experiments. They conclude with some observations on implementation issues and further research suggestions.

They claim that their theoretical model and management strategies can be implemented to specific improvement program settings, by proper adaptation of the model structures and calibration of parameters. Several managerial parameters in their models are set at some reasonable values that serve their research purpose. In further applied research and actual studies, these constants must be well estimated by data analysis and interviews and also tested by sensitivity analysis. They also suggest that their models can also be turned into interactive simulation games, microworlds and larger learning laboratories so as to provide a platform for organizational learning programs. On a more general note they assert that their models may provide useful starting points for different research projects on goal setting, performance measurement, evaluation and improvement.

Doyle et al. (Chapter 15) asserts that measuring change in the mental models of the participants in systems thinking and system dynamics interventions is unavoidable if the relative effectiveness of different interventions in promoting learning about complex dynamic systems is to be assessed. In this chapter, they describe the general features of rigorous methods designe to measure change in mental models of complex dynamic systems and provide a detailed example of the design, implementation, and analysis of one such method.

They claim that the experimental results produced by this process have in the present case been generally encouraging about the effectiveness of system dynamics interventions in promoting learning. Although the intervention tested was quite modest, involving simply a single play of a management flight simulator and related preparatory and debriefing sessions comprising about 5 hours of time spread out over a two week period, several statistically significant changes in participants' mental models resulted, including: an increase in the size of subjects' mental models, a change in the content of mental models toward the expert model of the problem posed by the intervention, and an increase in the degree of feedback thinking contained in subjects' models, they asset. They hypothesize that longer interventions that focus on building systems thinking or system dynamics modeling skills would elicit even stronger changes in mental models.

On the interpretation of their results they urge a great deal of caution: the paradigm of controlled laboratory experimentation does not allow the luxury of drawing grand conclusions from a single experiment, but instead relies on systematic replication that inches toward the truth one small step at a time. They highlight several important limitations and future research avenues: (i) for example, the study demonstrated positive effects due to the intervention, but does not address the possibility that some other type of intervention, unrelated to system dynamics, might be even more effective, (ii) the study also does not address what aspects of the intervention are important for producing the observed results: for example, were the changes in mental models due primarily to subjects' experience with the management flight simulator or to the information presented during the debriefing session?, and (iii) the limited time frame covered by the intervention did not offer the opportunity to assess the stability of the measured changes in mental models: it is entirely possible that the gains achieved by the intervention disappeared or at least decayed in the weeks and months after the intervention.

They claim that the described method of measuring the change in the mental models proved to be both practicable and capable of capturing and quantifying even subtle changes in mental models due to the intervention, and it can be adapted in a straightforward manner to resolve the above-stated questions as well as a wide variety of other questions related to the effectiveness of systems interventions in promoting learning. Finally, they suggest that the present work, which documents the messiness and sloppiness characteristic of mental models, as well as the problems faced by those who would attempt to measure and change them, will lead to an increased appreciation of the high degree of difficulty inherent in studying mental models of complex dynamic systems. Their work indeed contributes to the experimental approach based research work on dynamic decision making and learning in complex systems.

6. Future Directions on Decision Making in Complex Systems

We have discussed some generic issues of decision making in complex, dynamic systems in the previous section. In this section we propose some future directions of utilizing model based approaches to improve decision making in complex, dynamic tasks in the light of the collection of chapters in this volume. As discussed in Chapter 1, ILEs are computer simulation-based systems which use modeling and data and support decision making in an interactive fashion. Since management of the most real world systems essentially deals with integrated decision-making; decision making in complex, dynamic systems is a fertile area where ILEs can be used effectively.

The review in Chapter 1 shows that out 39 empirical papers studied, only 3 studies, departing from traditional dynamic decision making research focus on how poorly subjects perform in dynamic tasks, attempts to increase our understanding of the way in which expertise on decision making in complex tasks are acquired. These 3 studies attempt to unfold the process of decision making and make a case for the inclusion of (i) decisional aids in the form of human facilitation, and (ii) HCI design principles in any ILE model aimed at improving decision making in dynamic tasks. Contrary to MOF hypotheses, these studies suggest that the lack of emphasis on human facilitation and HCI design principles resulted, at least in part, in poor performance in dynamic tasks by people. Thus the use of decisional aids such as human facilitation and HCI design principles in an ILE show promise in improving the decision making capabilities of ILE users. This is encouraging but only future studies (e.g., on the effectiveness of various types of decisional aids and timing of these decisional aids (when to apply) in the context of a wider scale and scope of the dynamic tasks) could help make reliable predictions on the efficacy of these decisional aids in improving people's decision making and learning in complex, dynamic tasks.

It is thus observed that the future of experimental-based empirical research on the efficacy of modeling-based decisional aids in improving decision making in complex task is very fertile. In particular the relative effectiveness of the key components of an ILE—the decision maker, the decision task, and the decision making environment is not known. Table 1 of Chapter 1 and its subsequent analysis reveals a number of gaps in the way various characteristics of the decision makers impact their decision making performance. For example, task experience does help but only when the dynamic tasks have relatively simple structures (e.g., they have only few feedback loops and involve only first-order delays). But when the structure of the dynamic task has an increased level of complexity (e.g., due to multiple feedback loops, higher-order delays, and non-linear relationships between the variables of the task system), the impact of task experience and training is very limited. A series of future experimental research studies where the impact of same decisional aids, with some gradual and systematic changes in the complexity level of the dynamic task (e.g., increased number of feedback loops, higher-order time delays etc.) could be evaluated and should be on the research agenda of the researchers in decision making and learning with ILEs areas. The role of cognitive style of the decision makers is rarely evaluated in complex decision making research stream. For instance, do the high-analytical decision makers perform better in dynamic tasks than the individuals with low-analytical studies is highly relevant research question to be explored in future studies. Decision heuristics is another area where future studies could be developed and implemented. For example, in stable environments consistent decision strategies do help managers to make better decision. But in complex, dynamic tasks where environments are ever changing, the potential role of consistent decision strategies needs to be investigated through empirical experimental studies.

Decision makers, with ILEs and in the absence of human facilitation, face difficulties in assimilating the new knowledge with the existing mental models. Prior studies, as reviewed in Chapter 1, provide an insight into the effectiveness of human facilitation (in the form of debriefing) to decision making and learning with ILEs. Several gaps stills remain, (i) measures of effectiveness of debriefing lack a comprehensive framework, (ii) the effects of various forms of debriefing (e.g., outcome-oriented, process-oriented) are unknown, and (iii) no empirical evidence on how debriefing effects the decision making process is present. These issues together with the methodological issues on "goal dynamics" and "measuring change in mental models of the decision makers," highlighted in Chapter 14 and Chapter 15 of this volume, provide tremendous opportunities for researchers in complex decision making stream.

Finally, prior studies on dynamic decision making suffered due to restricted measures of task performance. Our review in Chapter 1 has shown the most of the empirical work in dynamic decision making and learning with ILEs consider "game score" as the measure of task performance. Given the promising development on how to measure expertise on dynamic decision making, as detailed by Spector (Chapter 2), future studies can explore the impact various decisional aids in the broader definition of performance measure of prototypic expertise.

5. Conclusions

This chapter provides a synthesis of all the previous chapters in this volume. It also discusses generic issues of better decision making in complex, dynamic environments. The chapter then highlights future directions of modeling based decisional aids in complex, dynamic tasks. In short, the future of modeling and simulation based decision support systems for complex domains is very bright. We highlighted some of the issues which should be targeted for further research and development of model-based decisional aids in complex systems.

Reference

Sterman, J. D. (1989). Modeling managerial behavior: Misperceptions of feedback in a dynamic decision making experiment. *Management Science*, **35**, 321–339.

CPSIA information can be obtained
at www.ICGtesting.com
Printed in the USA
LVHW011812041118
595903LV00012B/536/P

9 783540 736646